Efficiency and Sustainability in the Energy and Chemical Industries

Using classic thermodynamic principles as the point of departure, this new edition of a popular resource supplies the understanding and tools required to measure process efficiency and sustainability with much improved accuracy. Exploring the driving forces in the chemical and power industries, *Efficiency and Sustainability in the Energy and Chemical Industries: Scientific Principles and Case Studies, Third Edition* investigates why losses occur and explains how to reduce them. It focuses on the changing roles of refining and chemicals in industry and how the industry is transforming itself, and considers economics as a key enabler to look at technology choices and whether shareholder returns will be there.

- Includes new chapters on plastics recycling technologies and challenges, low carbon energy sources, the changing energy mix, and project economics, taxes, and subsidies.
- Illustrates techniques with wide-ranging case studies related to energy conversion, mining, and the chemical industries as well as examples and problems.
- Considers engineering layouts that reduce the environmental impact of chemical operations.
- Explains how to use energy analysis to accurately assess the quality and performance of chemical processes.
- Supplies quantitative tools for analyzing sustainability and efficiency.
- Investigates the challenges of the hydrogen economy and CO_2 and low carbon.
- Discusses plastics recycling, economics, and a changing energy mix.

Complete with the keys to quantification of process efficiency and sustainability, this cutting-edge book is the ideal guide for those engaged in the transition from fossil-based fuels to renewable and sustainable energy sources using low-waste procedures.

Green Chemistry and Chemical Engineering

Series Editor
Dominic C.Y. Foo
University of Nottingham, Malaysia

For more information about this series, please visit: www.routledge.com/Green-Chemistry-and-Chemical-Engineering/book-series/CRCGRECHECHE

Efficiency and Sustainability in the Energy and Chemical Industries

Scientific Principles and Case Studies

Third Edition

Krishnan Sankaranarayanan

CRC Press
Taylor & Francis Group
Boca Raton London New York

CRC Press is an imprint of the
Taylor & Francis Group, an **Informa** business

Designed cover image: Shutterstock

Third edition published 2024
by CRC Press
6000 Broken Sound Parkway NW, Suite 300, Boca Raton, FL 33487–2742

and by CRC Press
4 Park Square, Milton Park, Abingdon, Oxon, OX14 4RN

CRC Press is an imprint of Taylor & Francis Group, LLC

© 2024 Krishnan Sankaranarayanan

First edition published by CRC Press 2004
Second edition published by CRC Press 2010

ISBN: 978-1-032-30298-0 (hbk)
ISBN: 978-1-032-30301-7 (pbk)
ISBN: 978-1-003-30438-8 (ebk)

DOI: 10.1201/9781003304388

Typeset in Times
by Apex CoVantage, LLC

*To our children and grandchildren who may witness
the emergence of a sustainable society*

*To my father K. Sankaranarayanan (1934–2022),
who inspired us all to do our best and make a positive impact*

Contents

PART II   Thermodynamic Analysis of Processes

PART III   Case Studies

Chapter 9   Energy Conversion 103

    9.1   Introduction 103
    9.2   Global Energy Consumption 104
    9.3   Global Exergy Flows 106
    9.4   Exergy or Lost Work Analysis 108
    9.5   Electric Power Generation 108
            9.5.1   Steam Plants 109
            9.5.2   Gas Turbines 109
            9.5.3   Combined Cycle 110
            9.5.4   Nuclear Power 110
            9.5.5   Hydropower 113
            9.5.6   Wind Power 113
            9.5.7   Solar Power 113
            9.5.8   Geothermal Energy 114
    9.6   Coal Conversion Processes 114
            9.6.1   Fixed or Moving Beds 114
            9.6.2   Suspended Beds 114
            9.6.3   Fluidized Beds 115
            9.6.4   Thermodynamic Analysis of Coal Combustion 115
            9.6.5   Discussion 116
            9.6.6   Coal Gasification 118
    9.7   Thermodynamic Analysis of Gas Combustion 120
            9.7.1   Exergy In 120
            9.7.2   Air Requirements 120
            9.7.3   Exergy Out 122
            9.7.4   Efficiency 124
            9.7.5   Discussion 124
    9.8   Steam Power Plant 125
    9.9   Gas Turbines, Combined Cycles, and Cogeneration 127
            9.9.1   Gas Turbines 127
            9.9.2   Thermodynamic Analysis of Gas Turbines 128
            9.9.3   Combined Cycles, Cogeneration, and Cascading 129
            9.9.4   Example 130
    9.10   Concluding Remarks 130

    References 132

Chapter 10 Separations 133

    10.1   Introduction 133
    10.2   Propane, Propylene, and Their Separation 133
            10.2.1   Single-Column Process 134
            10.2.2   Double-Column Process 134
            10.2.3   Heat Pump Process 135
    10.3   Basics 135

Contents

Series Editor Introduction

GREEN CHEMISTRY AND CHEMICAL ENGINEERING

A BOOK SERIES BY CRC PRESS/TAYLOR & FRANCIS

Towards the late 20th century, the development of various environmentally friendly processes, techniques, and methodologies saw significant growth in the scientific community. The main driving force for such growth was due to the rising awareness of sustainable development, more stringent environmental regulation, and increasing costs of raw materials and waste treatment. After several decades of development, we now broadly term these environmentally friendly processes, techniques, and methodologies as *green/clean technologies*.

In the 21st century, the global community has experienced many extreme weather events such as prolonged drought, extreme heat, tornadoes, and wildfires. The scientific community believes that these extreme weather events are closely linked to climate change, and they are expected to increase in frequency and intensity in the future. Following the Paris Agreement and Glasgow Climate Pact, there is now an international commitment to limit the rise of global temperature to well below 2°C by end of this century and to pursue efforts to limit temperature increase to 1.5°C above pre-industrial levels. Hence, it is believed that green/clean technologies will have a much bolder role to play in combating the global climate change in the coming years.

It is also worth mentioning that the United Nations Sustainable Development Goals (UNSDG) that were launched in 2015 define 17 important goals to transform the world by 2030. It is believed that some of these goals may be addressed with the development of green/clean technologies. These include:

- **Goal 6—Clean Water and Sanitation:** Ensure access to water and sanitation for all
- **Goal 7—Affordable and Clean Energy:** Ensure access to affordable, reliable, sustainable and modern energy
- **Goal 12—Responsible Consumption and Production:** Ensure sustainable consumption and production patterns
- **Goal 13—Climate Action:** Take urgent action to combat climate change and its impacts

The *Green Chemistry and Chemical Engineering* book series by CRC Press/Taylor & Francis focuses on the subset of green technologies dedicated to address the "2E" agenda, i.e., *environment* and *energy*. It involves the development of materials (e.g., catalysts, nanomaterials), methodologies (e.g., process optimization, footprint reduction, artificial intelligence) and processes (e.g., waste treatment) that will bring forth solutions to address pressing problems such as:

- Greenhouse gas management and reduction
- Sustainable water production

- Wastewater treatment and recycling
- Circular economy and waste reduction
- Renewable energy
- Sustainable use of energy resources

I am hopeful that this *Green Chemistry and Chemical Engineering* series will serve as a de facto source of reference materials and practical guides for academics and industrial practitioners looking to advance the discipline and aims of green chemistry and chemical engineering.

Dominic C. Y. Foo
Centre for Green Technologies
University of Nottingham Malaysia

Preface

Climate change, energy transition, decarbonization, recycling, or in short, sustainability, seem to be terms which are very much in vogue these days. Indeed, many companies are pledging to reducing their CO_2 emissions, using plastics as feedstock, or bio fuels, and transitioning away from fossil fuels.

The debate regarding human activity, CO_2 emissions, and the observed freak weather events, and climate change seems to be essentially over, and the public is demanding action to ensure a more sustainable society that has high quality of life and is in better equilibrium with the environment.

There have been high profile summits and treaties, such as the 1997 Kyoto Protocol, 2015 Paris Agreement, and United Nations Climate Change Conference (COP26) at Glasgow, Scotland, United Kingdom.

However, the focus on energy transition is hardly new. For some of us, the energy crisis of the 1970s and 1980s still rings loud in our memories. The crisis was of political origin, and not of real shortage, but highlighted the global connectivities of the energy markets, and elicited a response in many countries to focus on increasing energy efficiency at home and in industry, and by reducing the dependency on liquid fossil fuels from the Middle East, which were the dominant producers in those days.

In retrospect, it sparked a massive shift to coal as an alternative energy source. Massive research and development programs were initiated to make available clean and efficient coal utilization and more easily handled materials as gaseous and liquid conversion products. Large integrated multinationals played an important role in these initiatives as they considered energy, not oil, as their ultimate business. In retrospect, the geopolitics were the key driver, with energy security as the primary motive.

However, it is unfair to state that this was the sole driver, as there was growing concern about the environment. With the industrial society proceeding at full speed with mass production and consumption, and the uplift from poverty of large portions of the world population by industrialization, the world became fully aware that this was accompanied by mass emission of waste. Air pollution, water pollution, deterioration of the soil, and so forth became topics that started worrying us immensely.

The irreversibility of our domestic and industrial activities seemed to ask a price for remediation that could become too high, if not for the present generations, then for the later ones. This insight developed a sense of responsibility that went beyond political, national, or other specific interests and seemed to be shared by all aware world citizens. Environmental legislation started becoming more and more important, and partly triggered by the activities of the Club of Rome, computer simulations showed the possible limits to growth for a growing world population with limited resources.

The 1987 Brundtland Report [1] emphasized our responsibility for future generations, and the need for "sustainable development." This showed the emergence of critical trilemma—with economic growth, need for resources, and care for the environment in a delicate balance. The sun as a renewable source of energy became more and more prominent, as exemplified in Japan's massive New Sunshine Program, and the emergence of "green chemistry," a development to fulfill our needs for chemicals in a sustainable way.

The desire to write this book was born more than two decades ago by Professors Jakob de Swaan Arons and Hedzer van der Kooi, who co-authored the first two editions of this book. The mere suggestion of the "sustainability," however, was met with ridicule in those days. However, at the request of Dutch industries, Jakob and Hedzer started their studies on efficiency in the mid-'90s, and sustainability around the same period, once a large multinational approached them with a question on how sustainable they were.

During a sabbatical in Japan in 1997, Jakob started writing the first draft, following a wish by the late Professor Kazuo Kojima of Nihon University to produce a book with him on exergy, a dominant thermodynamic property in the efficiency studies. Unfortunately Professor Kojima had to withdraw because of his declining health. But Jakob received great support from Hedzer and me, and the first edition of our book was the result and was published in New York seven years later, in 2004. The second edition with Krishnan as first author appeared in 2010.

Transitions, and the success of any change, is the result of the confluence of several factors. Some of the important ones are stakeholders (do they want it?), economics (is it cheaper?), feasibility (is it possible?), and playing field (legislation). It seems the societal drive is there, industry is mobilizing itself, but it is important that the other factors are satisfied as well.

This edition will add economics as a key enabler to look at technology choices and whether the share holder returns will be there. It is not simply sufficient that a technology is better, has lower CO_2 emissions, etc., if it is too expensive. The cost, in a sense, is also affected by the playing field, and whether there is legislation that makes certain choices more attractive, and whether there is, for example, a transparent global CO_2 tax. A local CO_2 will not have the desired effect and will simply stop all investment in that locality! Public policy, through taxes, and subsidies therefore is intimately tied to the economics.

A new chapter on low carbon and ways to decarbonize will be discussed, following some of the recent trends in industry. A new chapter on plastics recycling will also be added, and will outline the various technologies that are out there, and also the challenges facing plastics recycling. Finally, the changing roles of refining and chemicals in industry and how the industry is transforming itself, with different products (plastics, chemicals and energy).

REFERENCE

1. Brundtland, G.H., *Our Common Future, The World Commission on Environmental Development*, Oxford University Press: Oxford, UK, 1987.

About the Author

Krishnan Sankaranarayanan received his MSc at Delft University of Technology, the Netherlands and his PhD at Princeton University, New Jersey. At Delft, he did an extensive study of the energy efficiency of the polyolefin industry, for which activity DSM acted as host. After 12 years at ExxonMobil he joined SABIC, where he worked on a variety of technology developments, circular economy, and energy transition. He is currently leading Digital Transformation for the technology organization and is based in Houston, Texas. He also obtained an MBA at GWU in 2012.

Part I

Basics

Learn the fundamentals of the game and stick to them.

—Jack Nicklaus, golf legend

In Chapter 1, we look at some of the history leading to this book. In Chapter 2, we pay a renewed visit to thermodynamics. We review its essentials and the common structure of its applications. In Chapter 3, we focus on so-called energy consumption and identify the concepts of work available and work lost. The last concept can be related to entropy production, which is the subject of Chapter 4. This chapter shows how some of the findings of nonequilibrium thermodynamics are invaluable for process analysis. Chapter 5 is devoted to finite-time finite-size thermodynamics, the application of which allows us to establish optimal conditions for operating a process with minimum losses in available work.

DOI: 10.1201/9781003304388-1

Part I

Basics

Learn the fundamentals of sound and tuning...

In this chapter, we...

1 Introduction

Some years ago, we were teaching an advanced course in thermodynamics to process engineers of a multinational industry. Subjects included phase equilibria, the thermodynamics of mixtures, and models from molecular thermodynamics applied to industrial situations. Some participants raised the question whether some time could be spent on the subject of "the exergy analysis of processes." At that time this was a subject with which we were less familiar because energy-related issues fell less within the scope of our activities. We fell back on a small monograph by Seader [1] and the excellent textbook by Smith et al. [2], who dedicated the last chapter of their book not so much to exergy but to the *thermodynamic* analysis of processes. Concepts such as ideal work, entropy production, and lost work were clearly related to the efficient use of energy in industrial processes. The two industrial examples given—one on the liquefaction of natural gas, the other on the generation of electricity in a natural gas-fired power station—lent themselves very well not only for illustrative purposes but also for applying the exergy concept and exergy flow diagrams [3,4]. The latter concepts appealed to us because of their instrumental and visual power in illustrating the fate of energy in the processes (Figure 1.1).

After this experience in industry, we started to include the subject in advanced courses to our own chemical engineering students at Delft University of Technology. A colleague had pointed out to us that the design of a process is more valuable if the process has also been analyzed for its energy efficiency. For mechanical engineers, who were traditionally more engaged in energy conversion processes, this was obvious; for chemical engineers, until then more concerned with chemical conversion processes, this was relatively new. The subject grew in popularity with our students because it became more and more obvious that the state of the environment and energy consumption are closely related and that excessive energy consumption appeared to be one of the most important factors in affecting the quality of our environment.

In performing such an analysis, either for industry, or out of our own curiosity, we became more and more aware of the very important role that the second law is playing in our daily lives and how the thermodynamics of irreversible processes, until then for us a beautiful science but without significant applications, appeared to have a high "engineering" content. Atkins' statement that the second law is the driving force behind all change [5] had a lasting impact on us, as much as Goodstein's suggestion [6] that the second law may well turn out to be the central scientific truth of the twenty-first century. We discovered the importance of the relation between results from classical, engineering, and irreversible thermodynamics as we have tried to make visible in what we like to call "the magic triangle" behind the second law (Figure 1.2).

Later, when we were struck by the observation that complex industrial schemes and life processes or living systems have much in common, our attention was again

DOI: 10.1201/9781003304388-2   3

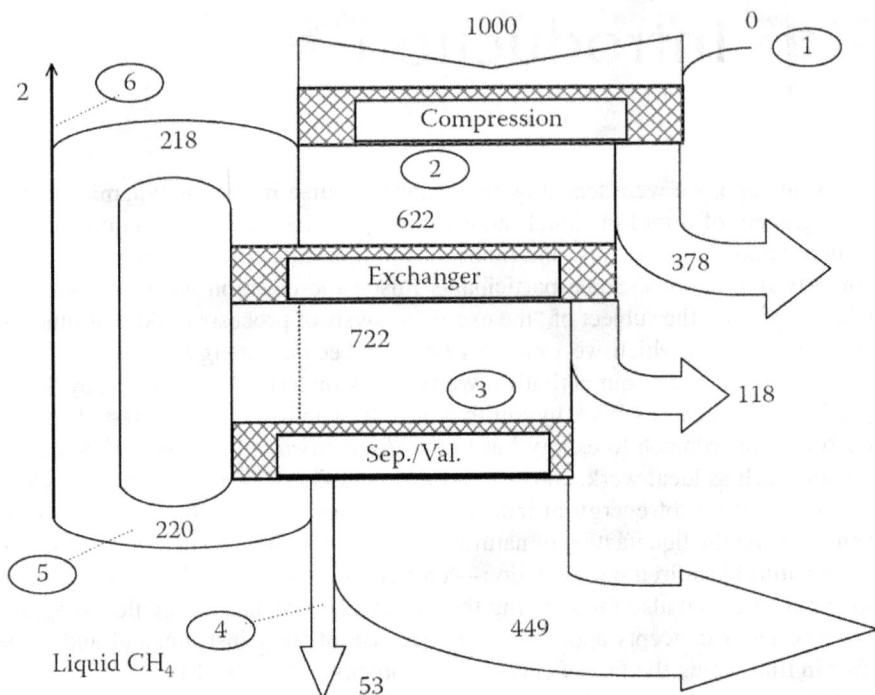

FIGURE 1.1 Grassmann diagram for the Linde liquefaction process of methane. One thousand exergy units of compression energy result in 53 exergy units of liquid methane. The thermodynamic efficiency of this process is 5.3%. The arrowed curves, bent to the right, show the losses in the various process steps.

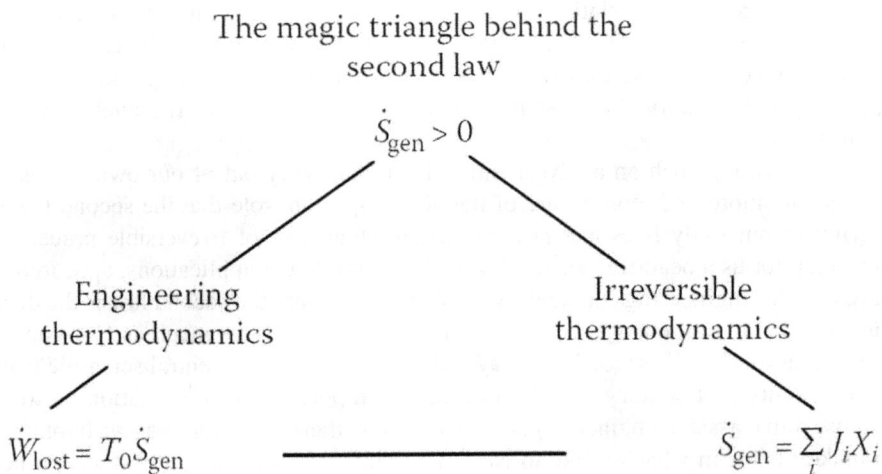

FIGURE 1.2 The magic triangle behind the second law. The relation between results from classical, engineering, and irreversible thermodynamics (see Chapter 4).

attracted to the meaning of the second law and the role of entropy production. This led us to the topic of energy flow in biology and the invaluable monographs by Schrodinger [7] and Morowitz [8]. This part of our education had come in a timely fashion, as became apparent when, more and more often, the words "sustainability" and "sustainable development" were brought in relation with the efficient use of energy. We were forced to see our analysis in the light of these concepts and to make efforts to extend our analysis to indicate, in quantitative terms, the extent to which processes or products are not only efficient but also contribute to sustainability. Once again we were stimulated by ideas and questions from colleagues within multinational industries. All these elements and influences can be found in this book and its structure.

Part I of this book, "Basics" (Chapters 2 through 5), reviews the main results of classical thermodynamics and identifies the important concepts of ideal work, lost work, and entropy generation, from using and combining the first and second laws for flowing systems. Having identified these concepts, we further interpret them in everyday technical terms by using the main results of irreversible thermodynamics. After reviewing possible ways to minimize the work lost, we conclude this part by giving attention to the thermodynamic cost of performing a process in finite time and space.

Part II, "Thermodynamic Analysis of Processes" (Chapters 6 through 8), discusses the thermodynamic efficiency of a process and how efficiency can be established and interpreted. A very useful thermodynamic property, called *exergy* or available work, is identified that makes it relatively easy to perform and integrate the environment into such an analysis. Some simple examples are given to illustrate the concept and its application in the thermodynamic or exergy analysis of chemical and nonchemical processes.

Part III, "Case Studies" (Chapters 9 through 12), takes these illustrations a bit further, namely, by demonstrating the analysis for some of the most important processes in industry: energy conversion, separations, and chemical conversion. Chapter 12 briefly discusses the concept of life cycle analysis, which aims to compare the consolidated inputs and outputs of a process or a product "from the cradle to the grave" [9], and its extension to include the minimization of process irreversibilities in a so-called exergetic life cycle analysis [10].

Part IV, "Sustainability" (Chapters 13 through 22), deals with the topics of sustainable development, efficiency, and sustainability in the chemical process industry and a very topical topic, carbon dioxide (CO_2). The sense and nonsense of green chemistry and biofuels is expounded in this part as well, followed by solar energy conversion and musings on hydrogen in the final chapter of this part.

Chapter 23 contains the author's thoughts on what the future may hold.

REFERENCES

1. Seader, J.D. *Thermodynamic Efficiency of Chemical Processes, Industrial Energy-Conservation Manual 1*, Gyftopoulos, E.P. (ed.), MIT Press: Cambridge, MA, 1982.
2. Smith, J.M.; Van Ness, H.C.; Abbott, M.M. *Introduction to Chemical Engineering Thermodynamics*, 4th edn., The McGraw-Hill Companies Inc.: New York, 1987.

3. Sussman, M.V. *Availability (Exergy) Analysis, A Self Instruction Manual*, Mulliken House: Lexington, MA, 1985.
4. de Swaan Arons, J.; van der Kooi, H.J. Exergy analysis, adding insight and precision to experience and intuition. In *Precision Process Technology. Perspectives for Pollution Prevention*, Weijnen, M.P.C.; Drinkenburg, A.A.H. (eds.), Kluwer Academic Publishers: Dordrecht, 1993, pp. 83–113.
5. Atkins, P.W. Educating chemists for the next millennium. *ChemTech* 1992, 390–392.
6. Goodstein, D. Chance and necessity. *Nature* 1994, *368*, 598.
7. Schrodinger, E. *What Is Life?* Cambridge University Press: Cambridge, 1980.
8. Morowitz, H.J. Energy flow in biology. In *Biological Organisation as a Problem in Thermal Physics*, OxBow Press: Woodbridge, CT, 1979.
9. Ayres, R.U.; Ayres, L.W.; Martinas, K. Exergy, waste accounting, and life-cycle analysis. *Energy* 1998, *23*, 355–363.
10. Cornelissen, R.L. *Thermodynamics and Sustainable Development*. PhD thesis, Twente University, Enschede, 1997.

2 Thermodynamics Revisited

In this chapter, we briefly review the essentials of thermodynamics and its principal applications. We cover the first and second laws and discuss the most important thermodynamic properties and their dependence on pressure, temperature, and composition, being the main process variables. Change in composition can be brought about with or without the transformation of phases or chemical species. The common structure of the solution of a thermodynamic problem is discussed.

2.1 THE SYSTEM AND ITS ENVIRONMENT

In thermodynamics, we distinguish between the system and its environment. The system is that part of the whole that takes our special interest and that we wish to study. This may be the contents of a reactor or a separation column or a certain amount of mass in a closed vessel. We define what is included in the system. The space outside the chosen system or, more often, a relevant selected part of it with defined properties, is defined as the environment.

Next, we distinguish between closed, open, and isolated systems. All are defined in relation to the flow of energy and mass between the system and its environment. A closed system does not exchange matter with its environment, but the exchange of energy (e.g., heat or work) is allowed. Open systems may exchange both energy and matter, but an isolated system exchanges neither energy nor mass with its environment.

2.2 STATES AND STATE PROPERTIES

The system of our choice will usually prevail in a certain macroscopic state, which is not under the influence of external forces. In equilibrium, the state can be characterized by state properties such as pressure (P) and temperature (T), which are called "intensive properties." Equally, the state can be characterized by extensive properties such as volume (V), internal energy (U), enthalpy (H), entropy (S), Gibbs energy (G), and Helmholtz energy (A). These properties are called "extensive" because they relate to the amount of mass considered; once related to a unit amount of mass, they also become intensive properties.

The equilibrium state does not change with time, but it may change with location as in a flowing system where P, T, and other state properties can gradually change with position. Then we speak of a steady state. If the state temporarily changes with time, as in the startup of a plant, we call it a "transient" state.

If an isolated system is in a nonequilibrium state, its properties will usually differ from its equilibrium properties and it will not be stable. If such a system can absorb

DOI: 10.1201/9781003304388-3

local fluctuations, it is in a metastable state; otherwise, the system and state are called unstable (Figure 4.2).

2.3   PROCESSES AND THEIR CONDITIONS

Often our system of interest is engaged in a process. If such a process takes place at a constant temperature, we speak of an isothermal process. Equally, the process can be defined as isobaric, isochoric, isentropic, or isenthalpic if pressure, volume, entropy, or enthalpy, respectively, remains unchanged during the process. A process is called "adiabatic" if no heat exchange takes place between the system and its environment. Finally, a process is called "reversible" if the frictional forces, which have to be overcome, tend to zero. The unrealistic feature of this process is that energy and material flows can take place in the limit of driving forces going to zero; for example, in an isothermal process, heat can be transferred without a temperature difference within the system or between the system and its environment. In a real process, frictional forces have to be overcome, requiring finite "driving forces" as ΔP, ΔT, ΔG, or when driving forces are already present in the system, this leads to processes where spontaneously is given in to such forces in a spontaneous expansion, mixing process, or reaction. Such processes are called "irreversible" processes and are a fact of real process life. As we will see later, the theory of irreversible thermodynamics identifies the so-called thermodynamic forces, for example, $\Delta(1/T)$ instead of ΔT, and the associated flow rate—in this instance, the heat flow rate Q.

2.4   THE FIRST LAW

Thermodynamics is solidly founded on its main laws. The first law is the law of conservation of energy. For a closed system that receives heat from the environment, Q_{in}, and performs work on the environment, W_{out}, we can write

$$\Delta U = Q_{in} - W_{out} \qquad (2.1)$$

Heat and work are forms of energy in *transfer* between the system and the environment. If more heat is introduced into the system than the system performs work on the environment, the difference is *stored* as an addition to the internal energy U of the system, a property of its state. In a more abstract way, the first law is said to define the fundamental thermodynamic state property, U, the internal energy.

Equation 2.1, in differential form, can be written as

$$dU = \delta Q_{in} - \delta W_{out} \qquad (2.2)$$

The δ-character is used to indicate small amounts of Q and W because heat and work are not state properties and depend on how the process takes place between two different states.

If the process is reversible and the sole form of work that the system can exert on its environment is that of volume expansion, then $dW_{out}^{rev} = PdV$. If, in addition, the process is isobaric, $PdV = d(PV)$ and

$$dQ_{in}^{rev} = d(U + PV) = dH \tag{2.3}$$

The enthalpy H is defined as $H = U + PV$ and is a property of state derived from the fundamental property U. If heat is stored reversibly and isobarically in a system, it is stored as an increase in the system's H-value. H has been defined for our convenience; it has no fundamental meaning other than that, under certain conditions, its change is related to the heat absorbed by the system. It can be shown that the specific heat at constant volume and pressure, c_v and c_p, respectively, can be expressed as

$$c_v = \left(\frac{\partial U}{\partial T}\right)_V \text{ and } {}^\circ c_p = \left(\frac{\partial H}{\partial T}\right)_P \tag{2.4}$$

For process engineering and design, it is important to know how enthalpy is a function of pressure, temperature, and composition. The last variable is discussed later. It can be found in any standard textbook that the differential of H can be expressed as a function of the differential of T and P as follows:

$$dH = c_p dT + \left[V - T\left(\frac{\partial V}{\partial T}\right)_P\right]dP \tag{2.5}$$

The first term depends on what is sometimes called the caloric equation of state, describing how *intra* molecular properties, the properties within the molecules, are a function of the state variables. The expression in brackets requires the mechanical equation of state, which expresses the dependency of a property, for example, V on the *inter*molecular interactions, the interactions between molecules. Process simulation models usually contain information and models for both types of equation of state.

Most often, we are not interested in the "absolute" value of H, but rather in its change, $\Delta_{tr}H$. The subscript "tr" refers to the nature of the change. If the change involves temperature and/or pressure for a one-phase system only, no subscript is used for Δ. But in case of a phase transition, of mixing, or of a chemical reaction, the subscript is used and may read "vap" for vaporization, "mix" for mixing, or "r" for reaction, and so forth.

When a superscript is used as in $\Delta_{tr}H^\circ$, it indicates that the change in H is considered for a transition under *standard* pressure, which usually is chosen as 1 bar. In the case of chemical reactions, the superscript $^\circ$ refers to standard pressure and to reactants and products in their pure state or otherwise defined standard states such as infinitely dilute solutions.

Finally, we present the first law for open systems as in the case of streams flowing through a fixed control volume at rest [1]

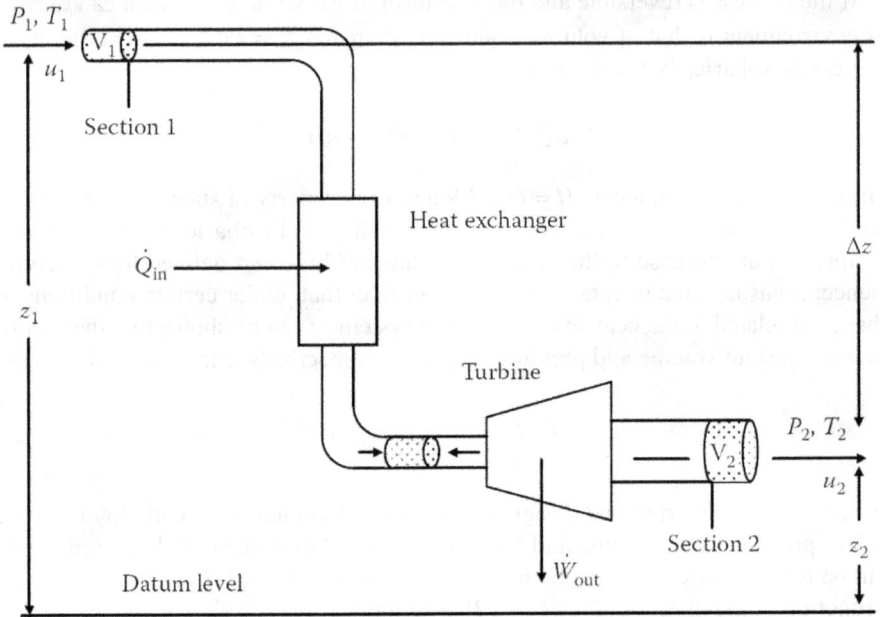

FIGURE 2.1 Changes in steady-state flow.

$$\left(\frac{dU}{dt}\right)_{cv} = \sum_{in}\dot{m}_i\left(h_i + \frac{u_i^2}{2} + gz_i\right)$$

$$-\sum_{out}\dot{m}_j\left(h_j + \frac{u_j^2}{2} + gz_j\right)$$

$$+\sum\dot{Q}_{in} - \sum\dot{Q}_{out} + \sum\dot{W}_{sh,in} - \sum\dot{W}_{sh,out}$$

(2.6)

For one ingoing and outgoing stream in the steady state (Figure 2.1), this equation simplifies into

$$\dot{m}\left(\Delta H + \frac{\Delta u^2}{2} + g\Delta z\right) = \dot{Q}_{in} - \dot{W}_{out}$$

(2.7)

where

$\dot{m}$ refers to the mass flow rate considered
u is the velocity of the flowing system
z is its height in the gravitational field

2.5 THE SECOND LAW AND BOLTZMANN

The second law is associated with the direction of a process. It defines the fundamental property entropy, S, and states that in any real process the direction of the process corresponds to the direction in which the total entropy increases, that is, the entropy change of both the system and environment should in total result in a positive result or in equation form

$$\Delta S + \Delta S_{environment} = S_{generated} \tag{2.8}$$

$$S_{generated} > 0 \tag{2.9}$$

In other words, every process generates entropy. The best interpretation, in our opinion, of this important law is given by adopting a postulate by Boltzmann:

$$S = k \ln \Omega \tag{2.10}$$

This equation expresses the relation between entropy and the thermodynamic probability Ω, where k is Boltzmann's constant. If in an isolated vessel, filled with gas, at $t = 0$ half of the molecules are nitrogen, the other half are oxygen, and all nitrogen molecules fill the left half of the vessel whereas all oxygen molecules fill the right half of the vessel, then this makes for a highly unlikely distribution, that is, one of a low thermodynamic probability Ω_0. As time passes, the system will evolve gradually into one with an even distribution of all molecules over space. This new state has comparatively a high thermodynamic probability Ω, and the generated entropy is given by standard textbooks give ample examples of how Ω can be calculated [1].

$$S_{generated} = S_{final} - S_{original} = k \ln \left(\frac{\Omega}{\Omega_0} \right) \tag{2.11}$$

Notice that the direction of the process and time have been determined: This has been called the *arrow of time* [2]. Time proceeds in the direction of entropy generation, that is, toward a state of greater probability for the total of the system and its environment, which, in the widest sense, makes up the universe. Finally, we wish to point out that an interesting implication of Equation 2.10 is that for substances in the perfect crystalline state at $T = 0\ K$, the thermodynamic probability $\Omega = 1$ and thus $S = 0$.

2.6   THE SECOND LAW AND CLAUSIUS

As the first law is sometimes referred to as the law that defines the fundamental thermodynamic property U, the internal energy of the system, the second law is considered to define the other fundamental property, the entropy S. Classical thermodynamics, via Clausius's thorough analysis [3] of thermodynamic cycles that extract work from available heat, has produced the relation between S and the heat added reversibly to the system at a temperature T:

$$dS = \frac{\delta Q_{in}^{rev}}{T} \tag{2.12}$$

This relation plays an important role in the derivation of the universal and fundamental thermodynamic relation

$$dU = TdS - PdV \tag{2.13}$$

which is often called the Gibbs relation. This relation is instrumental in connecting the changes between the most important thermodynamic state properties.

From this relation the differential dS can be expressed as the following function of the differentials of pressure and temperature:

$$dS = \frac{C_P}{T} dT - \left(\frac{\partial V}{\partial T} \right)_P dP \tag{2.14}$$

P, T, and also composition are the state variables most often used to characterize the state of the system, as they can be easily measured and controlled. As we show in Part II, Equations 2.5 and 2.14 are important to perform the thermodynamic analysis of a process $\Delta_r S_{298}^0$, which expresses the change in entropy of a reaction at 298 K and at standard pressure. The reaction is defined to take place between compounds in their standard state, that is, in the most stable aggregation state under standard conditions, like liquid water for water at 298 K and 1 bar. Analogous to Equation 2.6, for the first law for open systems, the second law reads and simplifies to Equation 2.8 for single flows in and out the control volume in the steady state.

$$\left(\frac{dS}{dt} \right)_{cv} = \sum_{in} \dot{m}_i S_i - \sum_{out} \dot{m}_j S_j$$
$$+ \sum_{in} \int \frac{\delta \dot{Q}_k}{T} - \sum_{out} \int \frac{\delta \dot{Q}_l}{T} + \dot{S}_{generated} \tag{2.15}$$

Finally, equilibrium processes can be defined as processes between and passing states that all have the same thermodynamic probability. On the one hand, these processes proceed without driving forces; on the other hand, and this is inconsistent and unrealistic, there is no incentive for the process to proceed. These imaginary processes function only to establish the minimum amount of work required, or the maximum amount of work available, in proceeding from one state to the other.

2.7 CHANGE IN COMPOSITION

So far our discussion of thermodynamic concepts has referred to systems that did not seem to change with composition, only with pressure P, temperature T, or the state of aggregation. Thermodynamics, however, is much more general than being limited to these conditions, fortunately, for changes in composition are the rule rather than the exception in engineering situations. Pure homogeneous phases may mix, and a homogeneous mixture may split into two phases. A homogeneous or heterogeneous mixture may spontaneously react to one or more products. In all these cases changes in composition will take place. This part of thermodynamics is usually referred to as "chemical" thermodynamics, and its spiritual father is Josiah Willard Gibbs [4]. It has been the merit of Lewis [5] to "decipher" Gibbs' achievements and to translate these into readily applicable practical and comprehensible concepts such as the Gibbs energy of a substance i, or the individual thermodynamic potential μ_i fugacity f_i, or activity a_i concepts now widely used in process design. The thermodynamic or chemical potential can be considered to be the decisive property for an individual molecular species transport or chemical behavior. It has been one of the main achievements of Prausnitz et al. [6] to be instrumental in quantifying thermodynamic properties for ready application by taking into account the molecular characteristics and properties of those molecules making up the mixture in a particular thermodynamic state. This branch of thermodynamics is often referred to as "molecular thermodynamics," and many consider Prausnitz as its most prominent founding father.

If a mixing process or chemical transformation is brought about, spontaneously or by applying work on the system, the process will take place with entropy generation:

$$S_{generated} > 0 \tag{2.9}$$

and the total entropy will tend to a maximum value that will be reached for the equilibrium state.

Indeed, for different molecules, which otherwise are nearly the same, such as isomers, or molecules of about the same size, polarity, or other properties, the thermodynamic probability of the mixed state at the same P and T is much larger than that of the respective pure states (in molar units):

$$\Delta_{mix}S = S_{generated} = R \ln \frac{\Omega_{mixed}}{\Omega_{separated}} \tag{2.16}$$

with $\Omega_{mixed} \ggg \Omega_{separated}$.

If the change is in composition only, at constant P and T, and confined to the system we wish to consider, for instance, in a mixer, separation column, or a reactor, then a system property G, the Gibbs energy, can be identified and has been defined as follows:

$$G \equiv U - TS + PV \tag{2.17}$$

It has the property that in time (t) and at constant pressure and temperature it tends to a minimum value that will be reached when the system has reached equilibrium (Figure 2.2);

$$G_{P,T} \to min \tag{2.18}$$

or

$$\frac{dG_{P,T}}{dt} \leq 0 \tag{2.19}$$

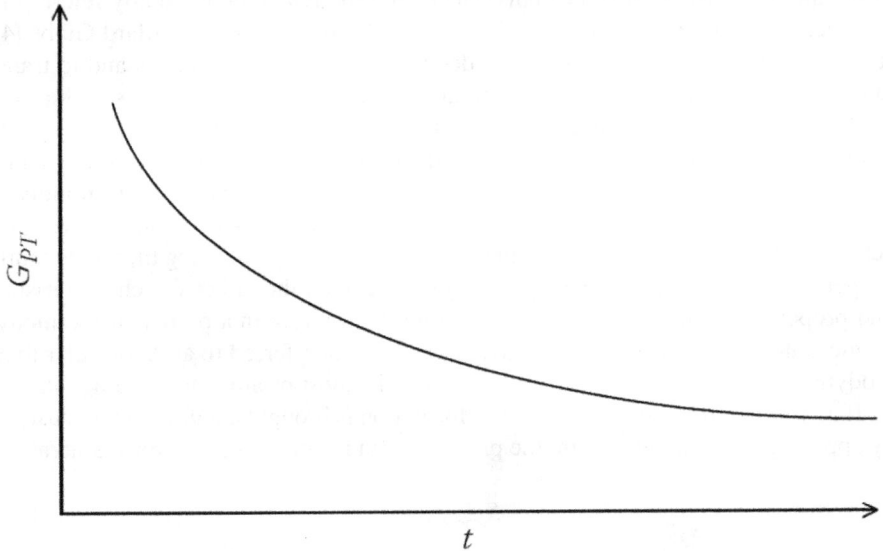

FIGURE 2.2   The Gibbs energy G on approaching equilibrium.

If we consider a chemical reaction that takes place in a homogeneous mixture, the equilibrium composition of the mixture can be found from the equality condition in Equation 2.19 if the dependence of G on the composition is known.

If the process of mixing takes place with negligible change of the internal energy U and volume V, we speak of ideal mixing and it can be shown then that for 1 mol of mixture

$$\Delta_{mix} S^{ideal} = -R \sum x_i \ln x_i \qquad (2.20)$$

in which

  x_i is the molar fraction of constituent i in the mixture
  R is the universal gas constant

For an ideal mixture, $\Delta_{mix} U^{ideal}$ and $\Delta_{mix} V^{ideal}$ are zero for mixing at constant P and T, and so $\Delta_{mix} G^{ideal}$ is given by

$$\Delta_{mix} G^{ideal} = RT \sum x_i \ln x_i \qquad (2.21)$$

Notice that $\Delta_{mix} H^{ideal}$ is also zero and thus, with Equation 2.3 in mind, ideal mixing at constant P and T will take place without heat effects.

For deviations from ideal mixing, the excess property M^E is defined as

$$M^E = \Delta_{mix} M - \Delta_{mix} M^{ideal} \qquad (2.22)$$

An important excess property is the excess Gibbs energy G^E. Many models have been developed to describe and predict G^E from the properties of the molecules in the mixture and their mutual interactions. G^E models often refer to the condensed state, the solid and liquid phases. In case significant changes in the volume take place upon mixing, or separation, the Helmholtz energy A, defined as

$$A \equiv U - TS \tag{2.23}$$

and its excess property A^E are the preferred choices for describing the process. This requires an equation of state that expresses the volumetric behavior of the mixture as a function of pressure, temperature, and composition. For the models most applied in practice for G^E and A^E, the reader is referred to Ref. [7].

Partial molar properties take a special place in the thermodynamics of mixtures and phase equilibria. They are defined as

$$\bar{M}_i \equiv \left[\frac{\partial(nM)}{\partial n_i} \right]_{P,T,n_j \neq i} \tag{2.24}$$

The best-known example is the partial molar Gibbs energy, better known as the earlier-mentioned thermodynamic potential μ. The thermodynamic potential of component i in a homogeneous mixture is

$$\mu_i \equiv \bar{G}_i = \left[\frac{\partial(nG)}{\partial n_i} \right]_{P,T,n_j \neq i} \tag{2.25}$$

An important condition for phase equilibria is

$$\mu_i' = \mu_i'' = \mu_i''' = \cdots \tag{2.26}$$

or in terms of the fugacity in mixtures

$$\hat{f}_i' = \hat{f}_i'' = \cdots \tag{2.27}$$

in which equations, the primes indicate the respective phases. Fugacity and activity (see the following) are directly related to the thermodynamic potential. The latter property has the dimension of Joules per mole, whereas f_i has the dimension of pressure and a_i is dimensionless.

The last equation, applied to a vapor-liquid equilibrium, reads

$$\hat{\phi}_i y_i P = \gamma_i x_i P_i^{\text{sat}} \phi_i^{\text{sat}} \exp \left| \frac{V_i^\ell \left(P - P_i^{\text{sat}} \right)}{AT} \right| \tag{2.28}$$

which simplifies to Raoult's law for ideal gas behavior, for which the fugacity coefficients $\hat{\phi}_i$, and $\hat{\phi}_i^{\text{sat}}$ are equal to 1, and ideal mixing in the liquid state, for which the activity coefficient γ_i equals 1, and the Poynting factor (the exponential in Equation 2.28), with V_i^ℓ the liquid-phase molar volume is approximately unity:

$$y_i P = x_i P_i^{\text{sat}} \tag{2.29}$$

The fugacity coefficient $\hat{\phi}_i$, can be calculated from a valid equation of state; the activity coefficient γ_i can be derived from an applicable G^E expression. The activity a_i is the product of γ_i and x_i.

The property known as the Gibbs energy G, and defined by Equation 2.17, plays an important role in describing on the one hand the transformation between phases where species stay the same but distribute differently over the phases present, such as vapor and liquid, and on the other hand in transformations where species change identity, the chemical reaction. Chemical reactions and phase transformations both proceed in the directions that fulfill Equations 2.18 and 2.19.

For a chemical reaction

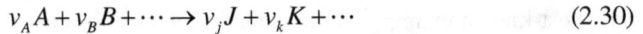

$$v_A A + v_B B + \cdots \rightarrow v_j J + v_k K + \cdots \tag{2.30}$$

in which v_i is the stoichiometric coefficient of species i, defined as positive for a product and negative for a reactant, it can be shown that progress of the reaction can be characterized by a reaction property, the so-called degree of advancement of reaction, which is defined as [1]

$$d\xi = \frac{dn_i}{v_i} \tag{2.31}$$

ξ and the chemical reaction velocity v_{chem} are related by

$$v_{\text{chem}} = \frac{d\xi}{dt} \tag{2.32}$$

Equilibrium is reached for (Figure 2.3)

$$\frac{dG}{d\xi} = 0 \tag{2.33}$$

and

$$v_{\text{chem}} = 0 \tag{2.34}$$

If G is known as function of composition, the position of the chemical equilibrium can be determined with the help of Equation 2.33, which is instrumental in finding

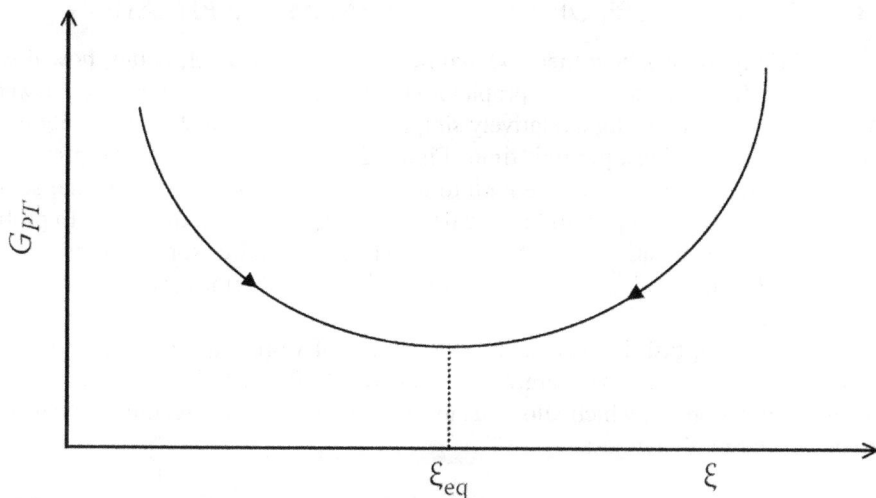

FIGURE 2.3   The Gibbs energy G and the degree of advancement of reaction ξ.

the equilibrium composition. It can be shown that chemical equilibrium is character-ized by the equation

$$\sum v_i \mu_i = 0 \tag{2.35}$$

From this, the chemical equilibrium constant at temperature T

$$K_T = \Pi \left(\frac{\hat{f}_i}{f_i^0} \right)^{v_i} \tag{2.36}$$

can be identified as

$$RT \ln K_T = -\Delta_r G_T^0 \tag{2.37}$$

The dependency on T is given by

$$\frac{d \ln K_T}{d(1/T)} = -\frac{\Delta_r H_T^0}{R} \tag{2.38}$$

Knowledge of these changes in standard Gibbs energy and enthalpy allows one to calculate the equilibrium composition and its variation with temperature.

2.8 THE STRUCTURE OF A THERMODYNAMIC APPLICATION

We now briefly discuss how thermodynamics can work for us or, better, how thermodynamics functions to solve a problem where it can help to provide the answer. We wish to illustrate this for a relatively simple problem: how much work is required to compress a unit of gas per unit time (Figure 2.4) from a low to a high pressure. Figure 2.5 schematically gives the path to the answer and the structure of the solution. In fact, the same steps will have to be taken to apply thermodynamics to problems such as the calculation of the heat released from or required for a process, of the position of the chemical or phase equilibrium, or of the thermodynamic efficiency of a process.

In our case, the task is to calculate the amount of work required to compress a gas at pressure P_1 and temperature T_1 to a pressure P_2. We turn to the first law in the version of Equation 2.7, which allows us to translate the original, technical, question into one of thermodynamics:

$$\dot{W}_{in} = \Delta H + \dot{Q}_{out} \tag{2.39}$$

We assume that the compression is adiabatic: it will take place without exchange of heat with the environment, $Q_{out} = 0$. So the first law tells us that W_{in} is known if we know the change in enthalpy of the gas. For this we need to know how the gas enthalpy is a function of pressure and temperature.

Assuming, for simplicity, that the gas behaves like an ideal gas, for which the enthalpy is not a function of pressure, we end up with the relation

$$\Delta H = c_p \left(T_2 - T_1 \right) \tag{2.40}$$

P_1, T_1, H_1

W_{in}

P_2, T_2, H_2

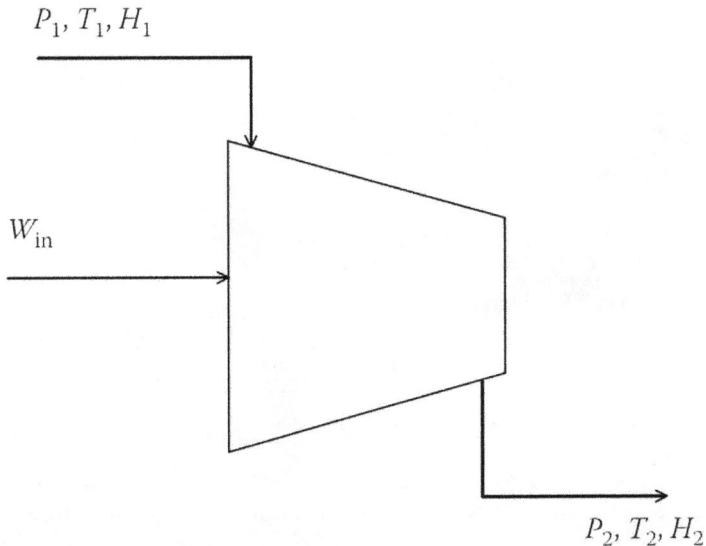

FIGURE 2.4   The compressor.

Required steps

Task $\longrightarrow$ 1. Translation

2. Fundamental relations

3. Models

4. Handleable relations

5. Retrieval, estimation basic data
   and parameters

6. Computation $\longrightarrow$ Completion

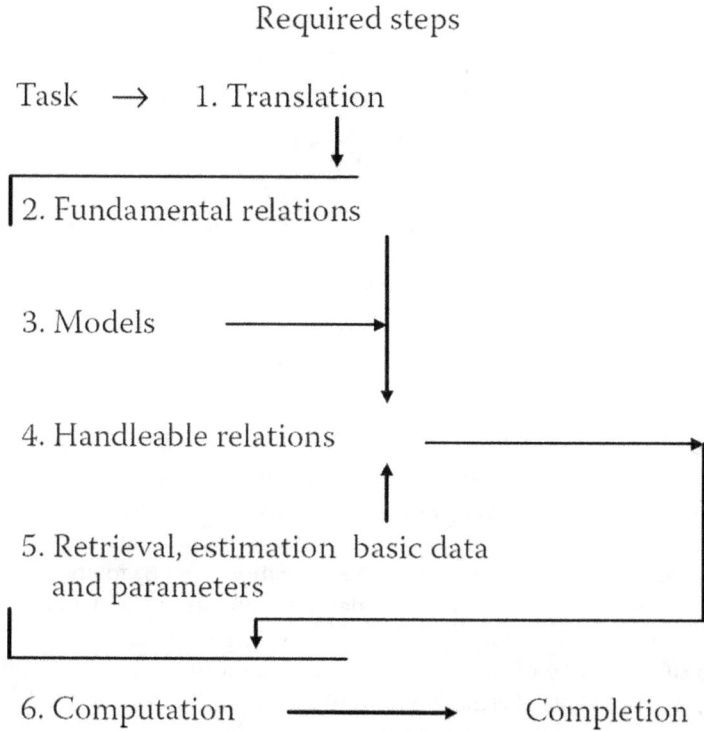

FIGURE 2.5 The structure of a thermodynamic application.

by which we have also assumed that the specific heat at constant pressure, c_p, is not a function of temperature. Unfortunately, we do not know T_2, and for that we turn to the second law. As the process takes place adiabatically, we write Equation 2.8 in the version

$$\Delta S \geq 0 \qquad (2.41)$$

First, we assume that the process takes place reversibly, and thus

$$\Delta_{rev} S = 0 \qquad (2.42)$$

With the assumptions that the gas behaves as an ideal gas and c_p is not a function of temperature, we may write

$$\Delta_{rev} S = c_p \ln \frac{T_2^{rev}}{T_1} - R \ln \frac{P_2}{P_1} \qquad (2.43)$$

These equations allow us to calculate T_2^{rev}, $\Delta_{rev} H$, and thus W_{in}^{rev}, the work required if the compression had taken place reversibly. But in the real process $\Delta S > 0$, and according to Equation 2.43, T_2 must be larger than T_2^{rev} and $\Delta H > \Delta_{rev} H$. Thus, $W_{in} > W_{in}^{rev}$. Usually, this is expressed in the compressor's efficiency

$$\eta \equiv \frac{W_{in}^{rev}}{W_{in}} < 1 \qquad (2.44)$$

When this efficiency is known, and for a specific compressor it usually is, W_{in} can be calculated and with this T_2. If we now turn again to Figure 2.5, the fundamental laws (step 1) were instrumental in translating the task, whereas the fundamental relations (step 2) that express how H and S are functions of P and T provide us with equations to proceed to the answer. Usually, the gas does not behave as an ideal gas and we need *models* (step 3) for what we earlier called the mechanical equation of state and the caloric equation of state. This will lead us to what we call *handleable*, as opposed to abstract, equations. The abstract equations do not allow us to calculate anything, but the "handleable" equations allow us to perform calculations if we know some numbers. Given certain basic data, from experiment or predicted, and associated parameters, such as for expressing c_p as a function of temperature (step 5), these equations allow us to perform the computation (step 6) to complete the task and come up with the answer to the original question. Steps 1 and 2 are extensively discussed in textbooks of chemical engineering thermodynamics [1], step 3 falls within the subdiscipline of molecular thermodynamics [6], and step 5 falls within that of the prediction of thermodynamic properties as a function of composition and based on the molecular structure of the mixture's constituents and their mutual interactions [7].

We conclude this example with the observation that the second law for real processes expresses that entropy generation is positive and that the implication is that the real amount of work required is larger than that calculated for the reversible or ideal(ized) process. This suggests a relation between entropy generation and excess work. This relation is of fundamental significance for the subject of this book, as we will demonstrate later.

REFERENCES

1. Smith, J.M.; van Ness, H.C.; Abbott, M.M. *Introduction to Chemical Engineering Thermodynamics*, 4th edn., McGraw-Hill: New York, 1987.
2. Blum, H.F. *Time's Arrow and Evolution*, Harper: New York, 1962.
3. Carnot, S. *Reflections on the Motive Power of Fire and Other Papers on the Second Law of Thermodynamics*, Clausius, R.; Clayperon, E. (eds.), Dover Publications: New York, 1960.
4. Gibbs, J.W. *Thermodynamische Studien*, Wilhelm Engelmann Verlag: Leipzig, 1982.
5. Lewis, G.N.; Randall, M. *Thermodynamics*, Pitzer, K.S.; Brewer, L. (eds.), McGraw-Hill: New York, 1961.
6. Prausnitz, J.M.; Gomes de Azevedo, E.; Lichtenthaler, R.N. *Molecular Thermodynamics of Fluid Phase Equilibria*, 3rd edn., Prentice Hall: Englewood Cliffs, NJ, 1999.
7. Reid, R.C.; Prausnitz, J.M.; Poling, B.E. *The Properties of Gases and Liquids*, 5th edn., McGraw-Hill: New York, 2001.

3 Energy "Consumption" and Lost Work

In this chapter, we show that it is not so much energy that is consumed but its quality, that is, the extent to which it is available for work. The quality of heat is the well-known thermal efficiency, the Carnot factor. If quality is lost, work has been consumed and lost. Lost work can be expressed in the products of flow rates and driving forces of a process. Its relation to entropy generation is established, which will allow us later to arrive at a universal relation between lost work and the driving forces in a process.

3.1 INTRODUCTION

We all know what is meant by energy consumption. Most of us pay energy bills and we accept that we are charged for consuming energy just as we are charged for consuming other things. But are we consuming energy? According to the first law of thermodynamics, energy cannot be created nor annihilated. Then what is it that we consume if it is not energy?

In a way the situation can be compared with consuming food. If we made a thorough analysis of food consumption, we would conclude that it is not its mass that we have consumed, as the mass balance is not affected. Nor is it the energy that we have consumed as a properly performed energy balance will show. This led Schrödinger [1] to his somewhat desperate question: "If it is not mass nor energy that we extract from food then what is it . . .?"

As should become clear from the following sections, it is not energy that we consume but its quality, by which is meant the extent to which energy is available for performing work. In the spontaneous combustion of natural gas, mass and energy are conserved, but the work stored and available in the chemical bonds of the gas will, to a large extent, get lost. By energy "consumption" we mean consumption of or decrease in available work. Loss of available work is called lost work, W_{lost}. As we shall show, lost work can, in thermodynamic terms, be identified as the product of the entropy generated and the absolute temperature of the environment T_0. This is expressed in the remarkable relation known as the Gouy-Stodola relation [2,3].

3.2 THE CARNOT FACTOR

Carnot allowed us to answer the following question: Which part of heat Q, available at a temperature $T > T_0$, can at most be converted into useful work? Provided the process is cyclic and conducted reversibly, the maximum amount of *work available* is given by

DOI: 10.1201/9781003304388-4

$$W_{\text{out}}^{\text{max}} = Q\left(1 - \frac{T_0}{T}\right) \qquad (3.1)$$

The factor $1 - (T_0/T)$ is often called the thermal efficiency, but we prefer to call it the Carnot factor.[1] For example, if heat is supplied at 600 K and the temperature of the environment is 300 K, the Carnot factor is 1/2. We could also say that in this instance the quality q of every Joule of heat is 1/2 J/J, if we wish to express that *at most* half of the Joule of heat supplied can be made available for useful work with respect to our environment at T_0:

$$q = \frac{W_{\text{out}}^{\text{max}}}{Q} = 1 - \frac{T_0}{T} \qquad (3.2)$$

So we define the quality of heat supplied at a temperature level $T > T_0$ as the maximum fraction available for useful work. Baehr [4] has called this part of heat the *exergy* of heat. The remaining part is unavailable for useful work and is called *anergy*. It is the minimal part of the original heat that will be transferred as heat Q_0^{min} to the environment. In Baehr's terminology, we could say that in this instance the ideal heat engine would achieve the following separation between useful and useless Joules:

$$Q = Q\left(1 - \frac{T_0}{T}\right) + Q \cdot \frac{T_0}{T}$$
$$= W_{\text{out}}^{\text{max}} + Q_0^{\text{min}} \qquad (3.3)$$
$$= \text{exergy} + \text{anergy}$$

Overall, the cycle takes up an amount of energy Q, produces an amount of work $W_{\text{out}}^{\text{me}}$, and releases an amount of heat $Q - W_{\text{out}}^{\text{max}} = Q_o^{\text{min}}$ at the temperature T_0 to the environment.

This equation very clearly expresses the quality aspects of heat and must have tempted Sussmann [5] to a statement much inspired by Orwell in his famous novel *Animal Farm* [6]: "All Joules are equal, but some Joules are more equal than others." But, in a more earnest sense, this observation can also be found in discussions on energy policy [7]: "The quality of energy (i.e., exergy)—and not only the quantity—as an objective and clear starting point, must be included in making policy choices."

Later it will become clear why this observation is important and how far-reaching the implication is of making use of the distinction between both quantity and quality of the various Joules involved in a process.

Equation 3.3 is an expression of the first law of thermodynamics for the separation in useful and useless energy of the energy from heat. Equation 3.1 is clearly not an expression of the first law but, as we shall see later, an implication of the second law. In this context, it is worth recalling Baehr's formulation of the first and second laws [4]:

1. The sum of exergy and anergy is always constant.
2. Anergy can never be converted into exergy.

This unusual and much less-known formulation of the main laws of thermodynamics serves the purpose of this book very well.

3.3 LESSONS FROM A HEAT EXCHANGER

Some important lessons from engineering thermodynamics can be learned from the thermodynamic analysis of a heat exchanger. For illustrative purposes we assume that the heat is exchanged between a condensing fluid at a temperature T_{high} and an evaporating fluid at a temperature T_{low}, with $T_{high} > T_{low} > T_0$ (Figure 3.1). The mass flow rates have been chosen such that within the exchanger all high-temperature fluid condenses and all low-temperature fluid evaporates. The heat exchanger is not supposed to exchange heat with the environment and thus operates adiabatically. For steady-state operation, if changes in kinetic and potential energy are small compared to the enthalpy changes, the energy balance can be written according to Equation 2.7 as

$$\dot{m}_{high}\Delta H_{high} + \dot{m}_{low}\Delta H_{low} = 0 \qquad (3.4)$$

The amount of heat exchanged and taken up by the evaporating fluid per unit of time, the heat flow rate $\dot{Q}$, is positive and is given by

$$\dot{Q} = \dot{m}_{low}\Delta H_{low} \qquad (3.5)$$
$$= -\dot{m}_{high}\Delta H_{high}$$

and this amount flows spontaneously from T_{high} to T_{low}. At the high temperature its *exergy*, or *available work* as Americans prefer to call it, is

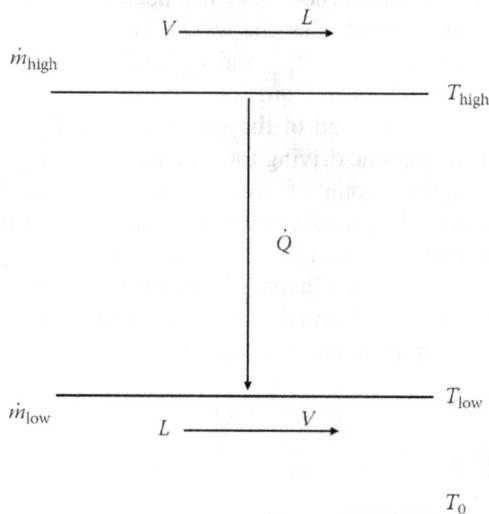

FIGURE 3.1 A heat exchanger in which the heat of condensation of a fluid at T_{high} is transferred to an evaporating fluid at T_{low}. We assume that $T_{high} > T_{low} > T_0$.

$$W_{high} = Q\left(1 - \frac{T_0}{T_{high}}\right) \tag{3.6}$$

At the low temperature its quality has decreased and the available work is now less:

$$W_{low} = Q\left(1 - \frac{T_0}{T_{low}}\right) \tag{3.7}$$

Thus, the exchange of heat has taken place with a rate of loss of available work:

$$\begin{aligned} \dot{W}_{lost} &= \dot{W}_{high} - \dot{W}_{low} \\ &= \dot{Q}\left[\left(1 - \frac{T_0}{T_{high}}\right) - \left(1 - \frac{T_0}{T_{low}}\right)\right] \\ &= \dot{Q} \cdot T_0 \left(\frac{1}{T_{low}} - \frac{1}{T_{high}}\right) \end{aligned} \tag{3.8}$$

or per unit mass instead of per unit time:

$$W_{lost} = Q \cdot T_0 \left(\frac{1}{T_{low}} - \frac{1}{T_{high}}\right) \tag{3.8a}$$

From Figure 3.2, in which the Carnot factor has been plotted against the amount of heat transferred, we can conclude that the work lost is represented by the enclosed area between the temperature levels T_{high} and T_{low} and the points of entry and exit.

The factor between brackets in Equation 3.8 can be identified as the "driving force" for heat transfer, but instead of the familiar $\Delta T = T_{high} - T_{low}$, our equation suggests that the thermodynamic driving force is $\Delta(1/T) = (1/T_{low}) - (1/T_{high})$, and thus Equation 3.8 shows that the amount of work lost is the product of T_0 and the product of (Q) and $\Delta(1/T)$, namely, the product of the flow rate and its driving force. This last product is one of the many that can be identified with the help of irreversible thermodynamics, as we will show in Chapter 4. Another interesting observation is the following. If a body or flow isothermally absorbs a positive amount of heat, Q_{in} at a temperature T, then its change in entropy is given by

$$\Delta S = \frac{Q_{in}}{T} \tag{3.9}$$

With this equation in mind, we can read the last part of Equation 3.8a as

$$W_{lost} = T_0 \left(\Delta S_{low} + \Delta S_{high}\right) = T_0 \Delta S \tag{3.10}$$

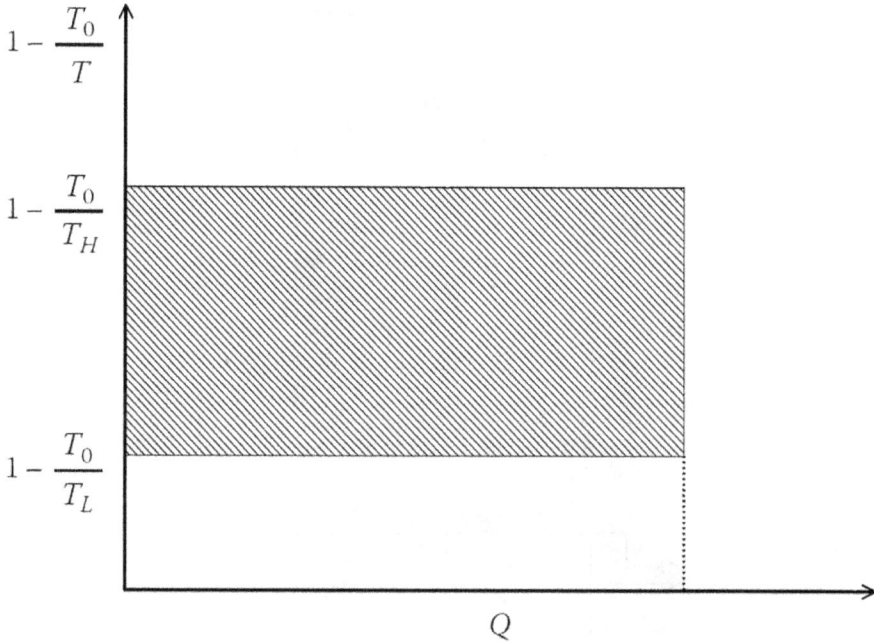

FIGURE 3.2 The enclosed area represents the amount of lost work in this plot of the Carnot factor against the heat transferred.

in which ΔS is the entropy change of the heat exchanger. As the heat exchanger operates adiabatically, there is no associated change in entropy of the environment, $\Delta S_0 = 0$, and thus the second law according to Equation 2.8 reads

$$S_{generated} = \Delta S \tag{3.11}$$

This allows us to write Equation 3.10

$$W_{lost} = T_0 S_{generated} \tag{3.12}$$

This remarkable, simple relation between the work lost and the entropy produced in a process dates back to close to the beginning of the twentieth century, when it was independently derived by Gouy [2] and Stodola [3].

This equation is not restricted to the process of heat exchange but instead has a universal validity. This is shown in the next section. The extent to which heat exchangers can contribute to the work lost in a process is clearly illustrated in Figure 3.3. Here we observe that, for the process of making ice, nearly 45% of the compressor work is lost in the evaporator and condenser of the ammonia refrigeration cycle. Not many process engineers associate lost work with heat exchange, but this example strikingly shows that they should.

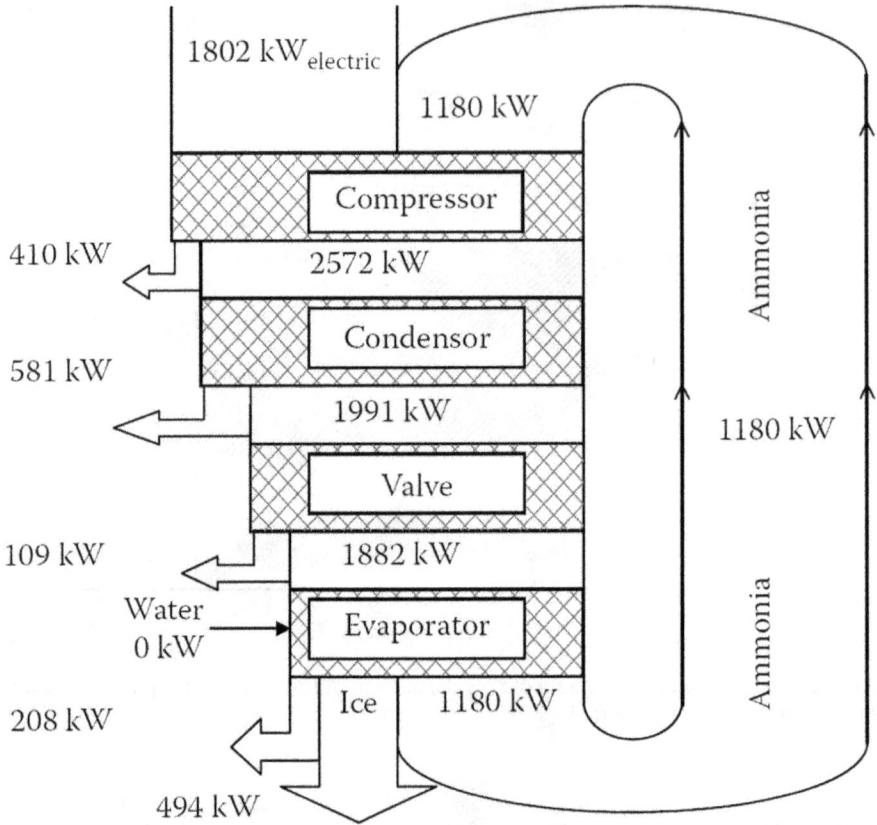

FIGURE 3.3 A refrigeration cycle to make ice from water with ammonia as the working fluid or energy carrier. Nearly 45% of the compression power is lost due to heat exchange in the evaporator and the condenser.

As mentioned earlier, this thermodynamic analysis suggests that $\Delta(1/T)$, not ΔT, is the driving force behind heat flow, contrary to everyday engineering practice. We may then write that to a first approximation

$$Q = k_H \cdot \Delta \frac{1}{T} \cdot A \qquad (3.13)$$

in which

k_H is the overall thermodynamic heat transfer coefficient
A is the surface of exchange

If we introduce this expression in Equation 3.8, we can write

$$W_{\text{lost}} = k_H \cdot A \cdot T_0 \left(\Delta \frac{1}{T} \right)^2 \tag{3.14}$$

This equation tells us that the amount of work lost in a heat exchanger is in the first instance proportional to the square of the driving force, and so if one wishes to be more economical with energy, the driving force should be made smaller. On the other hand, Equation 3.13 shows us that if we have to fulfill a certain heat transfer duty Q, the reduction of $\Delta(1/T)$ must be compensated either by an exchanger material with better heat conduction properties (a larger k_H) or with a larger surface for transfer. It is interesting how these simple equations express the economic need of optimizing between capital cost and the cost of energy. The powerful role of thermodynamics here becomes somewhat tempered by the role of economics. Both disciplines play a decisive role in the ultimate design of the heat exchanger.

3.4 LOST WORK AND ENTROPY GENERATION

We consider a steady flowing medium on which an amount of work is exerted, W_{in} (Figure 3.4). Heat is only exchanged with the environment at a temperature T_0. We assume that the flow is not undergoing significant changes in velocity nor in height and therefore neglect the macroscopic changes in kinetic and potential energy. The first law then reads according to Equation 2.39

$$W_{\text{in}} = \Delta H + Q_0 \tag{3.15}$$

The second law reads

$$S_{\text{generated}} = \Delta S + \Delta S_0 \tag{3.16}$$

We combine the equations, making use of the relation

$$\Delta S_0 = \frac{Q_0}{T_0} \tag{3.17}$$

and after eliminating Q_0 and ΔS_0, we arrive at the following equation:

$$W_{\text{in}} = \Delta H - T_0 \Delta S + T_0 S_{\text{generated}} \tag{3.18}$$

ΔH and ΔS express the changes in the flow's enthalpy and entropy from state 1 to state 2, for example,

$$\Delta H = H(P_2 T_2) - H(P_1 T_1) \tag{3.19}$$

The *minimum* amount of work to accomplish this change in conditions is apparently given by

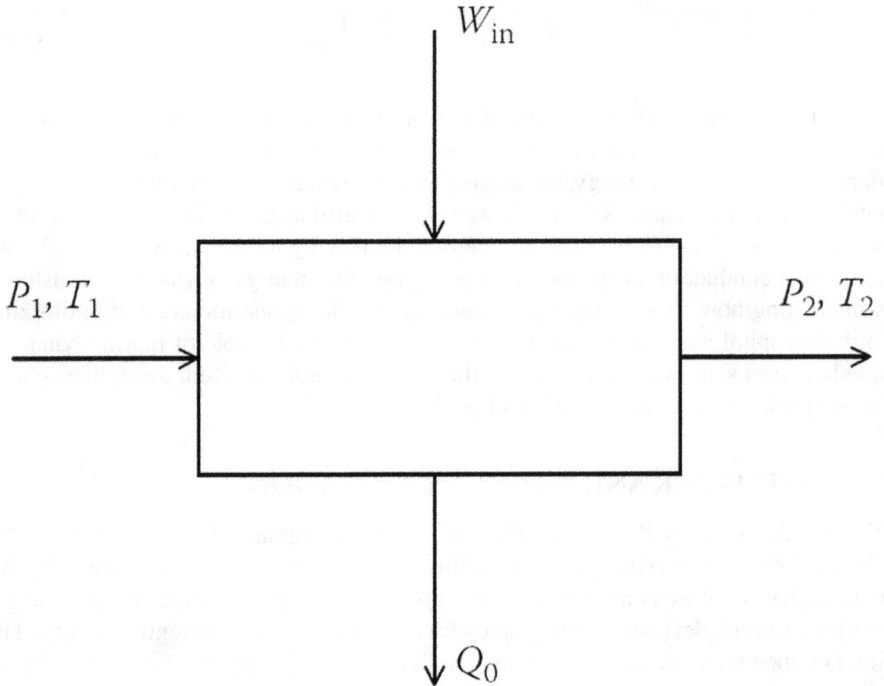

FIGURE 3.4 Work and heat are exchanged with the environment while a fluid is brought from condition 1 at P_1, T_1 to condition 2 at P_2, T_2.

$$W_{in}^{min} = \Delta H - T_0 \Delta S \tag{3.20}$$

The second law states that $S_{generated} > 0$; hence the real amount of work required must be larger:

$$W_{in} = W_{in}^{min} + T_0 S_{generated} \tag{3.21}$$

The amount of work lost in the process is defined as

$$W_{lost} \equiv W_{in} - W_{in}^{min} \tag{3.22}$$

and thus we arrive at the relation

$$W_{lost} = T_0 S_{generated} \tag{3.23}$$

This relation holds, of course, for the general exchange of work and heat between a flowing medium and its environment. The abstract formulation of the second law as in

$$S_{generated} > 0 \tag{3.24}$$

has found here a clear and nonambiguous translation into terms widely understood in the everyday world. After all, lost work in any process makes us rely more on energy resources such as coal, oil, natural gas, nuclear fuel, and so on. Knowing the prices of fuel, we can thus even translate entropy generation into terms of monetary units, but then the concept becomes as shaky as the value of a currency, and so we shall refrain from doing this. But the observation has to be made.

At this point the need arises to become more explicit about the nature of entropy generation. In the case of the heat exchanger, entropy generation appears to be equal to the product of the heat flow and a factor that can be identified as the thermodynamic driving force, $\Delta(1/T)$. In the next chapter we turn to a branch of thermodynamics, better known as irreversible thermodynamics or nonequilibrium thermodynamics, to convey a much more universal message on entropy generation, flows, and driving forces.

3.5 CONCLUSION

The quality of heat is defined as its maximum potential to perform work with respect to a defined environment. Usually, this is the environment within which the process takes place. The Carnot factor quantitatively expresses which fraction of heat is at most available for work. Heat in free fall from a higher to a lower temperature incurs a loss in this quality. The quality has vanished at T_0, the temperature of the prevailing environment. Lost work can be identified with entropy generation in a simple relation. This relation appears to have a universal value.

NOTE

1   The minimum amount of work, W_{in}^{min}, required to transfer to the environment the amount of "heat" leaked in from the environment at T_0 into a reservoir maintained at a temperature $T < T_0$ is $W_{in}^{min} = Q\left((T_0/T) - 1\right)$.

REFERENCES

1. Schrödinger, E.M. *What Is Life?* Cambridge University Press: Cambridge, 1944.
2. Gouy, G. Sur l'énergie utilizable. *Journal of Physics: Condensed Matter* 1889, *8*, 501.
3. Stodola, A. *Steam and Gas Turbines*, McGraw-Hill: New York, 1910.
4. Baehr, H.D. *Thermodynamik*, Springer-Verlag: Berlin, 1988.
5. Sussmann, M.V. *Availability (Exergy) Analysis*, 3rd edn., Mulliken House: Lexington, MA, 1985.
6. Orwell, G. *Animal Farm*, Penguin Longman Publishing Company: London, 1945.
7. Annual Report of the Dutch Electricity Producers (SEP), 1991.

4 Entropy Generation
Cause and Effect

In this chapter, we first introduce the principles of irreversible or nonequilibrium thermodynamics as opposed to those of equilibrium thermodynamics. Then, we identify important thermodynamic forces X (the cause) and their associated flow rates J (the effect). We show how these factors are responsible for the rate with which the entropy production increases and available work decreases in a process. This gives an excellent insight into the origin of the incurred losses. We pay attention to the relation between flows and forces and the possibility of coupling of processes and its implications.

4.1 EQUILIBRIUM THERMODYNAMICS

Equilibrium thermodynamics is the most important, most tangible result of classical thermodynamics. It is a monumental collection of relations between state properties such as temperature, pressure, composition, volume, internal energy, and so forth. It has impressed, maybe more so overwhelmed, many to the extent that most were left confused and hesitant, if not to say paralyzed, to apply its main results. The most characteristic thing that can be said about equilibrium thermodynamics is that it deals with transitions between well-defined states, equilibrium states, while there is a strict absence of macroscopic flows of energy and mass and of driving forces, potential differences, such as difference in pressure, temperature, or chemical potential. It allows, however, for nonequilibrium situations that are inherently unstable, out of equilibrium, but kinetically inhibited to change. The driving force is there, but the flow is effectively zero.

Some confusion may arise from the discussion of so-called reversible processes. Reversible processes take place in the limit of all driving forces going to zero. Most of us tacitly accept the terminology of "heat is isothermally transferred" only to become aware in daily engineering practice that there is no such thing as isothermal heat transfer. In daily engineering practice, heat transfer needs a temperature gradient, mass transfer needs a gradient in the thermodynamic potential, and chemical conversion needs a nonzero affinity between products and reactants. In fact, all heat exchangers, separation columns, chemical reactors, and so forth work by the grace of driving forces, which can be defined by thermodynamics, as we shall see. It is the virtue of irreversible thermodynamics to reconcile the results of equilibrium thermodynamics with the need to determine rates of processes encountered in process technology.

DOI: 10.1201/9781003304388-5

4.2 ON FORCES AND FLOWS: CAUSE AND EFFECT

In Section 3.3, we have shown that the entropy generation rate in the case of heat transfer in a heat exchanger is simply the product of the thermodynamic driving force $X = \Delta(1/T)$, the natural cause, and its effect, the resultant flow $J = \dot{Q}$, a velocity or rate. Selected monographs on irreversible thermodynamics, see, for example [1], show how entropy generation also has roots in other driving forces such as chemical potential differences or affinities.

Katchalsky and Curran [2], for example, consider a system separated from the environment by a rigid adiabatic wall. The system consists of two compartments 1 and 2, separated by a diathermal, elastic barrier that is permeable to one of the components in the system (Figure 4.1). It can be shown that the entropy generation rate is given by

$$\frac{dS_{gen}}{dt} = \frac{dQ_1}{dt}\left(\frac{1}{T_1} - \frac{1}{T_7}\right) + \frac{dV_1}{dt}\left(\frac{P_1}{T_1} - \frac{P_2}{T_7}\right) - \frac{dn_1}{dt}\left(\frac{\mu_1}{T_1} - \frac{\mu_2}{T_7}\right) \tag{4.1}$$

$$\dot{S}_{gen} = \dot{Q}_1 \cdot \Delta\left(\frac{1}{T}\right) + \dot{V}_1 \cdot \Delta\left(\frac{P}{T}\right) + \dot{n} \cdot \Delta\left(-\frac{\mu}{T}\right)$$

$$= \sum_i J_i X_i \tag{4.2}$$

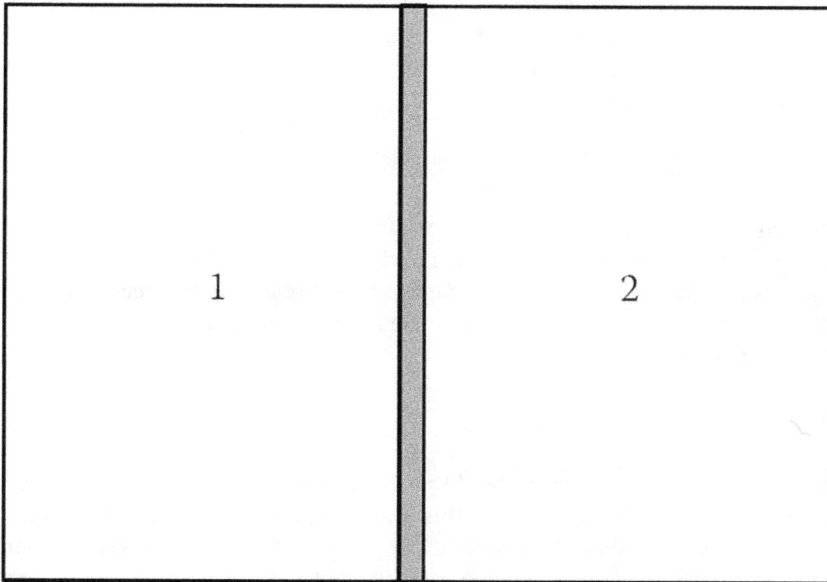

FIGURE 4.1   An isolated system separated in two compartments by a membrane.

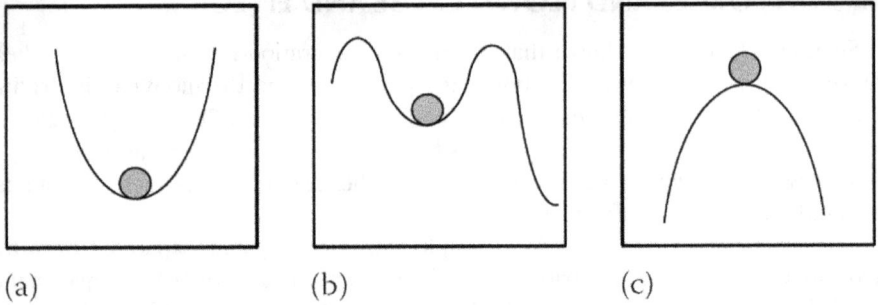

(a)   (b)   (c)

FIGURE 4.2   Examples of a stable (a), metastable (b), and unstable (c) system.

Here we have adopted the convention that J_i is the flow rate, a velocity, of heat, volume, and matter, and X_i is the corresponding affinity or driving force $\Delta(1/T)$, $\Delta(P/T)$, and $\Delta(-\mu/T)$. Irreversible thermodynamics smoothly transforms into equilibrium thermodynamics in the limiting case when the driving forces are going to zero. Without driving forces, there will be no macroscopic process and true equilibrium is established. The existence of driving forces does not automatically imply flows. Inhibition by either physical or chemical barriers may prevent the occurrence of flow, even if the driving force is not zero. These then are situations of instability. We speak of a *metastable* system if it is locally stable but globally unstable, and of an unstable system in the case of local instability (Figure 4.2) [3].

We have not dealt yet with another driving force better known as chemical affinity. We recollect that the condition for chemical equilibrium at a certain pressure P and temperature T is given by [4]

$$\sum v_i \mu_i = 0 \tag{4.3}$$

In this equation v_i and μ_i stand, respectively, for the stoichiometric coefficient and the thermodynamic potential of component i taking part in the reaction. The convention is that v_i is positive for a product and negative for a reactant. The reaction will proceed to the right if $\sum v_i \mu_i$ is negative and to the left if this sum is positive. It is therefore convenient to define the chemical affinity A as

$$A = -\sum v_i \mu_i \tag{4.4}$$

When A is positive, the reaction should proceed to the right. When $A = 0$, chemical equilibrium prevails. Formulated in this way, A suits our perception of chemical affinity perfectly. Another characteristic property is the degree of advancement of reaction ξ [4]. The chemical reaction velocity and the degree of advancement are related by

$$v_{ch} = \frac{d\xi}{dt} = \frac{1}{v_i}\frac{dn_i}{dt} \tag{4.5}$$

It can be shown that the contribution to the entropy generation rate $\dot{S}_{gen}$ due to the progress of a chemical reaction is given by

$$\dot{S}_{gen} = \frac{A}{T} \cdot v_{ch} \tag{4.6}$$

v_{ch} will be positive if A is positive; with v_{ch} approaching zero, A will approach 0; and with $v_{ch} = 0$, true chemical equilibrium has been established and $A = 0$. We have now identified a new driving force, A/T, which is an addition to the driving forces we have already encountered, $\Delta(1/T)$, $\Delta(P/T)$, and $\Delta(-\mu/T)$. The case of chemical affinity has stimulated some to call all driving forces affinities, but we will not adopt this convention, because we think that the word "affinity" should exclusively relate to chemistry.

When more reactions take place at the same time, the chemical contribution to $\dot{S}_{gen}$ is given by

$$\dot{S}_{gen} = \frac{A_1}{T_1} \cdot v_1 + \frac{A_2}{T_2} \cdot v_2 + \cdots \tag{4.7}$$

Each contribution to the total rate must be positive unless reactions are *coupled*. In the case of coupled reactions, one contribution may be negative (this reaction goes uphill) as long as the other coupled reaction is downhill and has a positive contribution large enough to make the sum positive. Later, we shall see that this case is one of the great challenges for the chemical industry to cut down on energy consumption. Living systems make extensive use of this ingenious principle [5] and thus serve as a splendid example to meet this challenge.

4.3   CAUSE AND EFFECT: THE RELATION BETWEEN FORCES AND FLOWS

Although irreversible thermodynamics neatly defines the driving forces behind associated flows, so far it has not told us about the relationship between these two properties. Such relations have been obtained from experiment, and famous empirical laws have been established like those of Fourier for heat conduction, Fick for simple binary material diffusion, and Ohm for electrical conductance. These laws are linear relations between force and associated flow rates that, close to equilibrium, seem to be valid. The heat conductivity, diffusion coefficient, and electrical conductivity, or reciprocal resistance, are well-known proportionality constants and as they have been obtained from experiment, they are called phenomenological coefficients L_{ii},

$$I_i = L_{ii} X_i \tag{4.8}$$

Newton's law is a special case, as Katchalsky and Curran point out [2]. Newton's law relates force to acceleration:

$$f = ma \tag{4.9}$$

This relation is restricted, however, to frictionless media. If frictional forces are included, and acceleration is assumed to be damped out, an equally linear relationship between velocity v and force f can be established as in Stokes' law, which clearly falls in the preceding category.

Equation 4.8 may be rewritten as

$$X_i = R_{ii}J_i \tag{4.10}$$

In this equation, just as in Newton's law adapted for friction, the reciprocal of the phenomenological coefficient L_{ii} has been introduced and acts as a friction coefficient, a resistance. Recalling the relations

$$\dot{W}_{lost} = T_0 \dot{S}_{gen} \tag{4.11}$$

and

$$\dot{S}_{gen} = \sum_i J_i X_i \tag{4.12}$$

a direct relation emerges between the work lost in the process and the frictions incurred:

$$\dot{W}_{lost} = T_0 \sum_i R_{ii} J_i^2 \tag{4.13}$$

This equation expresses the fact that in a process with the various flow rates J_i, work is continuously dissipated to overcome the barriers, the resistance, or the friction that all the processing such as heat and mass transfer and chemical conversion introduce. In Chapter 5, we refer to this as the result of the "magic triangle." There is no clearer way to illustrate the origin of the process's energy bill, nor a better way to calculate it. This relation also defines the challenge that in order to keep the energy bill as low as possible one should find, as Bejan calls it, "the path of least resistance" [6].

In 2009, Lems [7] proposed a new fundamental thermodynamic principle that leads to a universal and strictly thermodynamic relationship between flows and forces. This relationship applies to chemical reactions, diffusion, electrical conduction, and heat conduction, is nonlinear but shows linear behavior close to the equilibrium state. The linear approximation is usually well justified for diffusion, and heat and electrical conduction.

4.4 COUPLING

So far, we have discussed the relation between one driving force and what is called its conjugated flow. Experiments have empirically established many linear relations between flows and conjugated forces. But experiments also pointed to a possible extension of this linearity. Flow rate J_1 might as well be linearly related to driving force X_2 and flow rate J_2 to X_1, and so on. Such observations, according to Katchalsky and Curran [2], have been made, for example, for the "interference" of electrical and osmotic behavior, of mass and heat flow and of electricity and heat flow. This led Onsager [8], chemical engineer and Nobel laureate, to the formulation of the so-called phenomenological equations:

$$J_1 = L_{11}X_1 + L_{12}X_2$$
$$J_2 = L_{21}X_1 + L_{22}X_2$$

(4.14)

or

$$I_i = \sum_{n}^{k=1} L_{ik}X_k \, (i=1,2,\therefore \infty,n)$$

(4.15)

An alternative formulation may be given according to

$$X_i = \sum_{n}^{k=1} R_{ik}J_k \, (i=1,2..,n)$$

(4.16)

in which the coefficients R_{ik} represent generalized resistances or frictions. L_{ii} or R_{ii} are coefficients between the conjugated flow and force. L_{ik} and R_{ik} are so-called coupling or interference coefficients. Onsager [8] shows that if this type of coupling takes place, a simple relation between the coupling coefficients exists:

$$L_{ik} = L_{ki}$$

(4.17)

or

$$R_{ik} = R_{ki}$$

(4.18)

A rule of thumb for the validity of linear relationships is that processes should be slow and the thermodynamic states near equilibrium. But even then, not all flows can be coupled. Coupling is limited to certain cases. Casimir shows that coupling is only possible between driving forces of the same tensorial character, such as scalar, vectorial, and so forth. For more details, see Ref. [2].

An interesting question that immediately comes up is how coupling affects $\dot{S}_{gen}$ and $\dot{W}_{lost}$. Therefore, we recall that

$$\dot{S}_{gen} = \sum_{i=1}^{n} J_i X_i$$

(4.19)

which is an unconditional relation. Next, we assume a simple case of two pairs of conjugated flows and forces that are linearly related and coupled; thus

$$J_1 = L_{11}X_1 + L_{12}X_2$$
$$J_2 = L_{12}X_1 + L_{22}X_2$$

(4.20)

$$\dot{S}_{gen} = J_1 X_1 + J_2 X_2$$
$$= L_{11}X_1^2 + (L_{12} + L_{21})X_1 X_2 + L_{22}X_2^2 > 0$$

(4.21)

The second law requirement that $\dot{S}_{gen} > 0$ leads to the condition that

$$L_{11} > 0, L_{22} > 0 \qquad (4.22)$$

and

$$\left(L_{12}\right)^2 < L_{11}L_{22} \qquad (4.23)$$

This relation does not mention the sign of L_{12}. In ternary mixtures of two solutes diffusing in a solvent medium, cases are known where the coupling coefficients are negative and coupling can lead to some 25% reduction in the entropy generation rate. It has also been observed that matter can move against its thermodynamic potential gradient, thus causing a negative contribution to $\dot{S}_{gen}$. However, in such instances the positive contribution due to heat flow in the proper direction—namely, of lower temperatures—will more than compensate for that, thus obeying the second law.

Finally, we want to mention that Prigogine [1] has shown that for a coupled system of flows and associated driving forces, the rate of entropy generation is at a minimum when the system is in the nonequilibrium steady state; all non-steady states are associated with higher rates of entropy generation.[1]

4.5   LIMITED VALIDITY OF LINEAR LAWS

Linear laws appear to have a limited validity [7]. This may be experienced as impractical because in that case much more experimental evidence has to be collected for establishing the relation between J_i, and X_i, but as we shall see later, this is also the source for much more excitement in industrial practice and nature. The validity of linear laws for heat and mass transfer reaches as far as the phenomenological laws of Fourier and Fick reach. In the case of Fick's law, the range of validity may be shorter or longer because it is the thermodynamic potential difference, not the concentration difference, that is the driving force. Similarly, Fourier's law is based on temperature differences, but the true driving force is $\Delta(1/T)$ not ΔT, and again the true range of validity is decided by the validity of the linear relationship between the flow rate of heat and $\Delta(1/T)$.

An interesting case is presented by chemical reactions. For simplicity, we consider an ideal solution in which the molecules M and N participate in the reaction

$$M \rightleftarrows N \qquad (4.24)$$

Their steady-state concentrations are maintained such that the chemical affinity A, defined by Equation 4.4 ($A = \mu_M - \mu_N$), is positive and thus the chemical system is out of equilibrium. The reaction velocity v_{ch} is positive and given by

$$v_{ch} = \vec{v} - \vec{v} = \vec{k}[M] = \overleftarrow{k}[N] \qquad (4.25)$$

The rate of lost work is (Equations 4.6 and 4.11)

$$\dot{W}_{\text{lost}} = T_0 \dot{S}_{\text{gen}} = T_0 \cdot \frac{A}{T} \cdot v_{\text{ch}} \qquad (4.26)$$

We assume, again for simplicity, that $T = T_0$ and thus

$$\dot{W}_{\text{lost}} = A \cdot v_{\text{ch}} \qquad (4.27)$$

We now derive the relation between the chemical flow rate v_{ch} and its driving force A. Earlier we defined

$$A \equiv \mu_{\text{M}} - \mu_{\text{N}} \qquad (4.28)$$

and thus for the equilibrium situation

$$A_{\text{eq}} = \mu_{\text{M,eq}} - \mu_{\text{N,eq}} \\ = 0 \qquad (4.29)$$

Assuming that M and N are present in an ideal solution,

$$\mu_{\text{M}} - \mu_{\text{M,eq}} = RT_0 \ln\left(\frac{[M]}{[M]_{\text{eq}}}\right) \qquad (4.30)$$

with a similar expression for component N. Subtracting Equation 4.29 from Equation 4.28 results in

$$A = RT_0 \ln\left[\left(\frac{[N]}{(M)}\right)_{\text{eq}} \Big/ \left(\frac{[N]}{(M)}\right)\right] \qquad (4.31)$$

At equilibrium, the chemical velocity v_{ch} is zero and thus $\vec{v} = \bar{v}$.

This allows us to write

$$\vec{k}[M]_{\text{eq}} = \bar{k}[N]_{\text{eq}} \qquad (4.32)$$

and introducing this relation in Equation 4.31 we find that

$$A = RT_0 \ln\left(\frac{\vec{k}[M]}{\bar{k}[N]}\right) \\ = RT_0 \ln\frac{\vec{v}}{\bar{v}} \qquad (4.33)$$

in which we have assumed that we are close enough to equilibrium that $\vec{k}$ and $\bar{k}$ still have their equilibrium values.

From Equation 4.33 we find

$$\frac{\vec{v}}{\hat{v}} = \exp\left(\frac{A}{RT_0}\right)$$
(4.34)

Remembering that

$$v_{ch} = \vec{v} - \overleftarrow{v}$$

$$= \vec{v}\left(1 - \frac{\overleftarrow{v}}{\vec{v}}\right)$$
(4.35)

and introducing Equation 4.34 into Equation 4.35, we finally obtain the relation between v_{ch} and A:

$$v_{ch} = \vec{v}\left[1 - \exp\left(\frac{-A}{RT_0}\right)\right]$$
(4.36)

According to Equation 4.25, $\vec{v}$ is the molar rate of conversion of M into N. Equation 4.36 shows that if $A/RT_0 \ll 1$, then v_{ch} is linear in A: $v_{ch} = \vec{v}A/RT_0$. For increasing values of A/RT_0, the chemical velocity quickly approaches its "saturation" value, $v_{ch} = \vec{v}$, which is reached in the limit of $A/RT_0 \rightarrow \infty$, as can be seen from Figure 4.3.

Meanwhile, the rate of lost work is given by Equations 4.27 and 4.36

$$\dot{W}_{lost} = A \cdot \vec{v}\left[1 - \exp\left(-\frac{A}{RT_0}\right)\right]$$
(4.37)

To allow an easy interpretation of this equation, we introduce the dimensionless properties

$$y = \frac{\dot{W}_{lost}}{RT_0\vec{v}}$$
(4.38)

for the reduced lost work rate and

$$x = \frac{A}{RT_0}$$
(4.39)

for the reduced chemical affinity. Equation 4.37 then reads

$$y = x\left(1 - e^{-x}\right)$$
(4.40)

This equation has two interesting asymptotes (Figure 4.4). When $x \rightarrow 0$, that is, on approaching equilibrium, the chemical velocity depends linearly on the affinity (see text Equation 4.36) and the lost work rate changes with its square, $y \rightarrow x^2$. For $x \rightarrow \infty$,

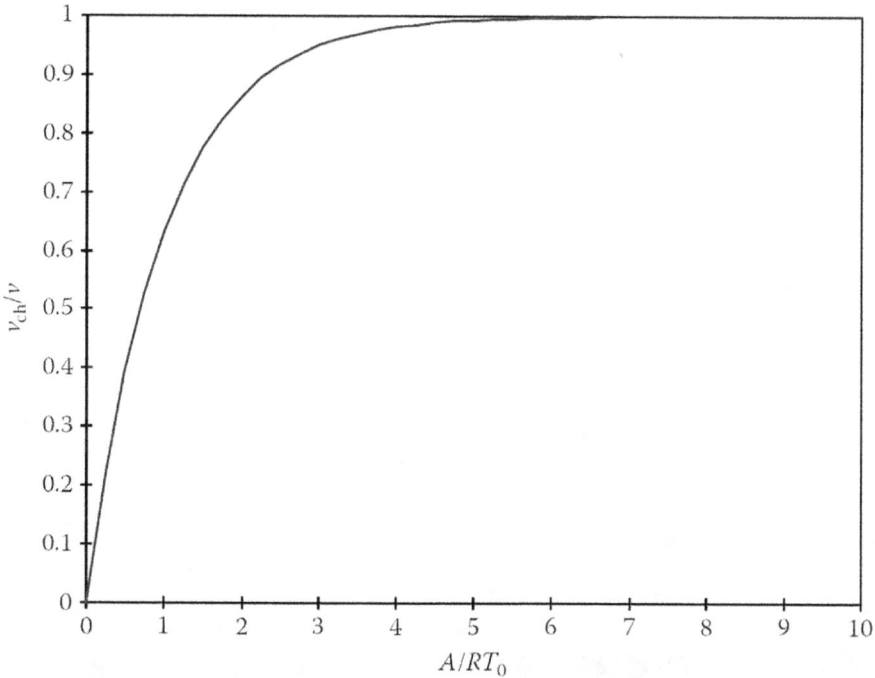

FIGURE 4.3 The reaction velocity as a function of the chemical affinity. The coordinates are in dimensionless units.

that is, giving in to a highly spontaneous and irreversible reaction, $y \to x$ and the lost work rate depends linearly on the affinity. These results can be read from Figure 4.4, where Equation 4.40 has been plotted as well as its two asymptotes. Surprisingly, outside the linear region, that is, from a reduced affinity of about 0.5, the lost work rate is lower than if the linear region extended over a larger range of affinities, although for larger affinities the lost work rate increases all the same but not as drastic as in the linear region. Perhaps this is unique for chemical reactions.

Now suppose it is possible to convert M into N via a number, n, of sequential reactions, of which we assume that they have equal rate constants.

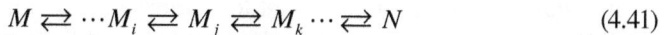

$$M \rightleftarrows \cdots M_i \rightleftarrows M_j \rightleftarrows M_k \cdots \rightleftarrows N \tag{4.41}$$

The overall affinity A remains the same, but each step i now has an affinity, on average, of $A_i = A/n$ for each i. In the steady state, the chemical reaction velocity v_{ch} is now

$$v_{ch} = \cdots v_i = v_j = v_k = \cdots \tag{4.42}$$

with

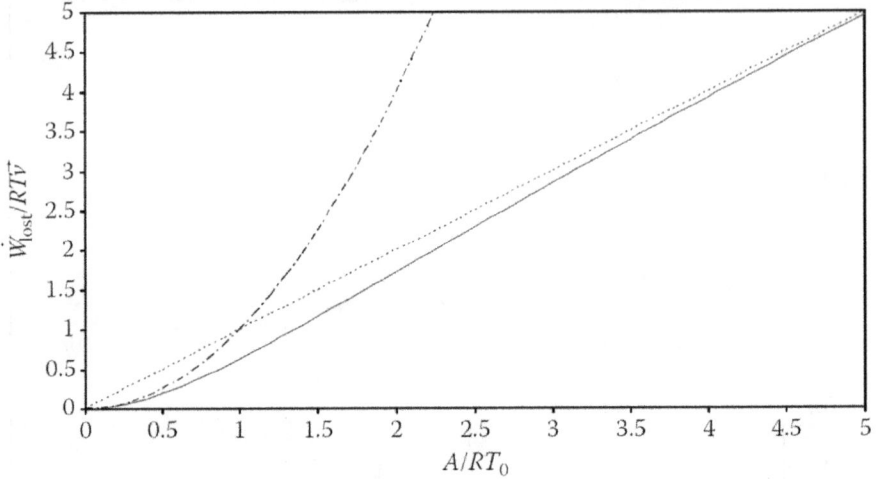

FIGURE 4.4 The rate of lost work as a function of the chemical affinity (in dimensionless units; see text). The solid line is Equation 4.37, whereas the dashed-dotted line is the parabolic asymptote for small A/RT_0 and the dotted line is the linear asymptote for large A/RT_0.

$$v_i = \bar{v}_i \left[1 - \exp\left(-\frac{A}{nRT_0} \right) \right] \qquad (4.43)$$

Equations 4.42 and 4.43 present a possible explanation for the frequent occurrence in biochemical networks of sequential reactions with many steps n. For an overall reduced (dimensionless) chemical affinity of, let us say, $A/RT_0 = 5$, the dimensionless chemical reaction rate $v_{ch}/\bar{v}$ is, according to Figure 4.3, close to 1, and the dimensionless lost work rate is the product: $5 \times 1 = 5$, as shown in Figure 4.4. In Figure 4.3, this product is represented by the area left of the vertical axis at $(A/RT_0) = 5$.

However, if the total affinity A is spread over, let us say, $n = 10$ reaction steps, then each individual reaction step corresponds, on average, to a reduced chemical affinity of $0.1A/RT_0 = 0.5$, with a reduced chemical velocity of 0.4, according to Figure 4.3. So, the same chemical "distance" is now covered with a much smaller velocity and the corresponding lost work rate has been reduced accordingly. Point C' in Figure 4.5 shows the reduced lost work per step, and Point C the total lost work for the multistep reaction. Point C in Figure 4.5 clearly illustrates this sharp reduction in lost work rate for $A/RT_0 = 5$ and $n = 10$ as compared to the value for $A/RT_0 = 5$ and $n = 1$ (Point D).

The lesson to be learned from this rather coarse and simplified analysis is that spontaneous reactions are costly and energy-inefficient. The spontaneous combustion of fossil fuels for example costs about 30% of the work available in the original fuel [4]. Instead, one should aim for bridging the distance in affinity by a limited number of coupled reactions, which are sequential and share a common intermediate.

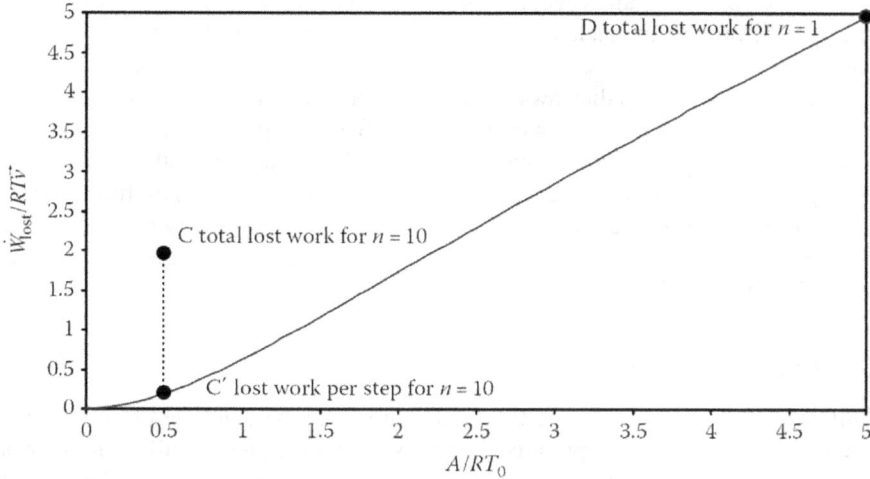

FIGURE 4.5 The rate of lost work as a function of the chemical affinity (in dimensionless units; see text).

Lehninger [5] gives some very clear examples of coupled, sequential reactions with a common intermediate in living systems. As an illustration, we consider the following reactions:

$$X \sim P + ADP \rightarrow X + ATP \qquad (4.44)$$

$$ATP + Y \rightarrow ADP + Y \sim P \qquad (4.45)$$

In reaction 4.44, $X \sim P$ is a biochemical molecule rich in chemical energy with respect to molecule Y through the attachment of the "energy-rich" phosphate group $\sim P$ Direct transfer of $\sim P$ to Y is not possible; it requires the common intermediate ATP, adenosine triphosphate. Adenosine diphosphate, ADP, takes over this $\sim P$ group first and forms ATP. Next, ATP transfers the $\sim P$ group to Y. In this reaction sequence, ATP is the common intermediate and acts as an energy carrier in the transfer of chemical energy from compound X to Y. Both reactions 4.44 and 4.45 make use of what we like to call the downhill-uphill principle: The downhill reaction from $X \sim P$ to $X + \sim P$ "pays" for the uphill reaction of ADP and $\sim P$ to ATP. The difference is the work lost in reaction 4.44, which may be an order smaller than the chemical energy transferred. The same analysis holds for reaction 4.45. In this way biochemical reactions seem to proceed less abruptly, less spontaneously, and more controlled. Chemical energy will have dissipated all the same, but in many small steps. In this way, an impressive complex, dynamic, and functional structure is sustained that stays out of equilibrium: the living system.

By the way, this is an example of what humans can learn from nature. The great distance in affinity between food molecules and their degradation products, which are ultimately recycled and recombined with the help of work available in sunlight, appears to be broken up in a lengthy sequence of reactions coupled by common intermediate molecules [5].

What will happen further away from equilibrium when the linear relationship between cause and effect breaks down is clear from the preceding example but for other instances is the subject of much research and speculation. With nonlinear relationships, the scope of phenomena becomes nearly unpredictable. Nonlinear dynamics [9,10] may well provide the clue to the phenomenon of macroscopic complexity [11], a rapidly expanding field of science, defined by some [12] as quickly becoming a field of "perplexity."

4.6 CONCLUSION

Equilibrium and nonequilibrium thermodynamics can be combined to give a complete thermodynamic description of a process or process step. Equilibrium thermodynamics allows us to calculate the changes in thermodynamic properties with the change in process conditions. Nonequilibrium thermodynamics allows us to calculate unambiguously the work that is lost associated with the process taking place. It relates this loss to the process's flows and forces driving these flows and identifies the various friction factors as a function of the relationship between flows and forces.

NOTE

1   This principle has been confirmed by Bejan and Tondeur (Bejan, A.; Tondeur, D. Equipartition, optimal allocation, and the constructal approach to predicting organization in nature. *Rev. Gén. Therm.* 1998, 37, 165–180.) in a highly original approach using the principle of equipartitioning in finite-time, finite-size thermodynamics. We refer for this treatment to Sections 5.3 and 5.4 in this book.

REFERENCES

1. Prigogine, I. *Thermodynamics of Irreversible Processes*, 3rd edn., Interscience Publishers: New York, 1967.
2. Katchalsky, A.; Curran, P.F. *Non Equilibrium Thermodynamics in Biophysics*, 5th edn., Harvard University Press: Cambridge, MA, 1981.
3. de Swaan Arons, J.; de Loos, Th.W. Chapter 5. Phase behaviour. In *Models for Thermodynamic and Phase Equilibria Calculations*, Sandier, S.I. (ed.), Marcel Dekker: New York, 1994.
4. Smith, J.M.; Van Ness, H.C.; Abbott, M.M. *Introduction to Chemical Engineering Thermodynamics*, 5th edn., McGraw-Hill: New York, 1996.
5. Lehninger, A.L. *Bioenergetics*, 2nd edn., Benjamin/Cummings Publishing Company: Meulo Park, CA, 1973.
6. Bejan, A., Personal communication. See also his book, *Entropy Generation Minimization*, CRC Press: Boca Raton, FL, 1996.
7. Lems, S. *Thermodynamic Explorations into Sustainable Energy Conversion. Learning From Living Systems*. PhD thesis, Delft University of Technology, Delft, 2009.
8. Onsager, L. Reciprocal relations in irreversible processes I. *Physical Review* 1931, *37*, 405 (Reciprocal relations in irreversible process II. *Physical Review* 1931, *38*, 2265).

9. Strogatz, S.H. *Nonlinear Dynamics and Chaos, with Applications to Physics, Biology, Chemistry and Engineering*, Addison-Wesley Publishing Company: Reading, MA, 1994.

10. Guckenheimer, J.; Holmes, P. *Nonlinear Oscillations, Dynamical Systems, and Bifurcations of Vector Fields* (Applied Mathematical Sciences), Vol. 42, Springer: New York, 1997.

11. Nicolis, G.; Prigogine, J. *Exploring Complexity*, W.H. Freeman: New York, 1989.

12. Horgan, J. From complexity to perplexity. *Scientific American*, June 1995, 104–109.

5 Reduction of Lost Work

This chapter establishes a direct relation between lost work and the fluxes and driving forces of a process. The Carnot cycle is revisited to investigate how the Carnot efficiency is affected by the irreversibilities in the process. We show to what extent the constraints of *finite size* and *finite time* reduce the efficiency of the process, but we also show that these constraints still allow a most favorable operation mode, the *thermodynamic optimum*, where the entropy generation and thus the lost work are at a minimum. Attention is given to the *equipartitioning principle*, which seems to be a universal characteristic of optimal operation in both animate and inanimate dynamic systems.

5.1 A REMARKABLE TRIANGLE

The second law in the form that suits us best states that in real processes, and thus in engineering practice, the production of entropy is positive:

$$\dot{S}_{gen} > 0 \tag{5.1}$$

By applying the first and second laws to processes in which heat and work are exchanged with the environment at P_0, T_0, we have shown before that this generated entropy is associated with a loss of work according to

$$\dot{W}_{lost} = T_0 \dot{S}_{gen} \tag{5.2}$$

On the other hand, irreversible thermodynamics has provided us with the insight that entropy generation is related to process flow rates like those of volume, $\dot{V}$, mass in moles, $\dot{n}$, chemical conversion, v_{ch}, and heat, $\dot{Q}$, and their so-called conjugated forces $\Delta(P/T)$, $-\Delta(\mu/T)$, A/T, and $\Delta(1/T)$. Although irreversible thermodynamics does not specify the relationship between these forces X and their conjugated flow rates J, it leaves no doubt about the identity of the thermodynamic driving forces X_i and the resultant entropy production rate:

$$\dot{S}_{gen} = \sum_i J_i X_i \tag{5.3}$$

The simple conclusion that we can now draw is that the work lost can be directly related to the process's flows and driving forces. By eliminating $\dot{S}_{gen}$ from Equations 5.2 and 5.3, we obtain

$$\dot{W}_{lost} = T_0 \sum_i J_i X_i > 0 \tag{5.4}$$

DOI: 10.1201/9781003304388-6

The magic triangle behind the
second law

$$\dot{S}_{gen} > 0$$

Engineering
thermodynamics

Irreversible
thermodynamics

$$\dot{W}_{lost} = T_0 \dot{S}_{gen}$$

$$\dot{S}_{gen} = \sum_i J_i X_i$$

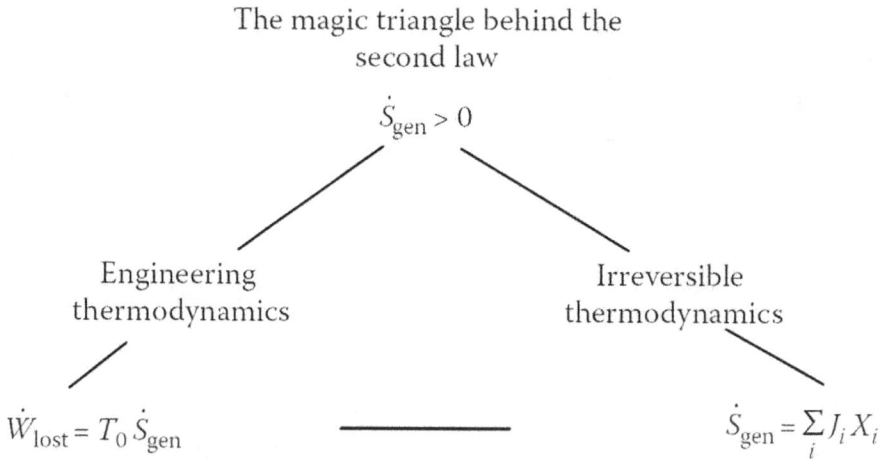

FIGURE 5.1 The magic triangle relating classical, engineering, and irreversible thermodynamics.

So another, more appealing, formulation of the second law is that in a real process available work, exergy, is always dissipated and correlated with the flows and driving forces of the process. This formulation has been represented in Figure 5.1. This figure illustrates how important insights from Clausius, Gouy and Stodola, and Onsager, or from classical, engineering, and irreversible thermodynamics, respectively, show a remarkable interrelationship. It is not without the second law that Equation 5.4 can be obtained, but once the entropy has been eliminated, this equation seems to be free from thermodynamics, with the exception perhaps of the definition of the thermodynamic forces. The equation expresses in a "down-to-earth" way the price that must be paid for energy "consumption" in terms of a process's flow rates and driving forces.

With X_i approaching zero, lost work will approach zero, but this result is not very realistic as flows will tend to become zero too. In practice, we deal with an equipment of finite size that we wish to operate in finite time. So, the question arises: With these constraints, what is the minimum amount of lost work? Thus, it is clear that minimization of lost work and thus of entropy production is a challenging subject with many aspects.

5.2 CARNOT REVISITED: FROM IDEAL TO REAL PROCESSES

We recall that the maximum amount of work available in a heat flow $\dot{Q}$ of a constant temperature $T_H > T_0$ is given by

$$\dot{W}_{out}^{max} = \dot{Q}_{in}\left(1 - \frac{T_0}{T_H}\right) \tag{5.5}$$

This perfect result can be achieved by operating a Carnot heat engine between the temperatures T_H and T_0 (Figure 5.2). In this operation, heat is isothermally transferred

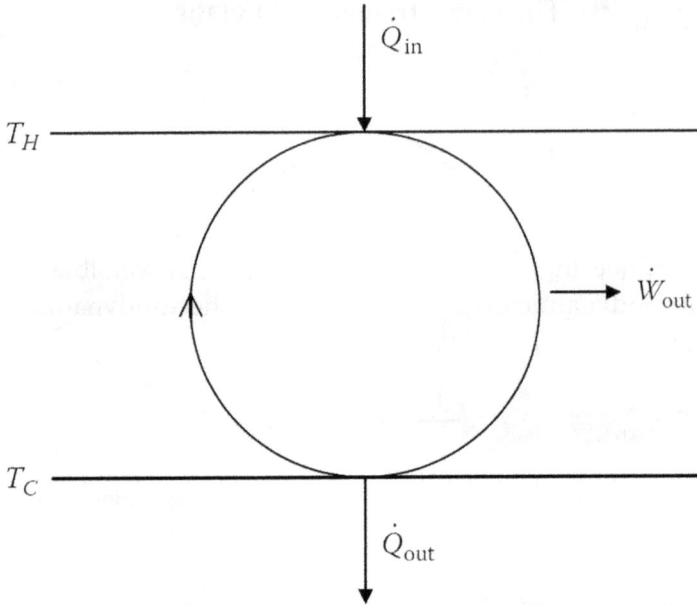

FIGURE 5.2   The Carnot engine operating between T_H and T_0.

from the heat source at a temperature T_H to a working fluid, whereas an amount of heat $\dot{Q}_{in} - \dot{W}_{out}^{max} = \dot{Q}_o^{min}$ is isothermally transferred from the working fluid to the environment at temperature T_0.

In reality, however, there is no such thing as isothermal heat transfer. After all, according to Equation 3.13, the transfer of a finite amount of heat requires a finite temperature difference ΔT, or more correctly, $\Delta(1/T)$. So the working fluid of the engine, strictly speaking, operates between T_{HC} and T_{0C}, rather than between T_H and T_0 (Figure 5.3). This operation implies the loss of power due to the lost work rate incurred between T_H and T_{HC} and T_{0C} and T_0, according to Equation 3.8:

$$\dot{W}_{lost} = \dot{Q}_{in} \cdot T_0 \cdot \Delta \frac{1}{T} \qquad (5.6)$$

The higher $\dot{Q}_{in}$ and $\dot{Q}_0$, the higher the required and associated temperature differences for heat transfer into the working fluid at high temperatures and out from the working fluid at low temperatures for heat exchangers with finite heat exchange area. If the assumption is made that these two losses during heat transfer are the *only* sources of irreversibility, of lost work in the Carnot cycle, we speak of *endoreversibility*, expressing that the Carnot cycle itself does not incur further losses in the power-producing and consuming devices of the power cycle.

The analysis of this endoreversible engine shows a remarkable result that was independently obtained by a number of researchers [1]. Two special limit cases can

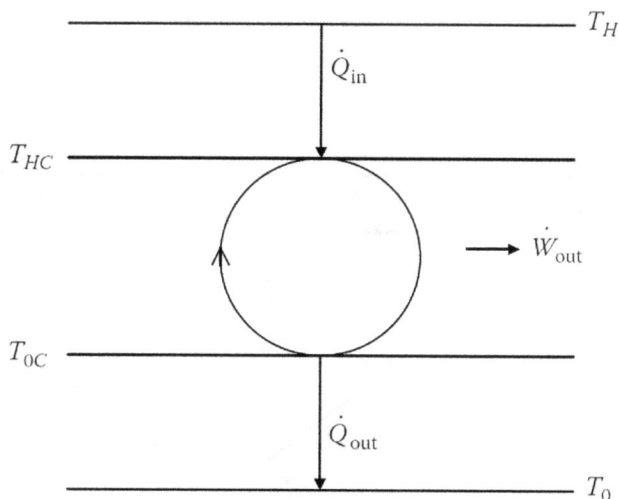

FIGURE 5.3 The endoreversible Carnot engine operating between T_{HC} and T_{0C}.

be identified. The first case is the limit where $\dot{Q}_{in} \to 0$ and thus $\dot{W}_{out} \to 0$, point 1 in Figure 5.4. Heat is thus infinitely slowly, reversibly, extracted from the heat source and transferred to the engine. The efficiency of the engine is the Carnot efficiency of an engine operating between T_H and T_0: $\eta = 1 - T_0/T_H$ because $T_{HC} = T_H$ and $T_{0C} = T_0$ in this limit case. The other limit case is point 2 in Figure 5.4, in which the rate at which heat is transferred into the cycle has reached a maximum and all its available work is dissipated between the temperature T_H and T_{HC} and T_{0C} and T_0 with $T_{HC} = T_{0C}$. Work originally available in the heat flow driving the engine is leaking away *through* the heat engine to the environment at T_0, whereas in the former limit, all available work was leaking away *outside* the engine to the environment.

Thus, there should be an optimum rate $\dot{Q}_{in}$, point 3 in Figure 5.4, for which the power output of the engine $\dot{W}_{out}$ has a maximum. This so-called maximum power is achieved for two optimal temperatures T_{HC}^{opt} and T_{0C}^{opt} related to the two original temperatures T_H and T_0 according to

$$\frac{T_{0C}^{opt}}{T_{HC}^{opt}} = \sqrt{\frac{T_0}{T_H}} \qquad (5.7)$$

The corresponding thermodynamic efficiency is given by

$$\eta = 1 - \frac{T_{0C}^{opt}}{T_{HC}^{opt}}$$

$$= 1 - \sqrt{\frac{T_0}{T_H}} \qquad (5.8)$$

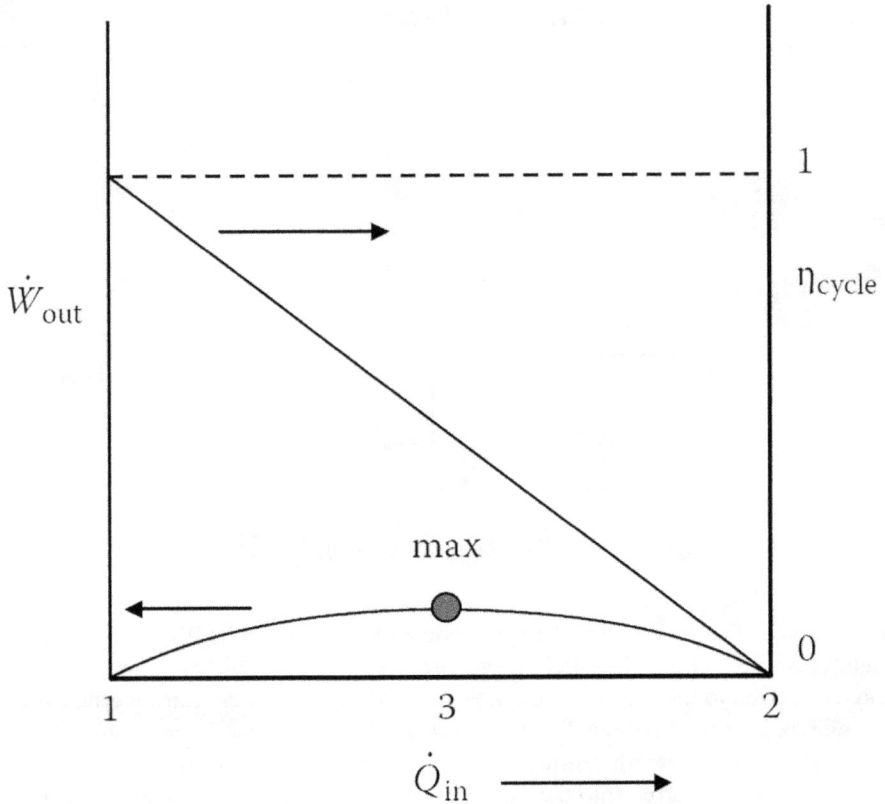

FIGURE 5.4   Power output $\dot{W}_{out}$ and efficiency η as a function of the heat flow rate $\dot{Q}_{in}$.

For a proof of this result, we refer to the original papers [1,2], but we can make the existence of these two optimal temperatures plausible by the following reasoning. Suppose we wish to introduce heat into the Carnot engine with a rate $\dot{Q}_{in}$. This fixes the upper temperature T_{HC} of the Carnot cycle, for example by the approximate relation

$$\dot{Q}_{in} = (kA)_H \left(T_H - T_{HC} \right) \tag{5.9}$$

in which k and A represent the overall heat transfer coefficient and heat transfer area, respectively. T_{HC}, in turn, fixes the lower temperature T_{0C} of the Carnot engine. After all, for a Carnot engine the heat taken in and the heat rejected are related according to

$$\frac{\dot{Q}_{out}}{\dot{Q}_{in}} = \frac{T_{0C}}{T_{HC}} \tag{5.10}$$

which in our case can be rewritten as

$$\frac{(kA)_L \left(T_{0C} - T_0 \right)}{(kA)_H \left(T_H - T_{HC} \right)} = \frac{T_{0C}}{T_{HC}} \tag{5.11}$$

It is obvious from Equation 5.9 that by fixing $\dot{Q}_{in}$, T_{HC} is fixed and that according to Equation 5.11 this fixes T_{0C}. The power of the engine is then given by

$$\dot{W}_{out} = \dot{Q}_{in} \left(1 - \frac{T_{0C}}{T_{HC}} \right) \tag{5.12}$$

Varying $\dot{Q}_{in}$ between the two extremes of points 1 and 2 should produce an optimal value $\dot{Q}_{in}^{opt}$ for which $\dot{W}_{out}$ is a maximum, as shown in Figure 5.4. This maximum power corresponds to optimal values for the temperatures between which the Carnot engine operates, T_{HC} and T_{0C}. These optimal values are related to T_H, the temperature at which heat is made available and T_0, the temperature of the environment that acts as the heat sink for the heat rejected by the engine. By maximizing $\dot{W}$ with respect to T_{HC}, we can find Equations 5.7 and 5.8.

Thus, for a heat engine operating reversibly between $T_0 = 288$ K and $T = 775$ K, the Carnot efficiency is 0.63, whereas the efficiency of the endoreversible engine considered at maximum power is 0.39, a result that shows remarkable agreement with the observed results for real power stations. Curzon and Ahlborn report in their original paper [2] on the efficiencies of three power stations in the United Kingdom, Canada, and Italy. The observed efficiencies were respectively 36%, 30%, and 16%, whereas calculation with Equation 5.8 gave respective values of 40°%, 28°%, and 17.5°%. The Carnot efficiencies were 64.1%, 48%, and 32.3%, respectively. The thermodynamic efficiency of the engine as a function of $\dot{Q}_{in}$ is also plotted in Figure 5.4 and moves from its Carnot value at $\dot{Q}_{in} = 0$ to the value zero for maximum heat throughput.

The impression may be given that operating at maximum power cannot take place at the most favorable conditions, namely, at the minimum entropy generation rate. But as Bejan subtly shows [3] the conditions for maximum power must be the conditions for the minimum entropy generation rate, completely in line with what one would expect from the Gouy-Stodola relation, Equation 3.12. In essence, we need to identify the contributions to the entropy generation rate. The first contribution is that from the heat exchanger at the high-temperature end of the Carnot engine:

$$\dot{S}_{gen,1} = \dot{Q}_{in} \left(\frac{1}{T_{HC}} - \frac{1}{T_H} \right) \tag{5.13}$$

The second contribution is that from the heat exchanger at the low-temperature end of the engine:

$$\dot{S}_{gen,2} = \dot{Q}_{out} \left(\frac{1}{T_0} - \frac{1}{T_{0C}} \right) \tag{5.14}$$

The third contribution is from the original heat flow $\dot{Q}_{source}$ after it has transferred part of its heat, $\dot{Q}_{in}$, to the engine. Suppose this flow $\dot{Q}_{source}$ is a saturated vapor, condensing at T_H to an extent that depends on the heat $\dot{Q}_{in}$ that has been drawn into the engine. After leaving the heat exchanger, the heat flow $\left(\dot{Q}_{source} - \dot{Q}_{in}\right)$ will exchange heat with the environment at T_0, resulting in the third contribution to the entropy generation rate, $\dot{S}_{gen,3} = \left(\dot{Q}_{source} - \dot{Q}_{in}\right)\left((1/T_0) - (1/T_H)\right)$. The total entropy generation is then

$$\dot{S}_{gen}^{total} = \dot{S}_{gen,1} + \dot{S}_{gen,2} + \dot{S}_{gen,3} \tag{5.15}$$

Minimizing $\dot{S}_{gen}^{total}$ with respect to T_{HC} or T_{0C} will result in the same optimal values for $\dot{Q}_{in}$, T_{HC}, and T_{LC}, and the same value for maximum power. This minimum value is, positioned at point 3, Figure 5.5, between the two extreme values of $\dot{S}_{gen}^{total}$ in points 1 and 2, respectively. In point 1, the engine operates so slowly that all the work available in the original heat flow is dissipated outside the engine: $\dot{S}_{gen,1}$ and $\dot{S}_{gen,2}$ are both zero and $\dot{S}_{gen,3}$ is at its maximum value. In point 2, again, all the originally available work is dissipated, the entropy generation rate in the heat exchangers is

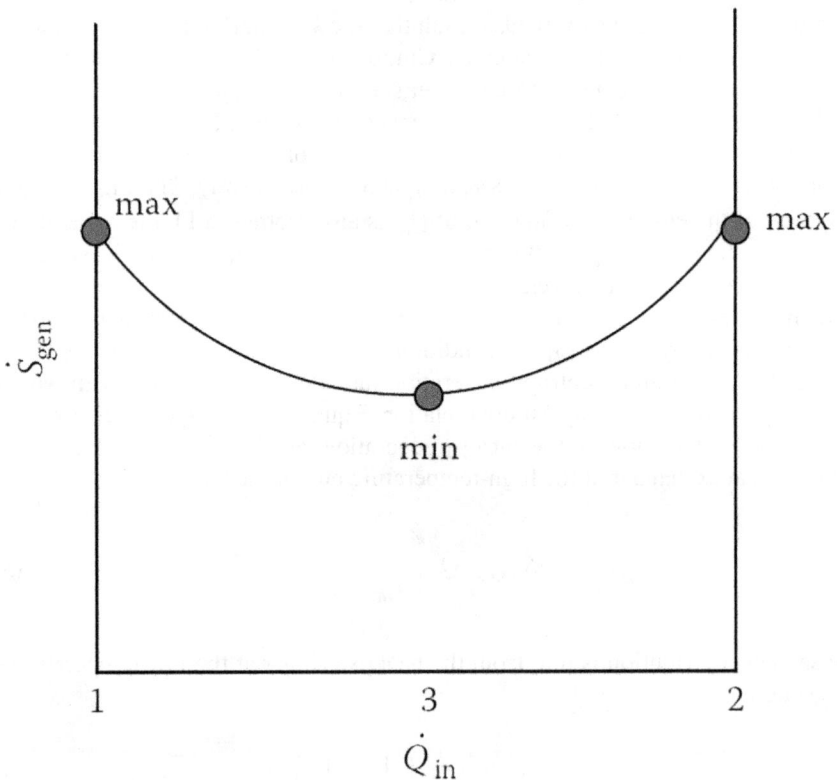

FIGURE 5.5   Entropy generation rate as a function of the heat flow rate.

maximum as $T_{HC} = T_{0C}$. In both 1 and 2 all heat originally available at T_H has been transferred to the environment, producing the maximum possible entropy generation.

In conclusion, an optimum heat flow rate $\dot{Q}_{in}^{opt}$ can be found (Figure 5.4) between a zero rate and a maximum rate of $\dot{Q}_{in}$. At zero rate, the engine operates at maximum, Carnot, efficiency but without the output of work because of the infinitely low rate. All the work available at the source leaks away into the environment, that is, *outside* the engine. At the maximum rate of $\dot{Q}_{in}$, the yield is zero again because all available work is leaking *through* and outside the system. For the optimum rate in between these rates, the power output is at a maximum and causing the *least* dissipation of work and the generation of entropy: at maximum power output, the entropy generation rate is at a minimum.

Finally, we want to point out that the concept of the endoreversible engine is a simplification that leads to a quick and clear insight into the role and thermodynamic cost of transfer processes in reducing the Carnot efficiency into smaller and more realistic values. But for finding the real optimum conditions, the concept of endoreversibility has to be sacrificed. This will complicate the matter to some extent but will allow for including all contributions to the lost work:

$$\dot{W}_{lost} = T_0 \sum_i J_i X_i \qquad (5.16)$$

thus, providing a complete insight into all sources of dissipation and their relative contributions.

Note: The transfer of heat, mass, and momentum seems to receive the most attention in textbooks on transport phenomena. However, the transfer of chemical energy, for example, by coupled reactions, of which bioenergetics provides many examples [4], is another interesting topic. We have shown [5] that living systems appear to operate as conductors for chemical energy, whereby, in a way of speaking, energy permeates the system to wherever it is required. Hill [6] launched the term "free energy transduction." From a closer look at this process, we can learn that via coupled chemical reactions, that is, the mechanism of downhill reactions driving uphill reactions, chemical energy is transported through the system. Every coupled reaction has, of course, its loss despite its, usually high, thermodynamic efficiency (>90%). In the end, all work originally available in food will be dissipated. In the meantime, the living system is kept in a dynamic, most often, steady state out of equilibrium with its environment, which is the very essence of life. His efforts to understand this from thermodynamic principles resulted in Schrodinger's famous monograph, *What Is Life?* [7].

5.3 FINITE-TIME, FINITE-SIZE THERMODYNAMICS

The analysis given in Section 5.2 is very revealing. Let us go over the highlights again. There is a source of heat available at a temperature T_H and at a fixed rate $\dot{Q}_{source}$. This fixes the amount of work available at the source per unit time. The largest fraction of this work that can be obtained as maximum power output, $\dot{W}_{out}^{max}$, is realized for an optimal value of $\dot{Q}_{in}$, the rate at which heat is transferred from the source to the endoreversible Carnot cycle. The other part, $\dot{Q}_{source} - \dot{Q}_{in}^{opt}$, leaks away to the environment, $\dot{Q}_{leakage}$. If $\dot{W}_{out}^{max}$ is the so-called rated power of the power station, it means that a finite amount of work has to be produced *in a finite time*. An optimal value of $\dot{Q}_{in}^{opt}$ takes care of this with $\dot{Q}_{in}^{opt} = (kA)_H \left(T_H - T_{HC}^{opt}\right)$. The optimal value for the upper temperature of the Carnot engine, T_{HC}^{opt}, follows from this relation, given the overall heat transfer coefficient k and the *finite size* of the heat exchanger, namely, the area of the heat transfer. The choice of other materials will of course affect the outcome.

We believe that finite-time, finite-size thermodynamics has a significant didactic value. The *unrealistic* limit of the Carnot or reversible cycle is replaced and a more *realistic* presentation of actual processes is obtained by combining the results of equilibrium thermodynamics with those of irreversible thermodynamics, which is more concerned with the rate and driving forces of processes.

5.4   THE PRINCIPLE OF EQUIPARTITIONING

In 1987, Tondeur and Kvaalen [8] drew attention to an important principle that they proved to be valid in the linear region of irreversible thermodynamics but that they expected to have a wider application. Discussing heat, momentum, and mass transfer, they stated that given a specified transfer duty and transfer area, the *total* entropy generation rate is minimal when the local rate of entropy generation is uniformly distributed, *equipartitioned*, along the space and time variables of the process. Some 10 years later, in 1998, Bejan and Tondeur [9] identified this principle as a universal design principle that accounts for the macroscopic organization in nature both in inanimate systems, such as rivers, and in animate systems, such as trees. They showed that the optimal performance of a finite-size system *with purpose* is always characterized by the equipartition of driving forces or even of the material required for process equipment, for example, heat exchangers, over the process. Bejan gives a series of examples of this remarkable principle [10].

To illustrate this principle, we choose the simplest possible example. Suppose the purpose of our process is to heat a flow of mass in *a finite time* such that its temperature increases with ΔT. This requires a heating duty of

$$I = \dot{m}c_p \Delta T \tag{5.17}$$

with $\dot{m}$ the mass flow rate and c_p the specific heat at constant pressure. For this process, we choose the heat exchanger as the equipment and identify between the heating and heated flow the local heat flux j and conjugate force $x = \Delta(1)/(T)$ (see Figure 5.6). We assume a linear relationship between flux and force: $j = kx$. The local entropy generation rate is thus $\dot{S}_{gen} = jx$. We may now write

$$I = \int_A j dA = k \int_A x dA \tag{5.18}$$

and

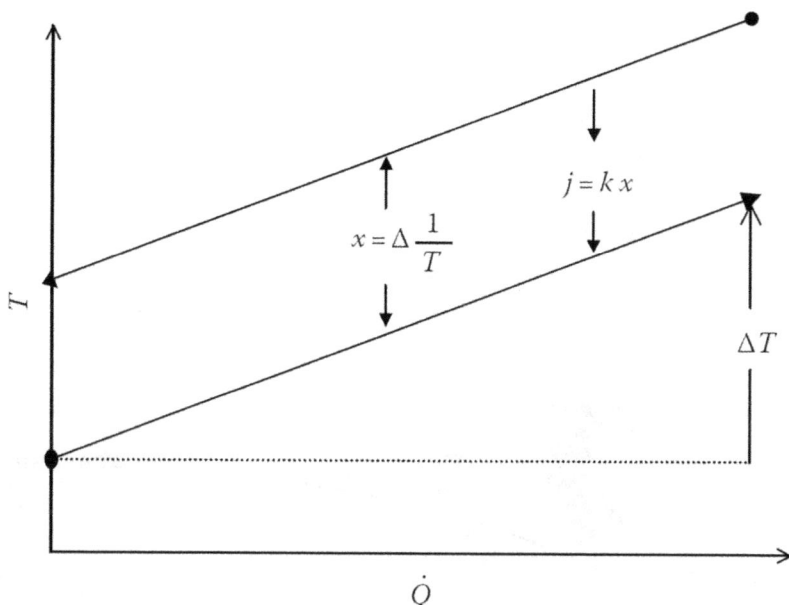

FIGURE 5.6 Local heat flux j and driving force $x = \Delta(1/T)$ in a heat exchanger. The temperature increase of the cold stream is ΔT.

$$\dot{S}_{gen} = \int_A \dot{S}_{gen} dA = k \int_A x^2 dA \qquad (5.19)$$

in which A represents the heat transfer area. We impose the *constraint* of *finite size* by putting A constant and ask ourselves what the optimal distribution is of the force x over the heat exchanger. Therefore, we seek the solution to the preceding equations in terms of the distribution of x that makes $\dot{S}_{gen}$ minimum. With the help of variational calculus [7], we find

$$x = \left(\Delta \frac{1}{T} \right)_{Local} = \text{constant} \qquad (5.20)$$

As a result, j and $\dot{S}_{gen}$ are constant as well. Local fluxes, forces, and entropy generation rates are equipartitioned or evenly distributed over the heat exchanger. In this way, we have obtained a realistic minimum for the lost work rate in this process, a minimum associated with the duty of the process, the finite size of the equipment, and the finite time in which the process has to be carried out. This is a splendid example of *thermodynamic optimization*, providing us with the optimal distribution of the driving force of the process for the most efficient operation under the set constraints. In a real heat exchanger it will be difficult to design for this distribution, but it can be approached in a countercurrent heat exchanger, which is far more efficient than the co- or cross-current heat exchanger. This is visualized in Figure 5.7, which compares the rate with which work is lost (represented by the enclosed areas; see Figure 3.2)

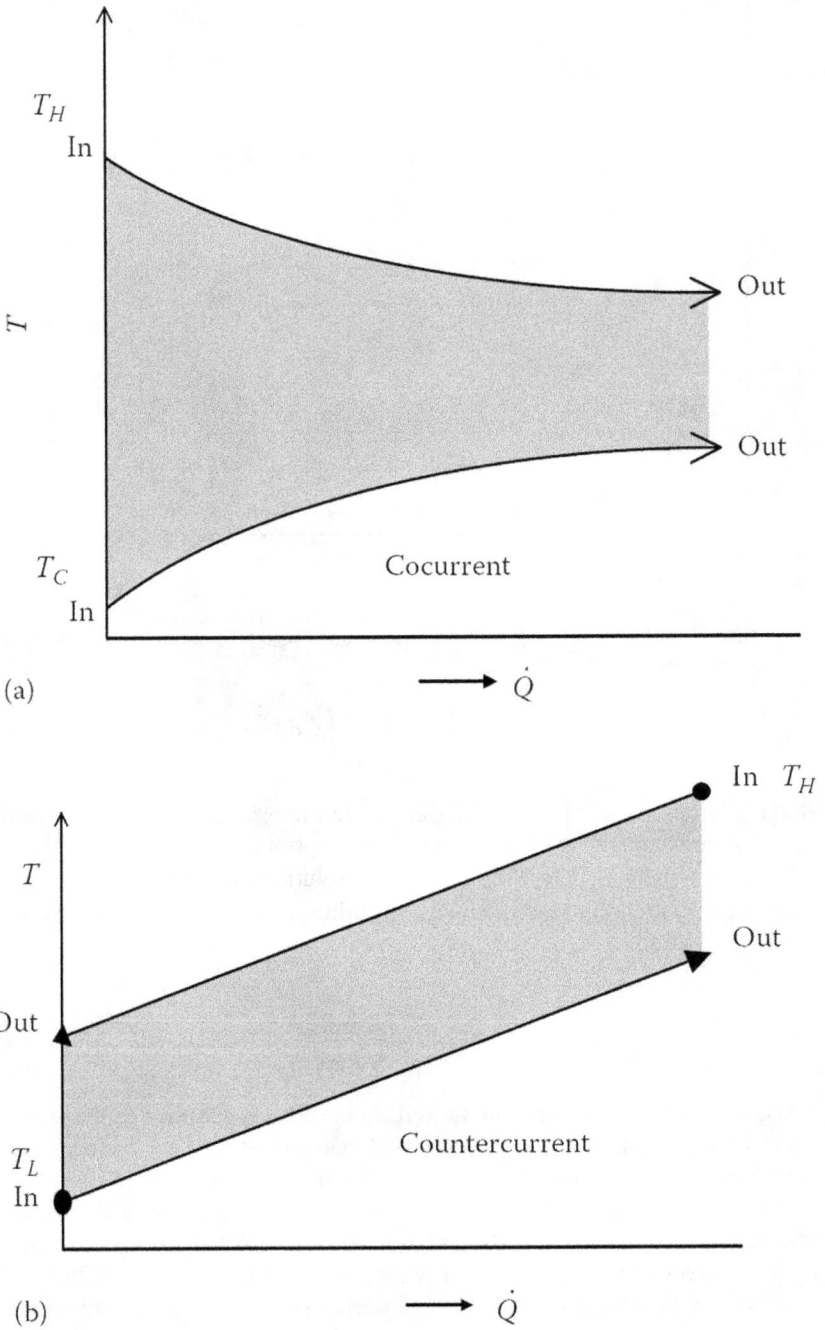

FIGURE 5.7 (a) Cocurrent versus (b) countercurrent heat exchange between the same inlet temperatures T_H and T_L.

in a countercurrent and a cocurrent heat exchanger with specified mass flow rates, specific heats, and inlet temperatures.

It is plausible from comparing these figures that the cocurrent configuration of the heat exchanger generates more entropy than the countercurrent configuration. Tondeur and Kvaalen [8] have generalized the theorem to the simultaneous transfer of mass, momentum, and heat, again under the restriction of the validity of linear laws and Onsager's relations of reciprocity. They also discuss how this theorem can be incorporated in the economic analysis of the process, but this subject is beyond the scope of this book and is left to the reader for further investigation. For a more thorough discussion and for some other illustrative examples of this principle, we refer to the original papers [8,9].

Bejan and Tondeur [9] make a number of other observations in their paper. One is that the relation between j and x is not necessarily linear. Another observation is that a similar analysis can show that the force x should be equipartitioned in time, which is another way of saying that the steady state is optimal. Prigogine gave an earlier proof of this principle [11]. The steady state is common in nature and often the favored state in industrial operation. It can be considered to be the "stable state" of nonequilibrium thermodynamics, comparable to the equilibrium state of reversible thermodynamics (see Figure 4.2). Of course, the latter is characterized by $\dot{S}_{gen} = 0$, whereas the former is characterized by a minimum value $\dot{S}_{gen}^{min}$, larger than zero. A disturbance of this stable and dynamic state will incur an increase of the entropy generation rate only to make it fall back to its minimum value. The stability condition therefore is

$$\frac{d\dot{S}_{gen}}{dt} < 0 \qquad (5.21)$$

5.5 CONCLUSION

The ideal, unrealistic, but basic limit of the thermodynamic efficiency of a process is that of the reversible process where all work available and entering the process is still available after the process. Work has simply been transferred from one carrier to another. Driving forces are infinitesimally small and the process is "friction-less": no barriers have to be taken. As a result, there is neither entropy generation nor loss of available work. The work requirements of the process can be accurately calculated from the thermodynamic properties of the equilibrium states that the process passes through.

For the establishment of the realistic limit, one has to take account of the rates of processes in which mass, heat, momentum, and chemical energy are transferred. In this so-called finite-time, finite-size thermodynamics, it is usually possible to establish optimal conditions for operating the process, namely, with a minimum, but nonzero, entropy generation and loss of work. Such optima seem to be characterized by a universal principle: equipartitioning of the process's driving forces in time and space. The optima may eventually be shifted by including economic and environmental parameters such as fixed and variable costs and emissions. For this aspect, we refer to Chapter 13.

REFERENCES

1. Bejan, A. Entropy generation minimization: The new thermodynamics of finite size devices and finite time processes. *Applied Physics Reviews* 1996, *79*(3), 1191–1218.
2. Curzon, F.L.; Ahlborn, B. Efficiency of a Carnot engine at maximum power output. *The American Journal of Physics* 1975, *43*, 22.
3. Bejan, A. *Entropy Generation Minimization*, CRS Press: Boca Raton, FL, 1996.
4. Lehninger, A.L. *Bioenergetics*, 2nd edn., W.A. Benjamin Inc.: Menlo Park, CA, 1973.
5. Lems, S.; van der Kooi, H.J.; de Swaan Arons, J. Thermodynamic optimization of energy transfer in (bio)chemical reaction systems. *Chemical Engineering Science* 2003, *58*(10), 2001–2009.
6. Hill, T.L. *Free Energy Transduction and Biochemical Cycle Kinetics*, Springer-Verlag: New York, 1989.
7. Schrödinger, E. *What Is Life?* Cambridge University Press: Cambridge, 2000.
8. Tondeur, D.; Kvaalen, E. Equifartition of entropy production. *Industrial & Engineering Chemistry Research* 1987, *26*, 50–56.
9. Bejan, A.; Tondeur, D. Equifartition, optimal allocation, and the constructal approach to predicting organization in nature. *Revue Géneral De Thermique* 1998, *37*, 165–180.
10. Bejan, A. *Shape and Structure, From Engineering to Nature*, Cambridge University Press: Cambridge, 2000.
11. Prigogine, I. *Introduction to Thermodynamics of Irreversible Processes*, 3rd edn., Interscience Publishers: New York, 1967.

Part II

Thermodynamic Analysis of Processes

He who would learn to fly one day must first learn to stand and walk and run and climb and dance. One cannot fly into flying.

—F. Nietzsche

Part II covers Chapters 6 through 8. In Chapter 6, the reader is introduced to the principles of the thermodynamic, or exergy, analysis of processes. Two new concepts are introduced: the concept of exergy (or available work), which enables and facilitates the integration of the environment into the analysis of the process, and the concept of the quality of the Joule, which helps to distinguish between "apples and pears" when one speaks about "energy." Chapter 7 concentrates on the chemical component of exergy. Chapter 8 presents some simple examples of the application of exergy analysis.

DOI: 10.1201/9781003304388-7

6 Exergy, a Convenient Concept

6.1 EXERGY

In view of the presentation of material in Part I, we can state that every task that we set ourselves in industry implies lost work. This holds for processes producing work equally well as for those requiring work. The challenge is to understand the nature of this lost work and from there to apply originality and intelligence to reduce it.

In the following, we deal with a steady-state flow process and consider a flow originally at ambient conditions, P_0 and T_0, and requiring work at the rate $\dot{W}_{in}$ to bring its conditions at P and T (Figure 6.1). In the process, heat is transferred to the environment at a rate of $\dot{Q}_{out}$.

As is justified for most situations in process technology, we ignore macroscopic changes in the kinetic and/or potential energy of the flow in this process. Applying the first law of thermodynamics for flow processes, we may write

$$\dot{W}_{in} = \dot{m}\Delta H + \dot{Q}_{out} \tag{6.1}$$

The second law for this process reads

$$\dot{S}_{gen} = \dot{m}\Delta S + \dot{S}_0 \geq 0 \tag{6.2}$$

in which $\dot{S}_0$ denotes the rate of the change in entropy of the environment.

Next, we wish to establish the minimum amount of work to accomplish the change in the flow's conditions from P_0, T_0 to P, T, that is, to bring about the corresponding changes in the state properties H and S of the flow, ΔH and ΔS, with

$$\Delta H = H_{P,T} - H_{P_0,T_0} \tag{6.3}$$

$$\Delta S = S_{P,T} - S_{P_0,T_0} \tag{6.4}$$

We will therefore rewrite Equation 6.1 as

$$\dot{W}_{in}^{min} = \dot{m}\Delta H + \dot{Q}_{out}^{min} \tag{6.5}$$

or even

$$\dot{W}_{in}^{min} = \dot{m}\Delta H + \dot{Q}_0^{min} \tag{6.6}$$

DOI: 10.1201/9781003304388-8

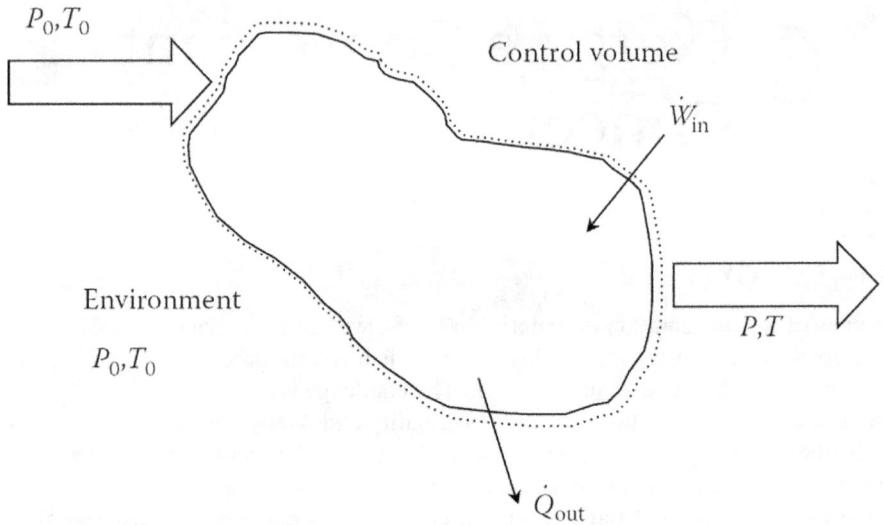

FIGURE 6.1 Work and heat exchange with the environment when a flow is brought from P_0, T_0 to P, T conditions.

By writing $\dot{Q}_0^{min}$ for $\dot{Q}_{out}^{min}$, we want to emphasize that the heat released to the environment must have no potential left, in order to assess the true minimum for $\dot{W}_{in}$. The minimum rate of input of work is associated with the minimum output of heat, as $\dot{m}\Delta H$ is fixed by the choice of the mass flow rate and the thermodynamic conditions of the initial state and the final state. Next, we combine Equations 6.2 and 6.6, replacing $\dot{Q}_0^{min}$ by applying the relation

$$\dot{Q}_0^{min} = T_0 \dot{S}_0 \tag{6.7}$$

and find

$$\dot{W}_{in}^{min} = \dot{m}\left(\Delta H - T_0 \Delta S\right) + T_0 \dot{S}_{gen} \tag{6.8}$$

The true minimum will, of course, require that the process is free of driving forces, such that

$$\dot{S}_{gen} = 0 \tag{6.9}$$

The minimum amount of work to bring about the required change in conditions appears then to be given by

$$\dot{W}_{in}^{min} = \dot{m}\left(\Delta H - T_0 \Delta S\right) \tag{6.10}$$

We can now define the property exergy, Ex, according to

$$\mathrm{Ex} \equiv \frac{\dot{W}_{in}^{min}}{\dot{m}}$$

$$= \left(H_{P,T} - H_{P_0,T_0}\right) - T_0\left(S_{P,T} - S_{P_0,T_0}\right) \tag{6.11}$$

as the amount of useful work confined in a unit of mass of the flow at conditions P and T with respect to the conditions of the environment. It is the maximum amount of work that a unit mass of flow can perform if it is brought reversibly to the conditions of the environment, and measures, in terms of work, to what extent this unit mass is out of equilibrium with the environment. We may rewrite Equation 6.11 as

$$Ex = \left(H - T_0 S\right)_{P,T} - \left(H - T_0 S\right)_{P_0,T_0} \tag{6.12}$$

In American literature [1], the term $H - T_0 S$ is often indicated as the availability function $B \equiv H - T_0 S$, which stems from the concept "available" work. This should not suggest, as it sometimes occurs in literature, that exergy is the same as availability. It is not, because from Equation 6.12 it follows that

$$Ex = B_{P,T} - B_{P_0,T_0} \tag{6.13}$$

At $P_0 T_0$, the exergy will be zero according to our definition, but this does not imply that $B_{P_0 T_0} = 0$. On the contrary, as

$$B_{P_0,T_0} \equiv \left(H - T_0 S\right)_{P_0,T_0} = G_{P_0,T_0} \tag{6.14}$$

the availability function at environmental conditions can be identified as the Gibbs energy at these conditions. As one can easily verify, this property is not determined as it contains enthalpy H and therefore the internal energy U as a property of unknown absolute magnitude. The zero level has often been arbitrarily chosen, albeit for good reasons. In the case of water, the melting point is the zero level for liquid water's enthalpy. With the exception for the conditions $P_0 T_0$, the availability expression at P and T

$$B = H - T_0 S \tag{6.15}$$

should not be mistaken for the value of the Gibbs energy G at P and T.

If the flow is brought from conditions P_1, T_1 to P_2, T_2 without a change in its composition, we may write

$$\Delta Ex = Ex_2 - Ex_1 \tag{6.16}$$

ΔEx is the minimum amount of work to bring about the indicated changes in thermodynamic conditions. The terms on the right-hand side of Equation 6.16 represent the maximum amount of work that the unit mass of flow can perform if the flow is reversibly brought back from states 1 and 2, respectively, to the pressure P_0 and temperature T_0 of the environment.

6.2   THE CONVENIENCE OF THE EXERGY CONCEPT

With the definition of exergy, it has become possible to assign to a stream a quality in terms of the potential to perform work. While lost work is dependent on the specific process that we perform to accomplish a certain goal and varies with the way the process is performed, exergy is, in contrast, an unambiguous property fully defined as a state property as soon as the conditions of the unit mass of flow and those of the environment are defined.[1] As exergy has been defined purely in terms of work, an exergy balance around a process, including its interaction with the environment, should result in an exergy loss that is precisely the amount of lost work; thus

$$\mathrm{Ex}_{in} - \mathrm{Ex}_{out} = \mathrm{Ex}_{lost}$$
$$= W_{lost} \tag{6.17}$$

Determination of the amount of lost work does not require the introduction of the concept of exergy. With the definition of Equation 6.10, the overall ΔH and ΔS decide on the amount of minimum work required to bring about the change in conditions. More work brought into the process than is strictly necessary according to Equation 6.10 must have been the work dissipated on overcoming the process "frictions."

Equation 6.10 led to the definition of exergy, whereas the same expression in Equation 3.20 does not. Both equations express the minimum amount of work to transform the conditions of a defined amount of mass from those in state 1 into those of state 2. But if we choose state 1 as that of the environment, the environment suddenly acts as the datum level for a property of the amount of mass considered. The simplest example we can think of is air. Ignoring for the moment other constituents than nitrogen and oxygen, air at environmental conditions $P_0 T_0$ is "powerless" to perform work for us. According to Equation 6.12 its exergy is indeed zero.

6.2.1   Out of Equilibrium with the Environment: What It Takes to Get There

In Chapter 2, we pointed out that in thermodynamics it is important to distinguish between the system and its environment. We discussed the state of equilibrium of a system as the state in which no driving forces in pressure, temperature, concentration, or otherwise are present, within the system or in its interaction with its environment. If, in contrast, we have a look at ourselves we must admit that we are living *out of equilibrium* with our environment. At this point we need to define what we mean by "the environment." By this we mean the nonliving, inorganic or dead, environment of, for example, oxygen in the air, the water in the seas and oceans, the sand and rocks on the beach, and so on. More in general, living systems, from the simplest microbe, to the largest animal or tree, are out of equilibrium with the dead environment, not so much in pressure, perhaps to some extent in temperature but most certainly in composition and structure. Thermodynamics allows us to characterize this state with at least two characteristic properties as we will show now. We will discuss the first parameter in this section, the second parameter in the next section.

We turn to Figure 6.2, which shows a specific mass, let us say 1 mol, of air that is out of equilibrium with the environment *only* with respect to its *temperature T*, which has been chosen to be larger than T_0, the temperature of the environment. As has been shown before in this chapter, Equation 6.11, thermodynamics allows us to calculate the minimum amount of work required to bring this mass out of equilibrium with the environment: $W_{required}^{min}$. $W_{required}^{min}$ is the first parameter to be considered as a measure for how far this mole of air is out of equilibrium with the environment. This value is known as *exergy*, as shown earlier, implying that this is the amount of work that this mole of air has at most available to perform work as it is brought back to the conditions of the environment. This is a physical example, but we can also take a chemical example. We consider 1 mol of methane, CH_4, the major component of natural gas, at pressure P_0 and temperature T_0 of the environment (Figure 6.3). This methane is out of equilibrium with the dead environment in *composition*. As we will show in Chapter 7, thermodynamics allows us to calculate the minimum work that is required to compose this mole of CH_4 from the constituents CO_2 and water in the air, releasing oxygen upon its formation

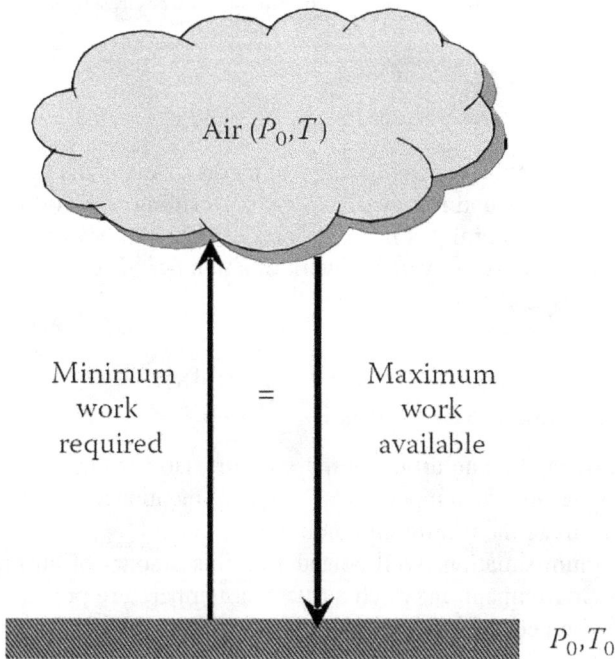

FIGURE 6.2 Physical exergy: out of equilibrium with the environment in temperature.

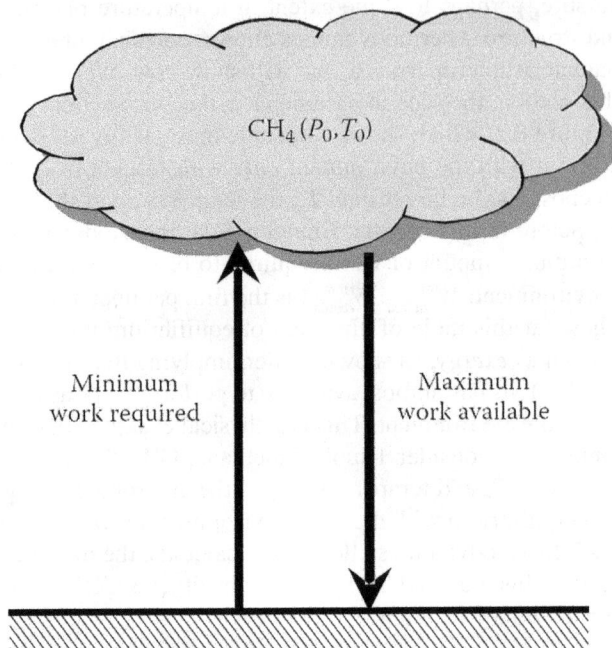

FIGURE 6.3 Chemical exergy: out of equilibrium with the environment in composition.

$$CO_2 + 2H_2O \rightarrow CH_4 + 2O_2 \qquad (6.18)$$

This work is called the *chemical exergy* of methane and is the maximum amount of work that this compound has available for performing work in the environment (Figure 6.3). Indeed, natural gas is an important fuel for a power station. Table 6.1 gives the chemical exergy or available work of a number of compounds: energy carriers, raw materials, and pure products.

6.2.2   Out of Equilibrium with the Environment: What It Takes to Stay There

In Figure 6.2, we studied an arbitrary mass of air that was out of equilibrium with its environment because its temperature $T > T_0$. We mentioned that thermodynamics allows us to calculate the minimum amount of work, $W_{required}^{min}$, that it takes to create this nonequilibrium situation. Well considered, this amount of air creates a structure within its environment, in which a different temperature prevails. Without this temperature difference, this amount of air would be unnoticed. Now that it differs in temperature, a spatial structure in temperature has been created. This structure can be made more complicated. Not only can we assume patterns of temperature in this sample of air, but equally patterns of pressure and nitrogen and oxygen concentrations. For each structure, we are able to calculate the minimum amount of work to

TABLE 6.1

Exergy Values for Various Compounds.

Fuels and Foods	Available Work (kJ/mol)
C (coal)	410
H_2 (fuel of the future?)	236
CH_4 (natural gas)	832
$-CH_2-$ (oil)	653
CH_3OH (methanol)	718
CH_xO_y (biomass)	490
CH_2O (carbohydrates)	504
$CH_2O_{0.1}$ (fat)	630

create it out of the air at the conditions of the environment. We have mentioned that this is the first parameter, of strictly thermodynamic nature, with which we can characterize the structured nonequilibrium state. But there is more to it. If we leave this structure to itself it will in no time spontaneously fall back to the equilibrium state within the environment. If the only difference with the environment is temperature $T > T_0$, heat will flow to the environment until the original state of homogeneous air at P_0 and T_0 has been restored. The work available in this sample of warm air, its exergy, has now been *dissipated*.

If we want to maintain the nonequilibrium state and prevent it from falling back to the equilibrium state of the environment, we will have to supply energy, or rather work, with at least the same rate as the rate with which the air sample is falling back to the equilibrium state at P_0 and T_0. So a second parameter to characterize the non-equilibrium state is the minimum power input, $\dot{W}_{maintenance}$, to keep the nonequilibrium state as it is. This input rate of work is required to compensate for the work lost in *dissipation* and corresponds to the minimum power input of an ideal heat pump that pumps heat leaked to the environment at T_0 back to the temperature $T > T_0$ (Figure 6.4).

6.2.3 DISSIPATIVE STRUCTURES

All work that is introduced per unit time into the heat pump is ultimately *lost work*, but with the advantage that the nonequilibrium state is maintained. Given the structured nature of the nonequilibrium state as discussed earlier, given the dissipation of the work input for maintenance, it will come as no surprise that Ilya Prigogine spoke of *dissipative structures*. Thinking about it, we will become aware that all living organisms are dissipative structures. By digesting one or the other source of food they succeed during their lifetime to stay out of equilibrium with their dead environment. But a dissipative structure does not necessarily need to be a living system. What to think of a chemical factory or an oil refinery in process? Such large structures are made up of many substructures such as distillation or extraction columns,

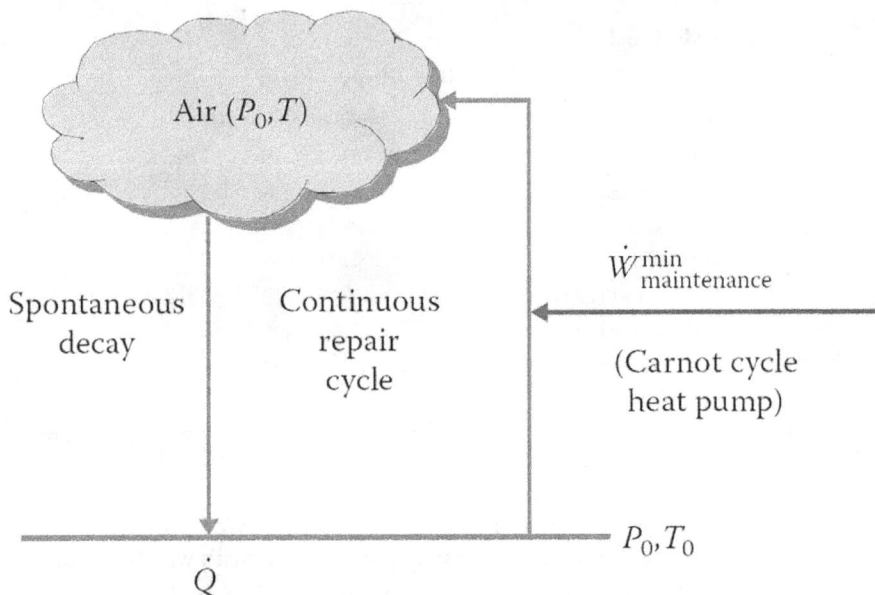

FIGURE 6.4　Maintenance of the nonequilibrium state.

reactors, and so on. Their periodic energy bills confirm that such working structures are dissipative structures too, but as they are not spontaneously reproducing themselves, they cannot be considered as living systems.

6.2.4　Physical and Chemical Exergy

Air at $P = 10$ bar and $T = T_0$ has, according to Equation 6.11, an exergy value different from zero because $H - H_0 = 0$ (ideal gas) and $S - S_0 = -R \ln P/P_0 = -R \ln 10$. Therefore, $Ex = RT_0 \ln 10$ and thus about 6 kJ/mol. Indeed, 1 mol of air at 10 bar should, potentially, be able to perform work for us living in an environment of 1 bar. This is *at most* the amount of work calculated and is called the exergy of air at 10 bar and ambient temperature. Reversely, it will require *at least* this amount of work to transform air from ambient conditions $P_0 T_0$ into air at 10 bar and ambient temperature. This is exactly the purpose of introducing the concept of exergy: expressing the amount of work available in an amount of mass, in rest or flowing, to perform work if it prevails at conditions different from those of the environment. This simple example may suggest that the exergy, at ambient conditions, of an amount of natural gas, for simplicity to be considered as methane only, is also zero. Indeed, applying Equation 6.11, we arrive at an exergy value of zero. But this calculation has only accounted for differences in *physical* conditions of methane with respect to the environment—pressure and temperature. Of course, as we discussed in Section 6.2.1, the *chemical* composition of methane is strikingly different from that of the environment, and that is, indeed, the reason why methane is so popular and is needed as

---

TABLE 6.2

Exergy Values of Methane (kJ/mol).

P (bar)	T (°C)	Ex_{ph}	Ex_{ch}	Ex
1	25	0.0	831.6	831.6
100	25	11.0	831.6	842.6
100	100	11.3	831.6	842.9

---

an "energy source." More correctly, this source of exergy is of chemical origin. By combustion with one component of the environment, oxygen, it can be converted into two other components of the environment, CO_2 and H_2O, and this reaction is widely used as the source of "energy." Therefore, we distinguish between physical exergy and chemical exergy as in the following equation

$$Ex = Ex_{phys} + Ex_{chem}^0 \qquad (6.19)$$

The first term on the right-hand side of this equation expresses the amount of work available due to differences in pressure and temperature with the environment. The second term, the chemical exergy, expresses the amount of work available due to the differences in *composition* with respect to the environment. The superscript in Ex_{chem}^0 expresses that the chemical exergy is considered at ambient conditions.

Table 6.2 lists exergy values for methane. It is clear from this table that methane carries an impressive amount of exergy as chemical exergy. Further, the table shows (1) the influence of increased pressure and temperature on the physical exergy and (2) that this latter contribution of exergy is nearly two orders smaller than the chemical contribution. Chemical exergy is the exclusive subject of Chapter 7.

6.3   EXAMPLE OF A SIMPLE ANALYSIS

Figure 6.5 summarizes the characteristics of an industrial process. Usually a raw material is required and "energy," the exergy or work available in an "energy" source or fuel, to produce the product. More often than not, the process also produces by-products and/or waste and cannot escape the second law that states that no real process can take place without entropy generation, which, as we have seen earlier, can be translated into the loss of the original work available or exergy.

The thermodynamic efficiency η of the process is then defined as the ratio of the available work leaving the process and the available work entering the process. Table 6.3 shows the chemical exergies of raw materials and products for some important processes.

We would now like to illustrate essentials of such an analysis and the role of the exergy concept with a simple example. We borrow this example from Sussman [2] because we can hardly think of a nicer and clearer illustration. Figure 6.5 illustrates how a stream of 1 kg/s of liquid water at 0°C is adiabatically mixed with a second

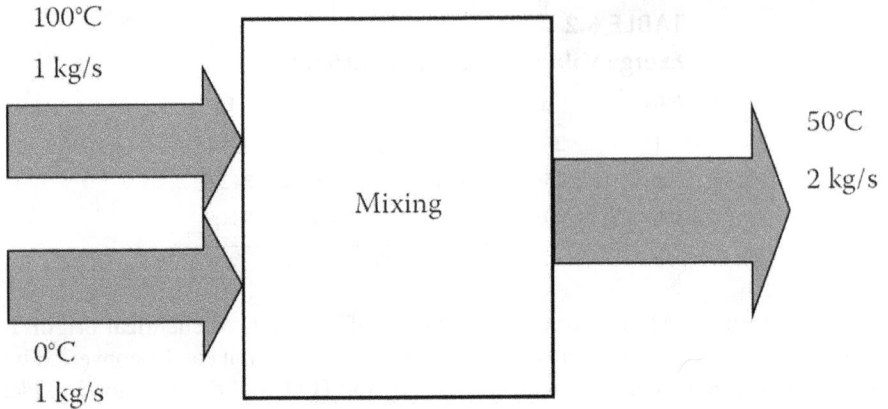

FIGURE 6.5 Flow diagram for the mixing of a hot and a cold stream.

TABLE 6.3
Sample Chemical Exergies of Raw Materials and Products.

Materials	Available Work (kJ/mol)		Raw Materials	Available Work (kJ/mol)
Si (silicon)	855	⇔	SiO_2	2–8
Al (aluminum)	888	⇔	Bauxite	<20
Cu (copper)	134	⇔	Copper oxide	<10
Fe (iron)	376	⇔	Iron ore	<20
Plastic	~653	⇔	Crude oil	653

stream of 1 kg/s of liquid water at 100°C to produce a stream of 2 kg/s of liquid water. The task at hand is to provide a thermodynamic analysis or exergy analysis of this process. The temperature of the environment is 25°C.

The analysis requires the calculation of three exergy flow rates, at 0°C, 50°C, and 100°C. As no heat or work is transferred between the considered system and its environment, the first law, Equation 6.1, yields that the overall enthalpy change is zero:

$$\dot{m}_3 H_3 - \dot{m}_1 H_1 - \dot{m}_2 H_2 = 0 \qquad (6.20)$$

The mass balance dictates that

$$\dot{m}_3 = \dot{m}_1 + \dot{m}_2 \qquad (6.21)$$

Multiplying Equation 6.21 with H_0, the enthalpy of liquid water at T_0, and subtracting the result from Equation 6.20, we obtain

$$\dot{m}_3 \left(H_3 - H_0 \right) = \dot{m}_1 \left(H_1 - H_0 \right) + \dot{m}_2 \left(H_2 - H_0 \right) \qquad (6.22)$$

As the change of enthalpy with pressure under these conditions can be neglected and, to some approximation, the specific heat of liquid water can be considered constant in this example, we can write

$$H_i - H_0 = c_p \left(T_i - T_0 \right)^{\circ} \ (i = 1,2,3) \qquad (6.23)$$

Introducing this relation into Equation 6.22 gives us the temperature T_3 of the flow leaving the mixer: $T_3 = 323.15$ K. The exergy flow rates can now be calculated from

$$\dot{E}x\,x_i = \dot{m}_i \left[\left(H_i - H_0 \right) - T_0 \left(S_i - S_0 \right) \right] \qquad (6.24)$$

The enthalpy term is taken care of by Equation 6.22, and the entropy term can be calculated from Equation 2.14 resulting in

$$S_i - S_0 = c_p \ln \left(\frac{T_i}{T_0} \right) \qquad (6.25)$$

where we have made the same assumptions as for arriving at the enthalpy term in Equation 6.23. The results are presented in the so-called Grassmann exergy flow diagram (Figure 6.6).

The diagram clearly shows the "value" of each flow in terms of its ability, or potential, to perform work. The diagram also shows that a considerable amount of the exergy flowing into the system is dissipated as a result of the mixing process.

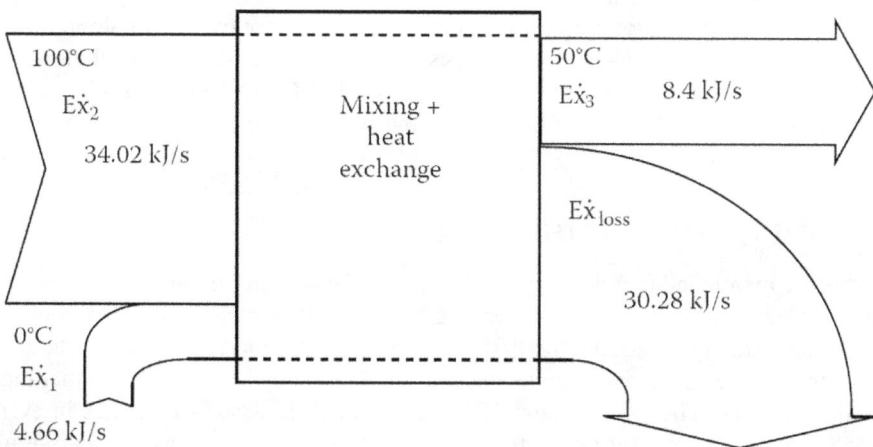

FIGURE 6.6 Exergy flow diagram or Grassmann diagram for the example of Figure 6.5.

However, exergy values are not required to calculate lost work. We could have calculated the same value from

$$\dot{W}_{lost} = T_0 \dot{S}_{gen} \tag{6.26}$$

with

$$\dot{S}_{gen} = \dot{m}_3 S_3 - \dot{m}_1 S_1 - \dot{m}_2 S_2 \tag{6.27}$$

which can be transformed into

$$\dot{S}_{gen} = \dot{m}_3 \left(S_3 - S_0 \right) - \dot{m}_1 \left(S_1 - S_0 \right) - \dot{m}_2 \left(S_2 - S_0 \right) \tag{6.28}$$

allowing the calculation of $\dot{S}_{gen}$ and thus of $\dot{W}_{lost}$ directly by making use of Equation 6.26. Although $\dot{W}_{lost}$ is directly related to the loss of energy resources, its value does not put the process in perspective. Exergy analysis emphasizes the value of the flows and thereby shows the efficiency of the process in terms of

$$\eta = \frac{\dot{E}_{out}}{\dot{E}x_{in}} = 0.216 \tag{6.29}$$

So nearly 80% of the original exergy is wasted in the process, which finds its origin in the temperature difference of 100°C between the flows entering the adiabatic mixer. This difference reduces to close to 0°C in the mixing process. Equation 3.8 explains the resulting work lost in the process. Reducing the loss is rather senseless in this case as the introduction of Carnot engines would not be worth the effort, so it seems. This observation gets another dimension, however, if one realizes that in a country like the Netherlands, nearly every citizen starts his or her day by taking a shower with water that is a mixture of tap water and water from a gas-fired boiler. The amount of exergy entering the process is then even (much) higher than in our example, reducing the efficiency to $\eta = 0.03$, or 3%. Losses on this scale warrant innovative measures, such as discussed in the next example in Section 6.5 and in Chapter 9.

6.4 THE QUALITY OF THE JOULE

Earlier we mentioned that the first law deals with the quantity of energy, and the second law deals with the quality of energy. Looking at the system *and* its environment at the same time, we see that the first law expresses that for a real process the total number of Joules involved remains unchanged. The second law expresses that their quality declines. The total amount of exergy—available work—dissipates in every process due to irreversibilities in the process. It would be nice if we could assign a quality to the Joule, expressing to what extent the Joule concerned has work available. This "quality" q with the dimension J/J should have a value $0 \leq q \leq 1$. We return to Figures 3.4 and 6.1 here and write

$$\dot{W}_{in} = \Delta\dot{H} + \dot{Q}_0 \qquad (6.30)$$

We rewrite this equation as

$$\dot{W}_{in} - \dot{Q}_0 = \Delta\dot{H} \qquad (6.31)$$

This equation expresses that the *net* energy transferred into the system is apparently stored in the system's enthalpy. So ΔH reflects the system's *energy* intake in its interaction with its environment. The same equation can be rearranged, by combining the first and second laws into

$$\dot{W}_{in} = \Delta\dot{H} - T_0\Delta\dot{S} + T_0\dot{S}_{gen} \qquad (6.32)$$

From this equation we established the earlier minimum for the amount of work, the ideal work, required to change the system's conditions from state 1 to 2 with $\Delta P = P_2 - P_1$ and $\Delta T = T_2 - T_1$ and thereby its thermodynamic properties $\Delta H = H_2 - H_1$ and $\Delta S = S_2 - S_1$:

$$\dot{W}_{in}^{min} = \Delta\dot{H} - T_0\Delta\dot{S} \qquad (6.33)$$

The real amount of work required exceeds this amount by

$$\dot{W}_{lost} = \dot{W}_{in} - \dot{W}_{in}^{min} = T_0\dot{S}_{gen} \qquad (6.34)$$

Equation 6.33 provides the definition of exergy if state 1 is chosen as the state at ambient condition, namely, $P_1 = P_0$ and $T_1 = T_0$: the minimum amount of work required to transfer the system from environmental conditions to those at P_2 and T_2. At these conditions, this is the maximum amount of work available for the reverse process. That is the valuable idea behind the exergy concept: to be able to assign to any process stream a value, its exergy, that expresses the confined work available in the stream. For the general change in state from P_0, T_0 to P, T, we can write the net energy input as

$$En_{in}^{net} = \Delta\dot{H} \qquad (6.35)$$

whereas the available work is given by

$$Ex = \Delta H - T_0\Delta S \qquad (6.36)$$

From this, it follows that every Joule that has entered the system has a quality defined as

$$q \equiv \frac{Ex}{E_{in}^{net}} = 1 - \frac{T_0\Delta S}{\Delta H} \qquad (6.37)$$

TABLE 6.4
The Quality of the Joule (Fraction Available for Work in J/J; $T_0 = 288.15$ K).

Chemical	~1
Hot combustion gases	0.6–0.8
Waste heat	0.2–0.3
Chimney gases	0.2
Hot water (100°C)	0.12
Warm water (50°C)	0.06

Table 6.4 lists the quality of the Joule for various energy carriers, from air in the environment to the Joule in hot combustion gases. The table illustrates that the chemical Joule brought in by methane has suffered in quality from the spontaneous reaction of burning the fuel to its combustion gases. It also illustrates the value of so-called waste heat if we take into consideration the quality of the Joule that we need for human comfort in the house via hot and warm water. We must emphasize that its meaning is lost as soon as the temperature falls below that of the environment. Nevertheless, in Chapter 9, which deals with the exergy analysis of energy conversion processes for producing work, we frequently use this concept for illustrative purposes.

At this point, it is useful to show how the second law can be related to the transformation of energy while making use of the "quality of the Joule" concept. Let us turn to Figure 6.7 in which the system is defined as contained within the rectangle and prevails in a *steady state*, that is, its properties do not change with time. Energy flows 1 and 2 enter the system, energy flows 3 and 4 leave the system, and the first law requires that

$$En_1 + En_2 = En_3 + En_4 \tag{6.38}$$

because energy is conserved. It is interesting to observe that if we would reverse the direction of the arrows, the same equation would be valid: the first law is not concerned with the direction of the process. Now suppose that we are able to assign a quality q to each flow of energy. By quality we mean a factor that indicates which fraction of energy is available for work (Figure 6.8). The second law states that the capacity of any total system to perform work is decreasing in time. So for the proper direction of the process

$$En_1 \cdot q_1 + En_2 \cdot q_2 > En_3 \cdot q_3 + En_4 \cdot q_4 \tag{6.39}$$

The work available in the inflowing streams is larger than the work available in the outflowing streams. In the process, *available work* is lost. The product of energy, En, and its quality with respect to our environment, q, is called *exergy*, Ex. So another formulation of the second law is that in a real process exergy is always, partly or

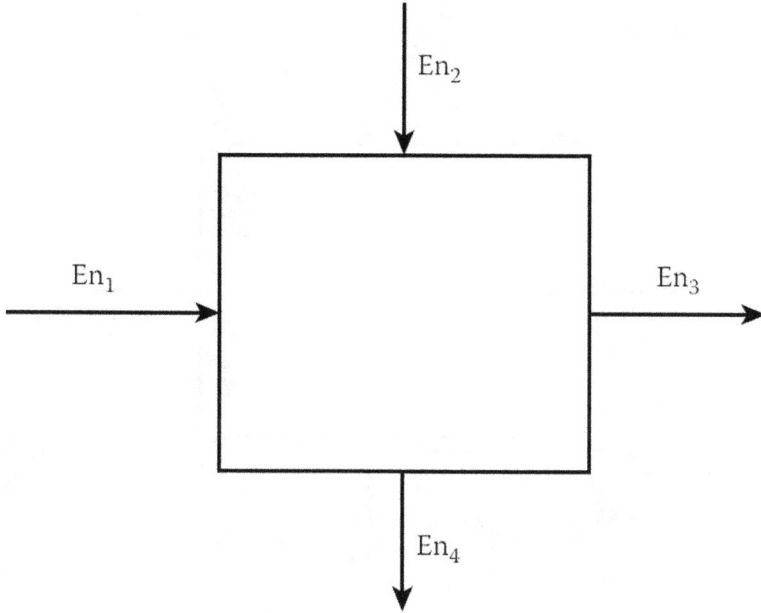

FIGURE 6.7 The first law in a steady-state flow process: energy is conserved, Equation 6.38.

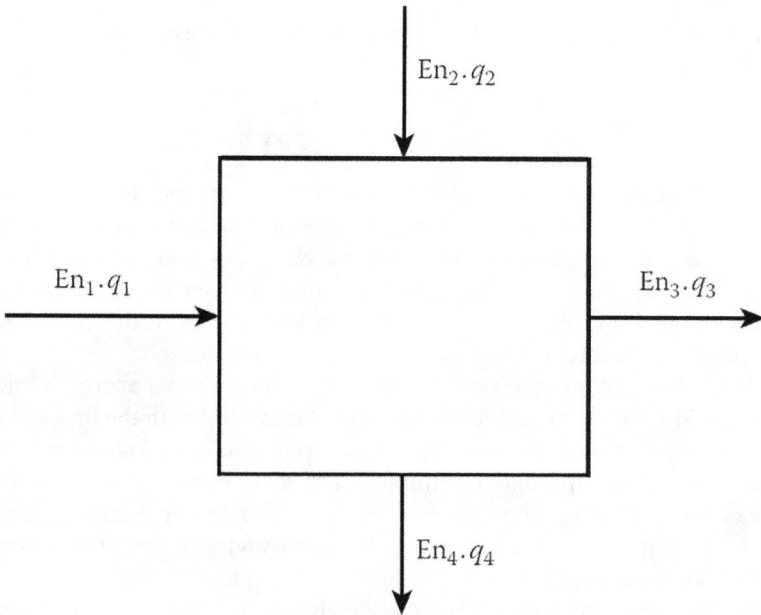

FIGURE 6.8 The second law states that the quality of energy is not conserved but diminishes Equation 6.39.

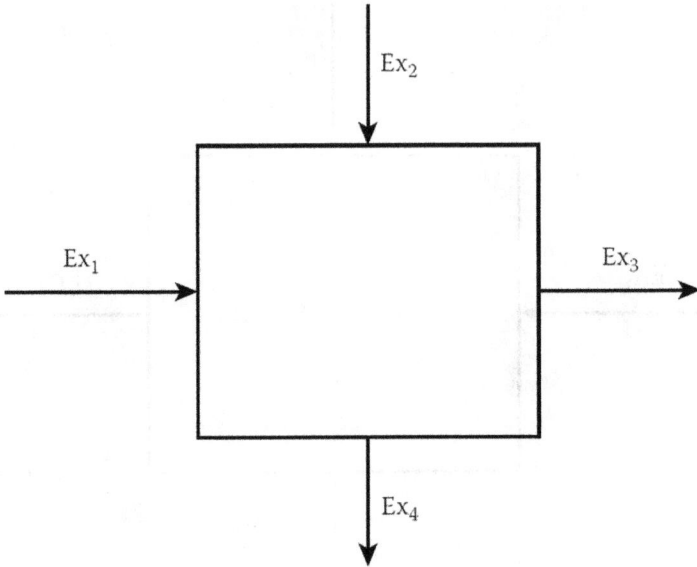

FIGURE 6.9 In a real process exergy is always lost, partly or completely: $Ex_1 + Ex_2 > Ex_3 + Ex_4$.

completely, lost, whereas only in an ideal reversible process exergy is conserved (Figure 6.9).

6.5 EXAMPLE OF THE QUALITY CONCEPT

Figure 6.10 pictures the case of a natural gas-fired power station. The hot combustion gases drive a combined cycle consisting of a gas turbine and a steam power plant. The power station "cogenerates," or produces, electricity and heat. The produced heat and electricity are used to upgrade the quality of water in the environment by raising its temperature. We wish to compare this process with the case where hot water is produced by direct combustion of the gas in a boiler.

The Joules brought into the system originate in the chemical energy of methane. We consider 10 J, $En_{in} = 10$, and follow the fate of these 10 J with the first and second laws. From process data, it follows that 5 J end up in electricity and 4 J in hot water. One Joule finds its way through the chimney. The 5 J of electricity increase the temperature of water of ambient conditions to that of hot water by means of a heat pump; based on the efficiency of this heat pump, 10 J are withdrawn from the environment and 15 J of hot water are the result.

Let us now follow this process in more detail with the second law and make use of the quality concept. To keep things simple, we make use of a simplified version of Equation 6.37

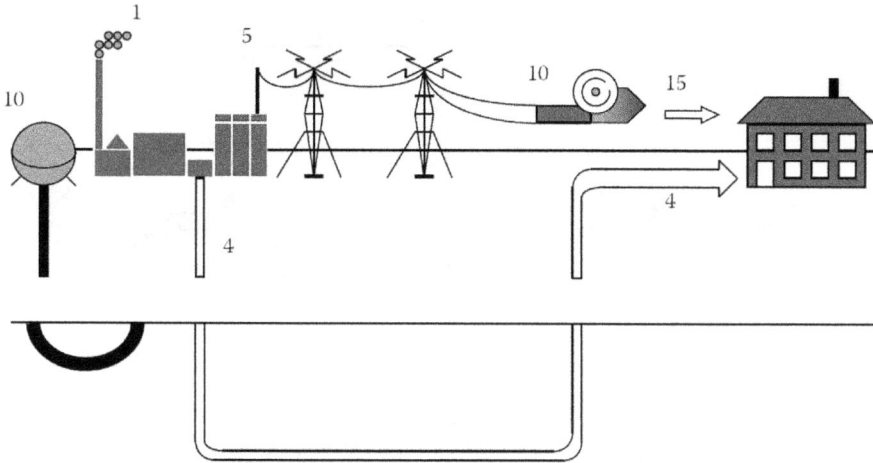

FIGURE 6.10 Natural gas-driven cogeneration power plant for the production of electricity and hot water.

$$q_i = 1 - \frac{T_0 \ln\left(T_i / T_0\right)}{T_i - T_0} \tag{6.40}$$

which results from Equation 6.37 by assuming c_p as a constant. By definition, the quality of Joules in electricity is unity, as theoretically, electricity can be fully converted into work. As we shall see in Chapter 7, the quality of Joules in natural gas is approximately unity as well. The quality of the Joules in the chimney gases and in the hot water has been calculated by means of Equation 6.40 and is found to be 0.20 and 0.12, respectively. This allows a simultaneous analysis of the energy and its quality in the alternative process: the production of hot water directly from the combustion of natural gas. For the latter, the exergetic efficiency can be calculated according to $(9 \times 0.12)/(10 \times 1) = 0.11$ because 10 J of fuel have been transformed into 9 J of hot water (see Table 6.4), and 1 J exits through the chimney. For the cogeneration process, the efficiency is calculated as $(19 \times 0.12)/(10 \times 1) = 0.23$, where we have neglected the exergy of the chimney gases, since these are not used to heat the water. Figure 6.11 summarizes the results.

The exergetic efficiency of the combustion has been calculated to be $\eta = 0.78$, or 78%. For this calculation, we used Equation 6.36, as we have assumed conditions given by Cengel and Boles [3] for the inlet gases of turbines, namely, 1150°C and 14 bar (see Chapter 9 for a similar calculation). This higher pressure will not affect ΔH. According to Equation 6.14 for the dependence of the entropy on pressure for ideal gases, this term has to be included. The exergetic efficiency of the cogeneration step is $(5 \times 1 + 4 \times 0.12)/(10 \times 0.78) = 0.70$ or 70% in which we have considered the exergy content of the chimney gases to be lost. The thermodynamic efficiency of the entire cogeneration plant is thus $0.78 \times 0.7 = 0.55$, or 55%, a result only achieved in the most

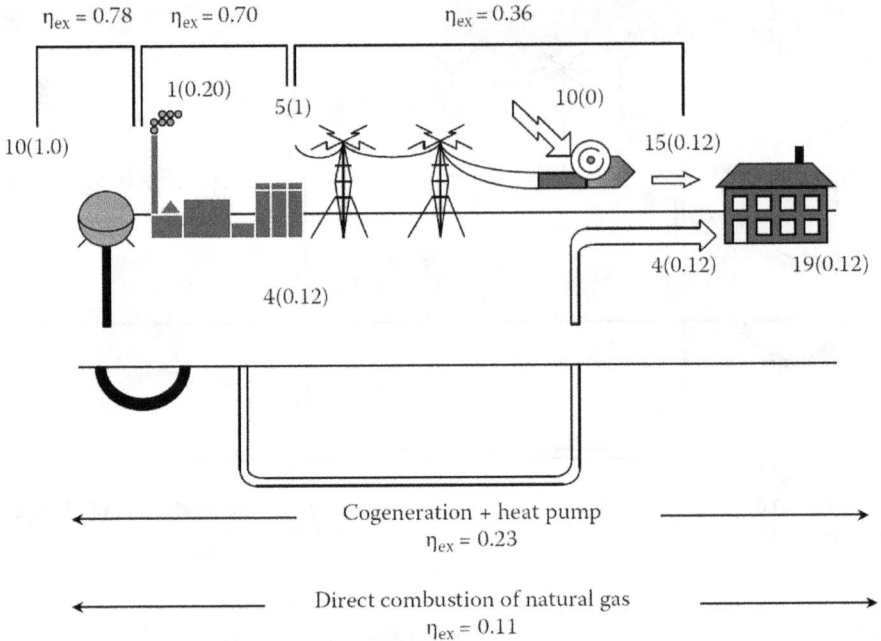

$\eta_{ex} = 0.78$ $\eta_{ex} = 0.70$ $\eta_{ex} = 0.36$

1(0.20)
5(1)   10(0)

10(1.0)   15(0.12)

4(0.12)   19(0.12)

4(0.12)

Cogeneration + heat pump
$\eta_{ex} = 0.23$

Direct combustion of natural gas
$\eta_{ex} = 0.11$

FIGURE 6.11 Thermodynamic analysis in terms of quality, quantity, and efficiency.

advanced power plants [3]. The thermodynamic efficiency of the heat pump has been computed from $(15 \times 0.12)/(5 \times 1) = 0.36$.

The exergetic efficiency of the cogeneration process including the heat pump, $\mu_{ex} = 0.23$, appears to be more than twice that of the direct combustion process, $\eta_{ex} = 0.11$. So, this is a 100% improvement in efficiency. We note that the exergetic efficiency of the heat pump is 0.36. If the quality aspect of the Joule were not considered and thus all Joules would be considered equal, which they are numerically but not in quality, then the efficiency of the heat pump could be interpreted as 15/5, or 300%. Such calculations can be very misleading, and therefore we believe that both the exergy and the quality concept are instrumental in improving communication among everybody involved and concerned. In this context, we wish to refer to the quote that we give in Chapter 3 on the quality and quantity of energy and the need to make this distinction in policy decisions.

6.6 CONCLUSION

Exergy is a convenient concept if one wishes to assign a quantitative quality mark to a stream or a product. This quality mark expresses the maximum available work or potential to perform work because of its possible differences in pressure, temperature, and composition with the prevailing environment. The physical exergy, Ex_{phys}, only accounts for the differences in *pressure* and *temperature*; the standard chemical exergy, Ex^0_{chem}, accounts for the difference in *composition* with the environment at the environment's pressure and temperature. Thus

$$Ex = Ex_{phys} + Ex^0_{chem}$$

(6.41)

The convenience of the exergy concept houses the possibility to discuss energy issues on a clear and quantitative basis, in particular if the exergy values assigned to process streams are combined with exergy losses incurred in the processes in which these streams participate, either actively or passively, as the next chapter illustrates.

NOTE

1 Other forms of exergy can be those associated with kinetic and potential energy, which we have neglected in this treatment.

REFERENCES

1. Seader, J.D. *Thermodynamic Efficiency of Chemical Processes*, MIT Press: Cambridge, MA, 1982.
2. Sussman, M.V. *Availability (Exergy) Analysis*, Mulliken House: Lexington, MA, 1980.
3. Cengel, Y.A.; Boles, M.A. *Thermodynamics*, McGraw-Hill: New York, 1994.

7 Chemical Exergy

In the last chapter, the concepts of exergy and physical exergy, in particular, were introduced. This chapter deals with three other important concepts, namely, exergy of mixing, chemical exergy, and cumulative exergy consumption, and their numerical evaluation.

7.1 INTRODUCTION

Recall that exergy values reflect the extent to which a compound or mixture is out of equilibrium with our environment. Examples are differences in pressure and temperature with the environment. Differences in temperature lead to heat transfer, while differences in pressure lead to mass flow. Chapter 6 shows that the *physical* exergy represents the maximum amount of work that can be obtained from a system by converting a system's pressure and temperature to those of our environment.

It appears, however, that when a system's physical exergy is zero and thus the system prevails in a state of thermomechanical equilibrium with the environment, it may still be out of equilibrium with that environment in other respects. The origin is to be found in the difference in the composition and nature of the components making up the system and the environment, respectively. These differences lead to values for the *exergy of mixing* and the *chemical exergy*. Earlier, we pointed out that though the physical exergy of methane is zero, its chemical exergy is not. Equally, pure nitrogen and oxygen have nonzero chemical exergies because their mole fraction in the environment is different from 1. The process of mixing plays an important role in the determination of the chemical exergy, so Section 7.2 will deal with the exergy of mixing.

7.2 EXERGY OF MIXING

To clarify the concept of the exergy of mixing, we give the example of pure oxygen at ambient conditions P_0 and T_0. Consider a system, for convenience chosen at P_0, T_0, isolated from the environment and consisting of two separate compartments containing oxygen and air, respectively. The two compartments, initially separated by an elastic diathermal barrier, and thus in mechanical and thermal equilibrium, are brought in contact with each other by removing the barrier. Oxygen and air will diffuse into each other, and eventually an equilibrium will be reached where oxygen and air have mixed into a homogeneous mixture. The initial condition of oxygen is apparently not one of complete equilibrium with the environment (i.e., with air) despite the equality in pressure (P_0) and temperature (T_0). The thermodynamic potential of pure oxygen is higher than that of oxygen in air at P_0 and T_0. On mixing of the components of air in their pure state to a homogeneous mixture, the thermodynamic potential of each component decreases. The associated change in exergy is

DOI: 10.1201/9781003304388-9

$$\Delta_{min}Ex = \Delta_{min}H - T_0\Delta_{min}S \qquad (7.1)$$

As the mixing process takes place at P_0, T_0, we may write

$$\Delta_{min}Ex_{P_0T_0} = \Delta_{min}G_{P_0T_0} \qquad (7.2)$$

For the calculation of the exergy value at P, T of a mixture, of a given composition, with respect to the exergy values of the pure components at P and T, the exergy difference is defined as

$$Ex_{min} \equiv \Delta_{min}Ex \qquad (7.3)$$

with values for $\Delta_{mix}H$ and $\Delta_{mix}S$ at the conditions P and T. Chapter 10 presents an example of the industrial distillative separation of the mixture of propane and propene, in which the exergy of mixing is very prominent.

7.3 CHEMICAL EXERGY

For the determination of a compound's chemical exergy value we need to define a reference environment. This reference environment is a reflection of our natural environment, the earth, and consists of components of the atmosphere, the oceans, and the earth's crust. If, at P_0 and T_0, the substances present in the atmosphere, the oceans, and the upper part of the crust of our earth are allowed to react with each other to the most stable state, the Gibbs energy of this whole system will have decreased to a minimum value. We can then define the value of the Gibbs energy for a subsystem, the "reference environment"—at sea level, at rest, and without other force fields present than the gravity field—to be zero as well as for each of the phases present under these conditions. It is a logical extension of these assumptions to define the thermodynamic potentials of each of the substances present in the different phases to have a value of 0 J/mol. With respect to this "reference environment," we then determine the thermodynamic potentials of all kinds of substances in all kinds of phases at P and T. From this "reference environment" it is not possible to obtain any work. Therefore, this state is also meaningful as a reference state for the determination of exergy values at P_0 and T_0. This finally leads to the definition $Exi (P_0, T_0) = \mu, (P_0, T_0)$ for the subsystem at sea level, at rest, and without the presence of any other force field than the gravity field. The concept of this "reference environment" is illustrated for a number of the so-called reference components.

7.3.1 REFERENCE COMPONENTS FROM AIR

Apart from differences in chemical concentration, or better, thermodynamic potentials, such as for oxygen, there can be other situations for being out of equilibrium with the environment at P_0, T_0. Consider, for instance, the material graphite. Graphite can spontaneously react with oxygen to form CO_2, but for kinetic reasons the reaction is very slow and graphite seems to be stable in our environment, although in the

TABLE 7.1

Partial Pressure of Various Components in Air.

Component	P_i (kPa)
N_2	75.78s
O_2	20.39
CO_2	0.0335
H_2O	2.2
D_2O	0.000342
He	0.000485
Ne	0.00177
Ar	0.906
Kr	0.000097
Xe	0.0000087

presence of oxygen it is metastable with respect to CO_2. As a result, it has a significant amount of *chemical* exergy available and can be considered as an important energy carrier because it is highly out of equilibrium with the environment.

In our environment, there are many substances that, like oxygen in our atmosphere, cannot further diffuse and/or react toward more stable configurations and may be considered to be in equilibrium with the environment. Neither chemical nor nuclear reactions can transform these components into even more stable compounds. From these components, we cannot extract any useful work, and therefore an exergy value of 0 kJ/mol has been assigned to them. This has been done for the usual constituents of air: N_2, O_2, CO_2, H_2O, D_2O, Ar, He, Ne, Kr, and Xe at $T_0 = 298.15$ K and $P_0 = 99.31$ kPa, the average atmospheric pressure [1]. Their partial pressures P_i in air are given in Table 7.1.

From these data, we can calculate the chemical exergy values of these components in the *pure* state at P_0 and T_0. Air at these conditions can, to a good approximation, be considered as an ideal gas; therefore, separation into its constituents will take place without a heat effect: $\Delta_{sep}H = 0$. And so, the only effect left in the exergy change of separation, $\Delta_{sep}Ex = -Ex_{mix}$ (see Equation 7.3), is that of the entropy of separation:

$$\Delta_{sep}Ex = \Delta_{sep}H - T_0\Delta_{sep}S$$
$$= T_0\Delta_{mix}S \tag{7.4}$$

As discussed in Chapter 2, the change in entropy associated with taking 1 mol of an ideal gas isothermally from pressure P_1 to a pressure P_2 is given by

$$\Delta S = -R\ln\left(\frac{P_2}{P_1}\right) \tag{7.5}$$

TABLE 7.2

Standard Chemical Exergy Values at P_0, T_0 of Various Components Present in Air.

Component	$\mathbf{Ex_{ch}^0}$ (kJ/mol)
N_2	0.72
O_2	3.97
CO_2	19.87
H_2O	9.49
D_2O	31.23
He	30.37
Ne	27.19
Ar	11.69
Kr	34.36
Xe	40.33

From this equation, we can show [2] that the standard chemical exergy at P_0 and T_0 of a pure component can be calculated from its partial pressure P_i in air with Equation 7.6:

$$Ex_{ch,i}^0 = RT_0 \ln\left(\frac{P_0}{P_i}\right) \tag{7.6}$$

The standard chemical exergy values for the main constituents of air as listed in Table 7.1 are given in Table 7.2.

Exergy values for the elements in their stable modification at $T_0 = 298.15$ K and $P_0 = 101.325$ kPa are called standard *chemical* exergy values Ex_{ch}^0. For the calculation of the chemical exergy value of all kinds of substances, the standard chemical exergy values of all elements are required.

7.3.2 Exergy Values of the Elements

The following example for graphite illustrates how the chemical exergy value for all other elements can now be calculated (Table 7.3). For the calculation of Ex_{ch}^0 of graphite, we make use of the reaction in which CO_2 is formed from the elements in their stable modification at P_0, T_0:

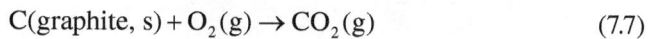

$$C(graphite, s) + O_2(g) \rightarrow CO_2(g) \tag{7.7}$$

TABLE 7.3
Standard Chemical Exergy Values of the Elements.

Element	Ex_{ch}^0 (kJ/mol)
Ag (s)	70.2
Al (S)	888.4
Ar (g)	11.69
As (s)	494.6
Au (s)	15.4
B (s)	628.5
Ba (s)	747.7
Bi (s)	274.5
Br_2 (1)	101.2
C (s, graphite)	410.26
Ca (s)	712.4
Cd (s,,)	293.2
Cl_2 (g)	123.6
Co (s_α)	265.0
Cr (s)	544.3
Cs (s)	404.4
Cu (S)	134.2
D_2 (g)	263.8
F_2 (g)	466.3
Fe (S_α)	376.4
H_2 (g)	236.1
He (g)	30.37
Hg (1)	115.9
I_2 (S)	174.7
K (s)	366.6
Kr (g)	34.36
Li (s)	393.0
Mg (s)	633.8
Mn (s_α)	482.3
Mo (s)	730.3
N_2 (g)	0.72
Na (s)	336.6
Ne (g)	27.19
Ni (s)	232.7
O_2 (g)	3.97
P (s, red)	863.6
Pb (s)	232.8
Rb (s)	388.6
S (s, rhombic)	609.6
Sb (s)	435.8
Se (s, black)	346.5
Si (s)	854.6

Element	Ex_{ch}^0 (kJ/mol)
Sn (s, white)	544.8
Sn (s)	730.2
Ti (s)	906.9
U (s)	1190.7
V (S)	721.1
W (s)	827.5
Xe (g)	40.33
Zn (s)	339.2

Source: Szargut, J. et al., *Exergy Analysis of Thermal, Chemical, and Metallurgical Process*, Hemisphere Publishing Corp., New York, 1988.

The corresponding change in standard Gibbs energy is called the standard Gibbs energy of formation of CO_2, $\Delta_f G_{298.15}^0$, and is defined as

$$\Delta_f G_{298.15}^0 \equiv \sum v_i \mu_{i,298.15}^0 \tag{7.8}$$

in which v is the so-called stoichiometric coefficient, denoted as positive for products and negative for reactants, and $\mu_i^0 = G_i^0$ is the standard thermodynamic potential or Gibbs energy for substance i. Equation 7.8 is based on the formation of 1 mol of the compound considered, in this instance 1 mol of CO_2. If we *define* the change in exergy in the same way

$$\Delta_f Ex_{298.15}^0 = \Delta_f G_{298.15}^0 = \sum_i v_i \mu_{i,298.15}^0$$
$$\equiv \sum v_i Ex_{ch,i}^0 \tag{7.9}$$

then the exergy of graphite can be calculated from

$$Ex_{ch,C(s)}^0 = -\Delta_f G_{298.15}^0 + 1 \cdot Ex_{ch,CO_2(g)}^0 - 1 \cdot Ex_{ch,O_2(g)}^0 \tag{7.10}$$

The values of $\Delta_f G_{298.15}^0$ for many compounds are listed in standard tables [2], and the value for CO_2 reads −394.359 kJ/mol. With the help of Table 7.2, which gives the standard chemical exergy values for CO_2 and O_2, Equation 7.10 allows the calculation of $Ex_{ch,C(s)}^0 = 394.359 + 19.87 - 3.97 = 410.26 kJ / mol$.

For the remaining elements, reference compounds have been chosen, as they occur in seawater or in the lithosphere, the earth's crust. An important aspect of this choice has been that the calculated exergy values of most compounds should be positive. Table 7.3 lists the standard chemical exergy values of the elements as presented in Szargut's well-known standard work [1]. Chapter 8 gives an example, the adiabatic combustion of H_2, to illustrate the use of these exergy values in an interesting application.

7.3.3   Chemical Exergy Values of Compounds

Table 7.3 is useful for the calculation of the standard chemical exergy values of compounds. We illustrate this for methane and start from its hypothetical formation reaction at standard conditions:

$$C(s) + 2H_2(g) \rightarrow CH_4(g) \tag{7.11}$$

Applying Equation 7.9 results in

$$Ex^0_{ch,CH_4(g)} = \Delta_f G^0_{298.15} + Ex^0_{ch,C(s)} + 2Ex^0_{ch,H_2(g)} \tag{7.12}$$

The first term on the right-hand side of this equation is the standard Gibbs energy of formation of methane, which is listed [2] as -50.460 kJ/mol and thus $Ex^0_{ch,CH_4(g)}$ can be calculated to be 831.6 kJ/mol. Chapter 9 illustrates the use of this exergy value in the analysis of a natural gas-driven power station.

In general, we can calculate the standard chemical exergy of a component j from the standard chemical exergy of its elements with the equation

$$Ex^0_{ch,j} = \Delta_f G^0_{j,298.15} + \sum v_i Ex^0_{ch,i} \tag{7.13}$$

We recall that the exergy of methane will be different for other values of P and T than P_0, T_0 and refer to Table 6.2 to demonstrate the influence of pressure and temperature on this exergy value. It is clear that the chemical contribution to the total exergy, $Ex = Ex_{phys} + Ex^0_{ch}$, in this case is dominant. At the same time, we should be aware that in a simple compression step, this contribution is irrelevant and should not be included in an exergy efficiency calculation. On the level of, let us say, 10 kJ/mol of physical exergy, the loss of 2.5 kJ/mol of exergy due to inefficiencies of the compressor results in a thermodynamic or exergetic efficiency of 75%. Had we included the 832 kJ of chemical exergy of methane, the thermodynamic efficiency would have been as high as 99.7%, which gives a completely blurred picture of the compressor's performance.

Finally, Table 7.4 gives the standard exergy values of a selected number of compounds that are relevant for the examples and topics presented in this book.

7.3.4   The Convenience of the Chemical Exergy Concept

In chemical thermodynamics, the reference components have been selected as the elements in their most common state at standard conditions, the standard state. These elements have been defined as having a zero standard Gibbs energy of formation. The standard Gibbs energy of formation of a compound is related to that of the elements from which it has been composed. Let us take liquid methanol, CH_3OH [1]. Its standard Gibbs energy of formation is -166.270 kJ/mol, a number that does not say very much other than that in the reaction

---

TABLE 7.4

Standard Chemical Exergy Values of Selected Compounds.

Substance	kJ/mol
CH_4 (g) "natural gas"	832
CH_3OH (g)	722
CH_3OH (l)	718
$-CH_{2-}$[a] "oil"	652
(CH_2O)[b] "biomass"	480
CO_2 (g)	20
SiO_2 (s, α quartz)	1.9
TiO_2 (s, rutile)	21.4
$Al_2O_3 \cdot H_2O$ (s) "bauxite"	200.8
Fe_2O_3 (s) "hematite"	16.5
NH_3 (g)	337.9
$CO(NH_2)_2$ (s) urea	689.0

[a] Crude oil on a per-carbon basis.

[b] Biomass (glucose) on a per-carbon basis.

$$C(s) + 2H_2(g) + \frac{1}{2}O_2(g) \rightarrow CH_3OH(l) \qquad (7.14)$$

the standard Gibbs energies at the left-hand side are zero and the standard Gibbs energy of reaction is also −166.270 kJ/mol. However, following the procedures as outlined in the preceding sections, we can calculate with Equation 7.13 the standard chemical exergy of liquid methanol to be −166.270 + 410.26 + 2 × 236.10 + 1/2 × 3.97 = 718.2 kJ/mol. This number is very meaningful, as it expresses the maximum amount of work available to us embodied in 1 mol of liquid methanol. We can then compare this with the value for methane and notice that the partial oxidation of methane to methanol has lowered the exergy value somewhat, from 832 to 718 kJ/mol. But methanol is in the liquid state, and this is an attractive feature for a transportation fuel. On the other hand, methanol has double the mass of methane, and so per unit of mass its available work or exergy is less than half. And last but not least, the efficiency of converting methane into methanol may be about 50%–60% (see Chapter 14) and much of the advantage of using methanol seems to have gone. Nevertheless, although, strictly speaking, the concept of exergy does not add anything in the fundamental sense, it certainly adds convenience, for example, for the discussion on the pros and cons of energy conversion such as in the earlier comparison of methane and methanol. This is one of the attractive features of the exergy concept that has made it so popular with many practitioners.

7.4 CUMULATIVE EXERGY CONSUMPTION

Suppose we deal with a process in which iron, Fe, has to be used as a reactant, for example, in a reduction reaction. The standard chemical exergy of Fe is 376.4 kJ/mol. If we wish to carry out a thermodynamic or exergy analysis of this process, this value is not appropriate. After all, to put the exergy cost of the product, for which Fe was needed as a reactant, in proper perspective, we need to consider all the exergetic costs incurred in order to produce this product all the way from the original natural resources—iron ore and fossil fuel in this example. The production of iron from, for example, the iron ore hematite and coal has a thermodynamic efficiency of about 30% [1], and therefore it is not 376.4 kJ/mol Fe that we need to consider but 376.4/0.3 = 1250 kJ/mol Fe. This value is called the cumulative exergy consumption (CExC) of Fe. It may well be that for proper analysis of the efficiency of the step consuming Fe to produce the product, we want to take the standard chemical exergy of Fe, but for the calculation of the CExC of the final product, we need to include the CExC of Fe. Chapter 14 discusses many examples where the exergy of the final product is compared with the CExC of the same product. Together these two values allow the calculation of the thermodynamic, exergetic, efficiency of a process yielding the product from natural resources. This is part of the subject of Chapter 14.

We recall that, without mentioning it, we touched upon the topic of cumulative exergy consumption before. In Chapter 6, we illustrate the application of the concept of physical exergy with the simple example of mixing liquid water of 100°C with that of 0°C. In that example, we first take the exergy value of hot water as 34 kJ/kg. But when this water has been produced from natural gas, its accumulated exergy consumption is calculated according to Table 6.4 to be 1/0.12 × 34 = 283 kJ/kg.

Finally, we consider another important contributor to the CExC of a product. We refer to the equipment being used in the process. This equipment also has to be manufactured from resources originally taken from the environment. This cumulative exergy consumption has to be discounted over the lifetime of this equipment and then added as a contribution to the cumulative exergy consumption of the product. Our experience is that this contribution is negligible for equipment that works continuously. For equipment performing with an irregular operation, such as a laundry machine at home, this contribution may be substantial and makes up a large part of the total exergy cost of the product.

7.5 CONCLUSION

The concept of *chemical exergy* has a distinct advantage over the standard Gibbs energy of formation. Whereas the latter is zero for the elements at standard conditions, the chemical exergy has a zero value for compounds or elements in equilibrium with, and as they occur, in our natural environment. Thus the standard chemical exergy of a compound clearly represents the amount of work available with respect to the environment in which we live and work. The chemical exergy can be simply calculated from the Gibbs energy of formation. The only difference between the two concepts is that their zero values are denned for different reference substances.

The chemical exergy of a molecule in a mixture is smaller than in its pure state, as it will require work to separate the mixture in its pure constituents, the exergy of separation. This exergy will be lost as the *exergy of mixing* when the pure constituents spontaneously form the mixture. The *total exergy* of a pure compound or element is therefore composed of three contributions: the *chemical exergy*, the *exergy of mixing*, and the *physical exergy*. The last element accounts for the fact that the molecule may be at different conditions of pressure and temperature than those of the environment, P_0 and T_0.

The concept of *cumulative chemical exergy consumption* is very useful and accounts for the fact that when a compound (e.g., ammonia) is introduced into a process, its chemical exergy has to be corrected for the exergy consumption accumulated since this compound was manufactured from its natural constituents (air and natural gas in the case of ammonia).

If the *thermodynamic efficiency* of a process step is calculated, the chemical exergies should be excluded from the calculation if the process step does not include chemical conversions. If it does, it may be appropriate to distinguish between the physical and the chemical efficiency, η_{phys} and η_{chem}, of the process step.

Finally, although the exergy concept is not strictly necessary for the calculation of the available work lost in the process, it is an extremely handy tool to calculate losses and efficiencies and for making a quick assessment of process options. Chapter 8 gives some simple illustrations, whereas Part III, "Case Studies" presents the results of integrated studies in the world of energy and chemical technology.

REFERENCES

1. Szargut, J.; Morris, D.R.; Steward, F.R. *Exergy Analysis of Thermal, Chemical, and Metallurgical Process*, Hemisphere Publishing Corp.: New York, 1988.
2. Smith, J.M.; Van Ness, H.C.; Abbott, M.M. *Introduction to Chemical Engineering Thermodynamics*, 5th edn., McGraw-Hill: New York, 1996.

8 Simple Applications

In this chapter, we present some examples of simple applications of exergy analysis. All applications refer to single process steps: the spontaneous expansion of a gas, the production of ice from water by cooling with evaporating ammonia in a heat exchanger, the compression of a gas, the vortex tube, the separation of a mixture, and the spontaneous combustion of hydrogen. This chapter will be followed by Part III, Chapters 9 through 12, which will deal with case studies of complete and integrated processes.

PROBLEM 8.1

The conditions of a natural gas reservoir are 30 MPa and 100°C. The gas, assumed to be pure methane, is spontaneously expanded to a pressure of 7 MPa (Figure 8.1). Assuming that this expansion is adiabatic, calculate the amount of work that is lost in the process, and express it as a fraction of the originally available amount of work per mole of gas in the reservoir. Carry out this calculation while making a distinction between the physical and chemical exergy of the gas.

Assume $t_0 = 20.00°C$.

Solution to Problem 8.1

The expansion of the natural gas is a spontaneous process and thus work must have been lost. According to the Gouy-Stodola relation, this lost work is related to the entropy production of the process.

$$W_{lost} = T_0 S_{gen} \tag{8.1}$$

The expansion has been assumed to be adiabatic, and thus the entropy generated equals the entropy increase of the gas, ΔS, as the entropy change of the environment, ΔS_0, can be set to zero because the process is adiabatic. The amount of lost work can now be calculated from the entropy values S_1 and S_2 of 1 mol of methane at the initial and final conditions, respectively. However, this requires knowledge not only of the final pressure P_2, which is known, but also of the final temperature T_2, which is unknown. Here, the first law helps us out. Applying Equation 2.39 and substituting zero for W_{in} and Q_{out}, we find $\Delta H = 0$ or $H_2 = H_1$. From the IUPAC data series number 16, dealing with methane [1], we find that the molar enthalpy and entropy at initial conditions are, respectively, 501.9 J/mol and −44.22 J/mol K. Thus, H_2 has the same value of 501.9 J/mol, but now at $P_2 = 7$ MPa and the unknown value of T_2. This allows us to find $T_2 = 336.13$ K and at the same time $S_2 = −32.88$ J/mol K from the same reference.

Thus, the lost work can be calculated and we find $W_{lost} = T_0(S_2 − S_1) = 3324$ J/mol. Note that this calculation did not require the application of the concept of exergy.

DOI: 10.1201/9781003304388-10

Simple Applications

P = 300 bar
t = 100°C

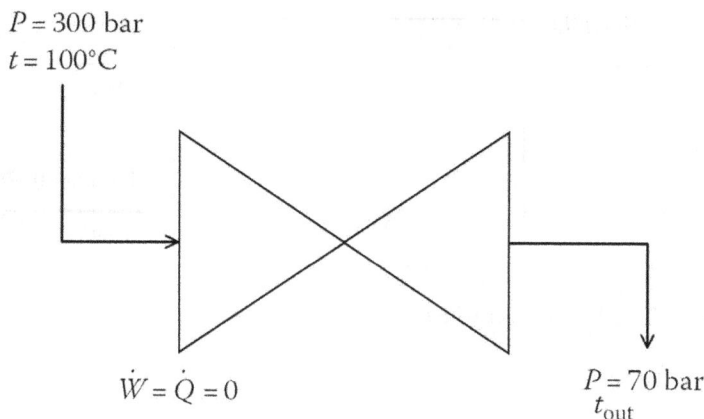

$\dot{W} = \dot{Q} = 0$

P = 70 bar
t_{out}

FIGURE 8.1 The adiabatic expansion of natural gas.

Next, we wish to calculate which fraction this lost work is of the work originally available in the gas. The chemical exergy of the gas, assumed to be methane, is significant, 831.65kJ/mol, but it should be excluded from the calculation because no chemistry is involved in the expansion step. The work available in the gas at initial and final conditions can be calculated from Equation 6.11:

$$Ex = (H - H_0) - T_0(S - S_0) \qquad (6.11)$$

From [1] we can find the values for $H_1, S_1, H_2, S_2,$ and H_0, S_0. They are 501.9 J/mol, −44.22 J/mol K, 501.9 J/mol, −32.88J/mol K, and −194 J/mol and −0.54 J/mol K, respectively. These last two values relate to the stable state of methane at $P_0 = 0.1000$ MPa and $T_0 = 293.15$ K, which is the gas phase. We calculate $Ex_1 = 13.501$ and $Ex_2 = 10.176$ kJ/mol. Of course, we should find the same amount of lost work as before, 3.325 kJ/mol, from

$$W_{lost} = Ex_1 - Ex_2 \qquad (8.2)$$

The fraction of nonchemical work available in the gas that has been lost in the expansion process can now be calculated from $W_{lost}/Ex_1 = 3.325/13.501 = 0.246$. If we had included the chemical exergy of the gas, this number would have been reduced to 0.00393, but as the expansion step is strictly nonchemical, this result is meaningless. Of course, the calculation of W_{lost} itself would not be affected as the chemical exergy would have to be included in both Ex_1 and Ex_2 and would drop out.

PROBLEM 8.2

Water at 20°C is cooled and frozen to ice at 0°C in a countercurrent process with evaporating ammonia (see Figure 8.2). The minimum temperature difference required for proper heat transfer is taken as 5°C between solid ice and liquid NH_3

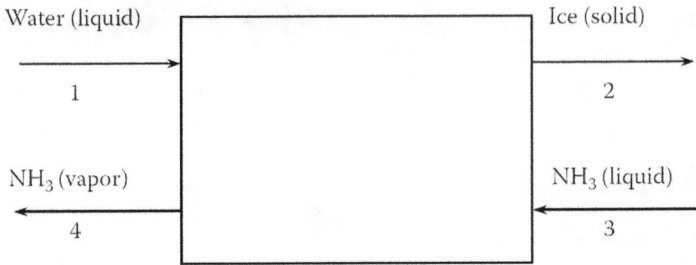

FIGURE 8.2   Freezing of water with the help of evaporating liquid ammonia.

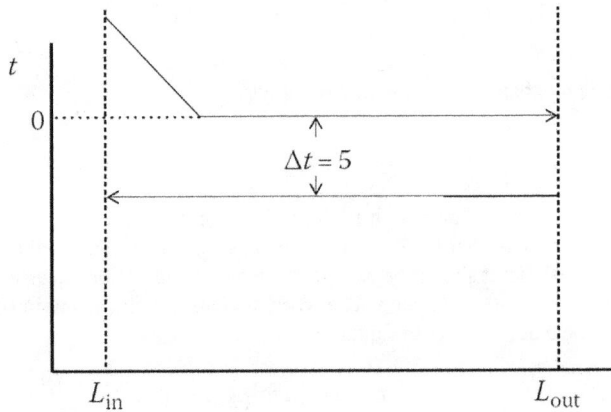

FIGURE 8.3   Temperature profiles in the heat exchange of Figure 8.2.

(Figure 8.3). Heat exchange takes place in equipment well isolated from the environment. Calculate

1. The exergy of 1 mol of ice
2. The mass flow rate of ammonia for the production of 1 mol/s of ice
3. The amount of work lost in this process step
4. The efficiency of this process step

Solution to Problem 8.2

1. We assume $t_0 = 20°C$. Next, we use the steam tables [2] to find the values for H_0^l, S_0^l at 20°C in state 1 and at 0°C in state 2 for *liquid* water, H_2^l and S_2^l. These are 1511J/mol, 5.339 J/mol K, −0.7 J/mol, and 0 J/mol K, respectively. From the heat and temperature of fusion, $\Delta_{fus} H = 6008$ J/mol and $T_{fus} = 273.15$ K, we can calculate the enthalpy and entropy at 0°C according to

$$H_2^s = H_2^l - \Delta_{fus} H$$

$$H_2^s = -0.7 - 6008 = -6008.7 \text{J} / \text{mol}$$

and

$$S_2^s = S_2^l - \frac{\Delta_{fus} H}{T_{fus}}$$

$$S_2^s = 0 - \frac{6008}{273.16} = -21.99 \text{J} / \text{molK}$$

Then we apply Equation 6.11 to calculate the exergy of 1 mol of ice

$$\text{Ex}_2^s = \left(H_2^s - H_0^l \right) - T_0 \left(S_2^s - S_0^l \right)$$

and find $\text{Ex}_2^s = 491.8 \text{J/mol}$, which is the exergy of 1 mol of ice at 0°C.

2. Next, we turn our attention to ammonia. What are the conditions of the evaporating ammonia? As the minimum temperature difference between ammonia and water is 5°C, $t_4 = t_3 = -5°C$ (Figure 8.3). State 3 is saturated ammonia in the liquid state, and state 4 is saturated ammonia in the vapor state. At this temperature, the saturated vapor pressure of ammonia is $P_{NH_3}^{sat} = 0.2951 \text{MPa}$. From the data compilation by Vargaftik et al. [3], we find for these conditions $H_4^v = 28.103$ and $H_3^l = 6.020 \text{kJ/mol}$. The heat exchanger operates adiabatically with respect to the environment; neither heat nor work is exchanged with the environment, thus the overall enthalpy change must be 0 according to the first law, Equation 2.39. Therefore,

$$\dot{n}_{NH_3} \left(H_4^v - H_3^l \right) = \dot{n}_{H_2O} \left(H_1^l - H_2^s \right) \tag{8.3}$$

As $\dot{n}_{H_2O} = 1 \text{mol/s}$, we can calculate for $\dot{n}_{NH_3} = 0.3405 \text{mol/s}$.

3. The amount of lost work, W_{lost}, can again be calculated from the Gouy-Stodola equation, Equation 3.12. This requires the values for $S_1^l = S_0^l$ and S_2^s for water and S_3^l and S_4^v for ammonia. These last values can again be found from Vargaftik et al. [3] and are 65.17 and 149.13 J/mol K, respectively. First we calculate $\dot{S}_{gen}$ from $\dot{S}_{gen} = \dot{n}_{H_2O} \left(S_2^s - S_1^l \right) + \dot{n}_{NH_3} \left(S_4^v - S_3^l \right) = 1.260 \text{J/Ks}$. Application of the Gouy-Stodola relation gives $\dot{W}_{lost} = 0.369 \text{kW/} \left(\text{mol ice} \right)$.

We could also have calculated $\dot{W}_{lost}$ by making use of the equation

$$\dot{W}_{lost} = E\dot{x}_{in} - E\dot{x}_{out} \tag{8.4}$$

with $\dot{E}x_{in} = \dot{n}_{H_2O} Ex_1^l + \dot{n}_{NH_3} Ex_3^l$ and $\dot{E}x_{out} = \dot{n}_{H_2O} Ex_2^s + \dot{n}_{NH_3} Ex_4^l$.
All enthalpy and entropy values are available to perform this
calculation. $Ex_1^l = (H_0 - H_0) - T_0 (S_0 - S_0) = 0\,\text{J/mol}$,
$Ex_3^l = 6,020 - 29,481 - 293.15(65.17 - 162.74) = 5,141\,\text{J/mol}$,
$Ex_2^s = 491.8\,\text{J/mol}$, and
$Ex_4^v = 28,103 - 29,481 - 293.15(149.13 - 162.74) = 2,612\,\text{J/mol}$.
We can then calculate $\dot{E}x_{in}$ and $\dot{E}x_{out}$.
These values are $\dot{E}x_{in} = 1 \cdot 0 + 0.3405 \cdot 5141 = 1750\,\text{W}$,
$\dot{E}x_{out} = 1 - 491.8 + 0.3405 \cdot 2612 = 1381\,\text{W}$. Applying Equation 8.4,
we again find $\dot{W}_{lost} = 0.369\,\text{kW}/(\text{mol ice})$.

4. To calculate the efficiency of this process step, we can compare $\dot{E}x_{out}$ and $\dot{E}x_{in}$.
   Defining the efficiency η as

$$\eta \equiv \frac{\dot{E}x_{out}}{\dot{E}x_{in}} = \frac{1381}{1750} \tag{8.5}$$

we find $\eta = 0.789$. Although the exchange of heat has taken place with an efficiency of
100%, the equipment is "well-isolated," the exchange of exergy has necessarily asked a
sacrifice in exergy due to the required temperature differences between the two flows,
which change from $20 - (-5) = 25°C$ at the entrance point to $5°C$ at the exit point of ice.
   A more reasonable definition of the efficiency is, however, the comparison of the
exergy change of the water and the exergy transferred from the working medium, in
our case ammonia. This efficiency is defined as

$$\eta \equiv \frac{Ex_2^s - Ex_1^l}{Ex_3^l - Ex_4^v} = \frac{491.8}{0.3405(5141 - 2612)} \tag{8.6}$$

We now find $\eta = 0.571$.

PROBLEM 8.3

A manufacturer of compressors wants to determine the efficiency of this piece of
equipment. In one experiment, the compression of 1 mol of gas per second from $P_1 = 5$
bar to $P_2 = 20$ bar requires a power input of $\dot{W}_{in} = 5.271\,\text{kW}$. During this experiment,
the temperature of the gas rises from $t_1 = t_0 = 20°C$ to $t_2 = 160.0°C$. What is the efficiency of the compressor according to the definition

$$\eta \equiv \frac{\dot{W}_{in}^{rev}}{\dot{W}_{in}} \tag{8.7}$$

The gas may be assumed to behave as an ideal gas, its c_p value may be considered at a constant value of 37.65 J/mol K and the gas constant $R = 8.314$ J/mol K. The compressor may be assumed to operate adiabatically.

Solution to Problem 8.3

The first law applied to this problem reads

$$W_{in} = \Delta H + Q_{out} \tag{8.8}$$

and as the compressor operates adiabatically $Q_{out} = 0$ J, so

$$W_{in} = \Delta H \tag{8.9}$$

The gas may be assumed to behave as an ideal gas, and its enthalpy is therefore only a function of temperature, not of pressure:

$$\Delta H = c_p(T_2 - T_1)$$
$$= 5271 J/mol \tag{8.10}$$

The second law reads

$$S_{gen} = \Delta S + \Delta S_0 \tag{8.11}$$

As the compressor operates adiabatically, the entropy change of the environment, $\Delta S_0 = 0$ J/K. The change in entropy of 1 mol of the ideal gas can be calculated from Equation 2.43.

$$\Delta S = c_p \ln \frac{T_2}{T_1} - R \ln \frac{P_2}{P_1} \tag{8.12}$$

and is found to be 3.173 J/mol K. According to Equation 8.11, this value is also the value for the generated entropy S_{gen}. The minimum amount of work is required when the process is carried out reversibly; this is in the limit as the driving forces are going to zero and thus with $\Delta S = S_{generated} = 0$J/K. From Equation 8.12, it follows elegantly that $T_2^{rev} < T_2$. This equation allows the calculation of $T_2^{rev} = 125.0°C$ by putting $\Delta S = 0$ and shows clearly that T_2^{rev} is indeed the minimum value for T_2. The more T_2 exceeds T_2^{rev}, the higher ΔS, S_{gen}, and W_{lost} and the lower the efficiency of the compressor will be. This also follows from combining Equations 8.7 and 8.9 for both T_2^{rev} and T_2 actually measured,

$$\eta_a = \frac{W_{in}^{rev}}{W_{in}} = \frac{c_p(T_2^{rev} - T_1)}{c_p(T_2 - T_1)} = \frac{W_{in} - W_{lost}}{W_{in}} \tag{8.13}$$

We find $\eta_a = 0.750$.

The work lost in the process is $W_{lost} = T_0 S_{gen}$ and is calculated to be 930 J/mol. If the thermodynamic efficiency is calculated from

$$\eta_b \equiv \frac{Ex_{out}}{Ex_{in}} = \frac{Ex_2}{Ex_1 + W_{in}} = \frac{Ex_2}{Ex_2 + W_{lost}} \tag{8.14}$$

we find $\eta_b = 0.899$. In Equation 8.14 we use that $W_{lost} = Ex_{in} - Ex_{out}$ with $Ex_{in} = Ex_1 + W_{in}$ and $Ex_{out} = Ex_2$. The latter value has been calculated from $Ex_2 = (H_2 - H_0) - T_0(S_2 - S_0) = c_p(T_2 - T_0) - T_0 c_p \ln T_2/T_0 - R \ln P_2^{/P}{}_0 = 8263$ J/mol.

The efficiency η_a calculated with Equation 8.13, and the one calculated with Equation 8.14, η_b, are different, which can be seen from the following alternative expressions:

$$\eta_a = \frac{W_{in} - W_{lost}}{W_{in}} \tag{8.15}$$

and

$$\eta_b = \frac{Ex_1 + W_{in} - W_{lost}}{Ex_1 + W_{in}} \tag{8.16}$$

The second efficiency, η_b, also accounts for the exergy of the entering flow, whereas the first efficiency, η_a, does not. W_{lost} is the same in both calculations. There is much to say for considering η_a as the value that expresses the performance of the compressor best. η_a expresses in a meaningful way the performance of the equipment whereas η_b expresses the quality of the process.

PROBLEM 8.4

A stream of gas of ambient temperature $t_1 = t_0 = 20°C$ and at a pressure of $P_1 = 0.8$ MPa is claimed to be separated adiabatically (Figure 8.4) into two equal flows of $t_2 = 70°C$ and $t_3 = -30°C$, respectively, both at $P_2 = P_3 = 0.1$ MPa. The gas may be assumed to behave as an ideal gas with a constant c_p value of 30 J/mol K.

1. Show that this process is possible.
2. Determine the exergetic efficiency of the process.

Solution to Problem 8.4

1. Whenever the question of "being possible" crops up, the second law offers relief. The first law shows that the overall enthalpy of the process does not change as no work is performed and no heat is exchanged with the environment. Based on 1 mol of gas splitting in two streams of 1/2 mol each, we can write

$$\Delta H = \frac{1}{2}(H_2 - H_1) + \frac{1}{2}(H_3 - H_1) = 0 \tag{8.17}$$

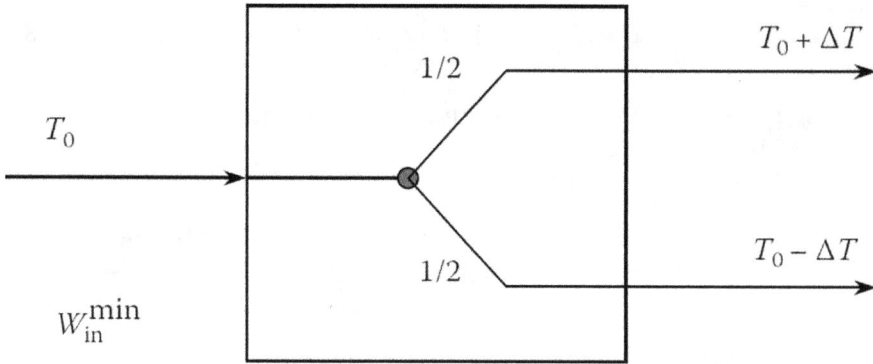

FIGURE 8.4 Adiabatic splitting of a compressed gas into two flows of different temperatures.

Equation 8.17 can be simplified in the case of ideal gases (with constant c_p values) to

$$\frac{1}{2}c_p\left(T_2 - T_1\right) + \frac{1}{2}c_p\left(T_3 - T_1\right) = 0$$

or can be written as

$$\left(t_2 - t_1\right) + \left(t_3 - t_1\right) = 0$$

the relation of which is satisfied indeed.

The second law according to Equation 8.11 simplifies to the equation $\Delta S = S_{gen} > 0$. The process is possible if the overall change in entropy of the streams is positive and indeed it is, as follows from applying Equation 8.12.

$$\Delta S = \frac{1}{2}c_p \ln\frac{T_2}{T_1} + \frac{1}{2}c_p \ln\frac{T_3}{T_1} - R\ln\frac{P_2}{P_1} \tag{8.18}$$

It is obvious from comparing the third term at the right side of the equation with the other two terms that it is the pressure difference between the initial and final states that more than compensates for the separation of the original stream at temperature T_0 in a stream at a higher and one at a lower temperature, a phenomenon that intuitively feels as "unlikely" and indeed is associated with a decrease in entropy as reflected by the two first terms right of the equal sign.

2. To explain the possibility of the process of splitting the original stream into one of a higher and one of a lower temperature, an exergy analysis is very revealing. If we assume that kinetic and potential energy contributions to the exergy values of the streams can be neglected, the exergy of the original stream is

$$Ex_1 = \left(H_1 - H_0\right) - T_0\left(S_1 - S_0\right) \tag{6.11}$$

$$Ex_1 = c_p\left(T_1 - T_0\right) - T_0\left(c_p \ln\frac{T_1}{T_0} - R\ln\frac{P_1}{P_0}\right) \tag{8.19}$$

and as $T_1 = T_0 = 293.15$ K, $P_1 = 0.8$ MPa, and $P_0 = 0.1$ MPa, Ex_1 becomes RT_0 ln $8 = 5068$ J.

The exergy of the split streams is

$$Ex_2 = \frac{1}{2}\left[c_p\left(T_2 - T_0\right) - T_0\left(c_p \ln\frac{T_2}{T_0} - R_0\ln\frac{P_2}{P_0}\right)\right] \tag{8.20}$$

which gives $Ex_2 = 57.5$ J
and

$$Ex_3 = \frac{1}{2}\left[c_p\left(T_3 - T_0\right) - T_0\left(c_p \ln\frac{T_3}{T_0} - R_0\ln\frac{P_3}{P_0}\right)\right] \tag{8.21}$$

resulting in $Ex_3 = 72.3$ J.

From comparing $Ex_1 = 5068$ J with $Ex_2 + Ex_3 = 130$ J, it is clear that the work available in the original single stream is more than enough to account for the work available in the split streams together, and we find an efficiency for our case of $\eta = (Ex_2 + Ex_3)/Ex_1 = 3\%$ and 97% is lost. Originally, this piece of equipment was described by Hilsch [4] and Ranque [5] and was named after them as the Hilsch-Ranque tube or the Vortex tube. In fact, a much larger separation in temperature T_2 and T_3 should be possible, as Ex_1 is considerably larger than $Ex_2 + Ex_3$. This is in line with the claimed efficiency of the modern version of the tube, the "Twister" tube, as this piece of equipment is now called [6]. The "Twister" tube is used to partially condense higher hydrocarbons from natural gas to bring the gas on specification such that condensation in the transport system is avoided.

PROBLEM 8.5

Air, assumed to be a mixture of 79 mol% nitrogen (N_2) and 21 mol% oxygen (O_2), is split into its pure components also at ambient conditions P_0, T_0 with $P_0 = 1$ bar and $T_0 = 293.15$ K (Figure 8.5). Under these conditions, air behaves as an ideal gas.

Calculate the minimum amount of work required for the separation of 1 mol of air.

Solution to Problem 8.5

The exergy of separation, Ex_{sep}, is given by

$$Ex_{sep} = 0.79Ex_{N_2}^0 + 0.21Ex_{O_2}^0 - 1\cdot Ex_{air}^0 \tag{8.22}$$

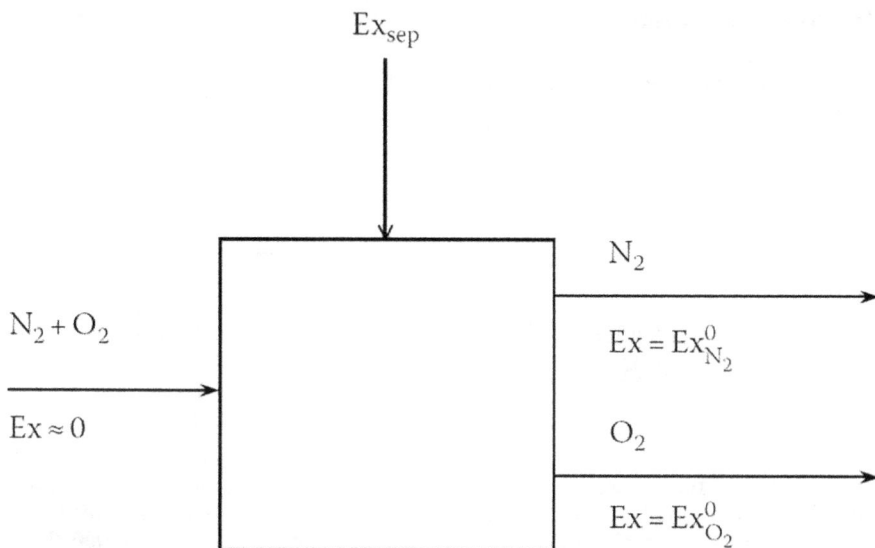

FIGURE 8.5   The separation of air into its main components.

$Ex^0_{N_2}$ and $Ex^0_{O_2}$ are the chemical exergy values of the pure components of air. The exergy value of air at standard conditions is given by

$$Ex^0_{air} = 0.79Ex^0_{N_2} + 0.21Ex^0_{O_2} + \Delta_{mix}H - T_0\Delta_{mix}S \qquad (8.23)$$

For 1 mol of an ideal gas mixture, $\Delta_{mix}H = 0$ J/mol and $\Delta_{mix}S = -R\Sigma x_i \ln x_i$, where x_i is the mole fraction of component i in the mixture. For air: $x_{N_2} = 0.79$ and $x_{O_2} = 0.21$. The exergy values of the pure components of air cancel against each other so Ex_{sep} can be simplified to $Ex_{sep} = -RT_0\Sigma x_i \ln x_i$. Substitution of the given mole fractions of T_0 and the gas constant R then give $Ex_{sep} = 1253$ J/mol.

This is the minimum amount of work required. In practice, the work requirement is much larger due to the inefficiencies of the real separation process. This is clearly illustrated in Chapter 10 for the separation by distillation of a mixture of propane and propene.

PROBLEM 8.6

Pure hydrogen gas at room pressure and temperature is adiabatically combusted with air. The combustion takes place with an amount of air that is 30% in excess of what is stoichiometrically required. Calculate the adiabatic flame temperature of the process, the work lost, and the thermo-dynamic efficiency of the process. Assume air to consist of a mixture of 79 mol% of N_2 and 21 mol% of O_2.

Solution to Problem 8.6

The equation for the stoichiometric combustion of hydrogen is

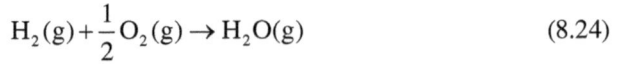

$$H_2(g) + \frac{1}{2}O_2(g) \rightarrow H_2O(g) \tag{8.24}$$

With 30% excess air, Equation 8.24 reads

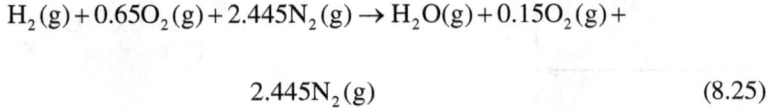

$$H_2(g) + 0.65O_2(g) + 2.445N_2(g) \rightarrow H_2O(g) + 0.15O_2(g) +$$

$$2.445N_2(g) \tag{8.25}$$

Assume that the hydrogen feed stream is 1 mol/s. The reactants are fed to the combustor at 298.15 K and 0.1 MPa, and the heat of reaction will be used to raise the temperature of the product mixture to its final value, the adiabatic flame temperature. The first law for this adiabatic process can be written as

$$\Delta\dot{H} = 0 = \left(\Delta_r \dot{H}_{298.15K}^0 + \dot{n}_{H_2O} \int_{298.15K}^T C_{P,O_2} dT + \dot{n}_{N_2} \int_{298.15K}^T C_{p,N_2} dT \right) \tag{8.26}$$

with $\dot{n}_{H_2O} = 1.0$, $\dot{n}_{O_2} = 0.15$, $\dot{n}_{N_2} = 2.445\,\mathrm{mol}/s$, and T is the final exit temperature of the products. We make the additional assumption that the gas is ideal. This allows us to use the ideal gas molar heat capacity values at constant pressure. The following expression for the temperature dependency of the heat capacity C_p^{ig} is used:

$$\frac{C_p^{ig}}{R} = A + BT + \frac{D}{T^2} \tag{8.27}$$

where the coefficients A, B, and D are given in Table 8.1.

The reaction enthalpy is $-241,818$ J/mol H_2 and is obtained from standard tables [2]. Substitution of this and Equation 8.27 into Equation 8.26 yields

$$0 = -241,818 + 8.314 \int_{298.15}^T \left(\sum_i \dot{n}_i A_i + \sum_i \dot{n}_i B_i T + \sum_i \dot{n}_i \frac{D_i}{T^2} \right) dT \tag{8.28}$$

From Table 8.1, we can readily compute $\sum_i \dot{n}_i A_i = 12.035\mathrm{s}^{-1}$, $\sum_i \dot{n}_i B_i = 2.976 \times 10^{-3}\,\mathrm{K}^{-1} \cdot \mathrm{s}^{-1}$, and $\sum_i \dot{n}_i D_i = 0.1848 \times 10^5\,\mathrm{K}^2 \cdot \mathrm{s}^{-1}$. Substitution yields

---

TABLE 8.1

Coefficients of Equation 8.27.

Species	A	B (K⁻¹)	D (K²)
N_2	3.280	0.593×10^{-3}	0.040×10^5
O_2	3.639	0.506×10^{-3}	-0.227×10^5
H_2O	3.470	1.450×10^{-3}	0.121×10^5

$$0 = -241{,}818 + 8.314 \left(12.035(T - 298.15) + \frac{1}{2} 2.976 \times 10^{-3} \left(T^2 - 298.15^2 \right) \right.$$
$$\left. - 0.1848 \times 10^5 \left(\frac{1}{T} - \frac{1}{298.15} \right) \right) \tag{8.29}$$

Solution of this algebraic equation yields $T = 2146$ K, which is the adiabatic flame temperature.

For this adiabatic combustion process, the entropy production is equal to the entropy change of the process and is given by

$$\dot{S}_{gen} = \Delta \dot{S} = -\Delta_{mix} \dot{S}_{air} + \Delta_r \dot{S}^0_{298.15} + \int_{298.15}^{T} \sum_i \dot{n}_i \frac{C_{P,i}}{T} dT$$
$$+ \Delta_{mix} \dot{S}_{products} \tag{8.30}$$

$$\dot{S}_{gen} = -3.095 \left(-R\Sigma x_i \ln x_i \right)_{air} + \frac{\Delta_r H_{298.15} 0 - \Delta_r G^0_{298.15}}{298.15}$$
$$+ R \int_{298.15}^{T} \left(\frac{12.035}{T} + 2.976 \times 10^{-3} + \frac{0.1848 \times 10^5}{T^3} \right) dT \tag{8.31}$$
$$+ 3.595 \left(-R\Sigma x_i \ln x_i \right)_{products}$$

The Gibbs energy of the reaction is $-228{,}572$ J/mol H_2 and is also obtained from the standard tables used before [2].

$$\dot{S}_{gen} = -3.095.4.273 - 44.427$$
$$+ 8.314 \left(12.035 \ln \frac{2146}{298.15} + 2.976 \times 10^{-3} (2146 - 298.15) \right.$$
$$\left. - \frac{0.1848 \times 10^5}{2} \left(\frac{1}{2146^2} - \frac{1}{298.15^2} \right) \right) \tag{8.32}$$
$$+ 3.595 - 8.314(0.2782 \ln 0.2782 + 0.04172 \ln 0.04172 + 0.6801 \ln 0.6801)$$

$$\dot{S}_{gen} = -13.23 - 44.43 + 244.06 + 22.44 = 208.8 \, W / K \qquad (8.33)$$

This yields for the lost work or lost exergy for this combustion reaction

$$\dot{W}_{lost} = \dot{Ex}_{lost} = T_0 \dot{S}_{gen} = 62.27 kW \qquad (8.34)$$

The exergy flowing in is the sum of the chemical exergy of hydrogen and the exergy value of air at standard conditions. This value is normally chosen to be 0 J/mol. In our case, air was considered to be dry air containing only 21 mol% oxygen and 79 mol% nitrogen. The exergy value of air can be calculated from

$$\dot{Ex}_{air} = \dot{n}_{air} \left(\sum x_i Ex_{chem.i} + RT_0 \sum x_i \ln x_i \right) \qquad (8.35)$$

$$\dot{Ex}_{air} = 3.095 \cdot (0.21 \cdot 3.97 + 0.79 \cdot 0.72$$
$$+ 8.314 \cdot 298.15 \cdot 10^{-3} \cdot (0.21 \ln 0.21 + 0.79 \ln 0.79)) = 0.4 kW \qquad (8.36)$$

$$\dot{Ex}_{in} = 236.1 + 0.4 = 236.5 kW \qquad (8.37)$$

The exergy flowing out of the system is then calculated from

$$\dot{Ex}_{out} = \dot{Ex}_{in} - \dot{Ex}_{lost} = 236.5 - 62.3 = 174.2 kW \qquad (8.38)$$

The thermodynamic efficiency is therefore $174.2/236.5 = 0.74$, or 74%. The price of the spontaneous combustion of H_2 is the loss of 26% of its original exergy. The remaining 74% is now available in the form of heat of its combustion products. As we see from Equation 8.33, the largest contribution to $\dot{S}_{gen}$ comes from the generation of thermal energy leading to a vast increase of the temperature. In a fuel cell, operating at a much lower temperature, this entropy generation can be reduced considerably.

REFERENCES

1. Angus, S.; Armstrong, B.; de Reuck, K.M. *International Thermodynamic Tables of the Fluid State-5, Methane* (IUPAC Data Series Number 16), Pergamon Press: Oxford, 1978.
2. Smith, J.M.; Van Ness, H.C.; Abbott, M.M. *Introduction to Chemical Engineering Thermodynamics*, 5th edn., McGraw-Hill: New York, 1996.
3. Vargaftik, N.B.; Vinogradov, Y.K.; Yargin, V.S. *Handbook of Physical Properties of Liquids and Gases Pure Substances and Mixtures*, 3rd edn., Begell House Inc.: New York, 1996, pp. 805–854.
4. Hilsch, R. The use of the expansion of gases in a centrifugal field as a cooling process. *Review of Scientific Instruments* 1947, *18*, 108.
5. Ranque, G. Experiments in a vortex with simultaneous exhaust of hot air and cold air. *Le Journal de Physique et le Radium* 1933, *4*, 1125.
6. www.pdbuchan.com/ranque-hilsch/ranque-hilsch.html.

Part III

Case Studies

I hear and I forget.
I see and I remember.
I do and I understand.

—Confucius

Chapters 9 through 12 demonstrate thermodynamic, or exergy analysis of industrial processes. First, Chapter 9 deals with the most common energy conversion processes. Then, Chapter 10 presents this analysis for an important industrial separation process, that of propane and propylene. Finally, Chapter 11 analyzes two industrial chemical processes involved in the production of polyethylene. Chapter 12 is included to discuss life cycle analysis, in particular its extension into exergetic life cycle analysis, which includes the "fate" or history of the quality of energy.

DOI: 10.1201/9781003304388-11

9 Energy Conversion

In this chapter, we explore how the exergy concept can be used in the analysis of energy conversion processes. We provide a brief overview of commonly used technologies and analyze the thermodynamic efficiency of (1) coal and gas combustion, (2) a simple steam power plant, (3) gas turbine, and (4) combined cycle and cogeneration. At the end of this chapter, we summarize our findings with some concluding remarks.

9.1 INTRODUCTION

The conversion of one form of energy into another has always been vital to the existence of man; man consumes food to liberate the chemical energy stored therein by means of oxidation. The discovery of fire by primitive man is a good example of transforming chemical energy present in the wood into heat and allowed man to consume cooked foods and ward off predators. Windmills, steam engines, hydroelectric plants, nuclear plants, and so on, all have a common purpose: the conversion of one form of energy into another.

Now, that energy can be neither created nor destroyed is a well-known statement of the principle of conservation of energy and is mathematically formulated in the first law of thermodynamics. Thus, if we speak loosely of "energy production," we do not mean its production from nothing, since this would violate the first law of thermodynamics, but simply the conversion of one form into another. It is this conversion that is the crux of human existence on earth today, and civilization as we know it depends entirely on various forms of energy conversion. The power outages in California in early 2001, and in the northeastern United States and Italy in 2003, the rapid rise of oil prices in 2008 and interplay with the world economies highlight the dependence of human society on energy. At present, the method of choice seems to be the use of chemical energy contained in fossil fuels.

Most of our available energy is obtained indirectly from chemical energy. In the steam turbine,[1] the generation of mechanical work proceeds through pathway I (Figure 9.1) and includes heat and electrical energies, whereas that of the internal combustion engine does not include electrical energy, which is pathway II [1–3].

Other pathways that involve technology to harness the power of the atom (nuclear fission) are also used, albeit not (yet?) as widespread as the aforementioned pathways. At present, anthropogenic nuclear fusion has not reached technological maturity and, in certain countries, nuclear energy has a negative image. Pathways that do not involve the generation of heat also exist, for example, the generation of mechanical work for milling processes from wind energy and propulsion from wind energy (in boats) have been around for centuries. The use of solar radiation to generate electricity in solar cells in satellites is also a good example, though using solar energy for terrestrial purposes has been increasing since then (see Chapter 17).

DOI: 10.1201/9781003304388-12

I. Chemical energy $\rightarrow$ Heat $\rightarrow$ Electrical energy $\rightarrow$ Mechanical work

II. Chemical energy $\rightarrow$ Heat $\rightarrow$ Mechanical work

FIGURE 9.1   Commonly used energy conversion pathways.

9.2   GLOBAL ENERGY CONSUMPTION

In the United States alone, approximately 8% of the GDP is used for the generation of energy, which amounts to $437 billion [2]. Since at least 1950, industry has been the largest energy-consuming sector in the U.S. economy, although its share has reduced from 47% in 1950 to 37% in 1990, and 30% in 2007 [4]. As Figure 9.2 shows, the other sectors are rapidly increasing and are projected to play an important role [4]. The breakup by energy source is as follows. In the United States, approximately 60% of the electricity is coal based, more than twice as much as the next largest source, nuclear power, at 17%, and natural gas, which makes up 12% [4]. Hydroelectric and alternative power sources such as geothermal energy generation take up only 10% and 1%, respectively. In terms of energy use by fuel, the list is topped by oil, which contributes to 48% of U.S. national energy spending, followed by natural gas, at 31%. Oil has the advantage that it has a high energy density (heating value), is liquid, and is therefore easy to transport and can be used in a variety of sectors (transportation, space heating, industrial heating, etc.). Oil, on the other hand, does have its problems. Its global distribution is uneven, leading to a concentration of wealth and power, and can lead to a struggle between those who have and those who have not. Furthermore, like all fossil fuels (e.g., coal) its production and use can have environmentally hazardous effects, including the emission of NO_x and SO_x and occasional oil spills. Natural gas has become an increasingly popular fuel due to its environmental cleanliness. Even though gas fields with sour gas are common, the sweetening process is relatively straightforward, as opposed to the "sweetening" of oil or coal.

The Energy Information Administration (EIA), an independent agency within the U.S. Department of Energy, predicted that over the next two decades, the world energy consumption will increase by approximately 40% (Figure 9.3) [4,5]. The predictions are that natural gas will remain the fastest-growing component of consumption across the globe, and the nuclear generation of electricity is expected to increase, peaking around 2015. The use of renewables as solar energy and biomass is expected to increase by 53%, and the largest share of world energy consumption will continue to be oil [5]. For example, if one considers Figure 9.4, which shows the primary energy use by fuel in the United States (trends are similar for the world), it is clear that conventional (fossil) energy sources will continue to play a role in the near future. This is well exemplified in Figure 9.5, which shows the world energy use by fuel. It is well known that the use of fossil fuels results in net emissions of CO_2, which in turn are alleged to contribute to enhanced atmospheric levels of this greenhouse gas. Although there is no scientific consensus, increased levels of CO_2 could contribute to global warming. It therefore makes sense to examine energy conversion processes for the development of energy conversion technologies. Figure 9.5 also shows

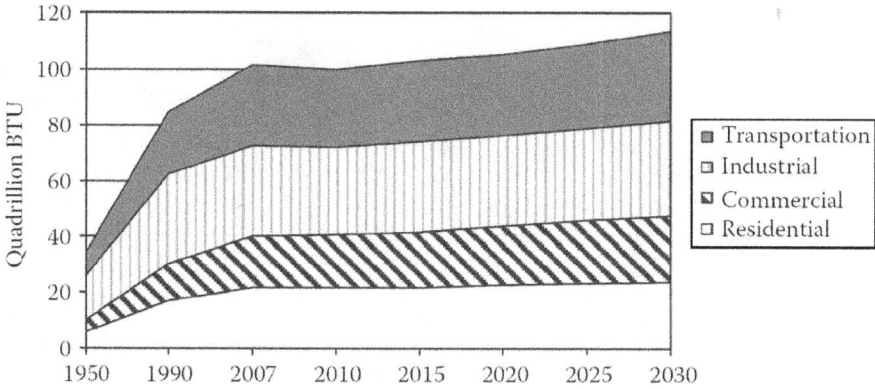

FIGURE 9.2 Energy use by sector in the United States. (With permission from U.S. Department of Energy, Energy Information Administration. Annual Energy Review 1991, DOE/EIA-0384(91), Washington, DC, June 1992.)

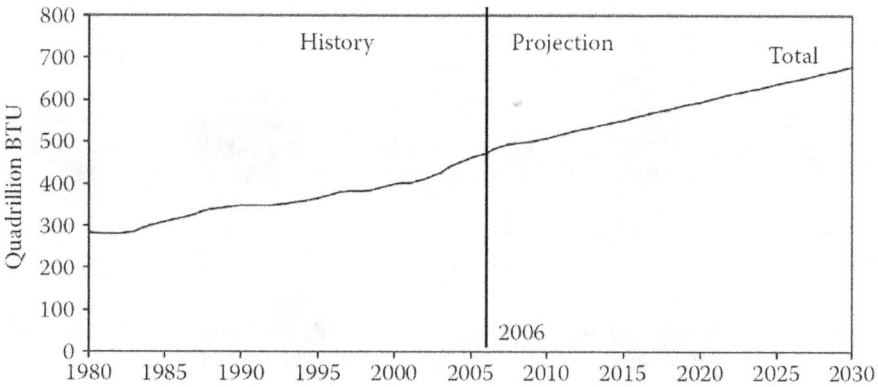

FIGURE 9.3 World energy demand. (Data up to 2006, projections after 2006.)

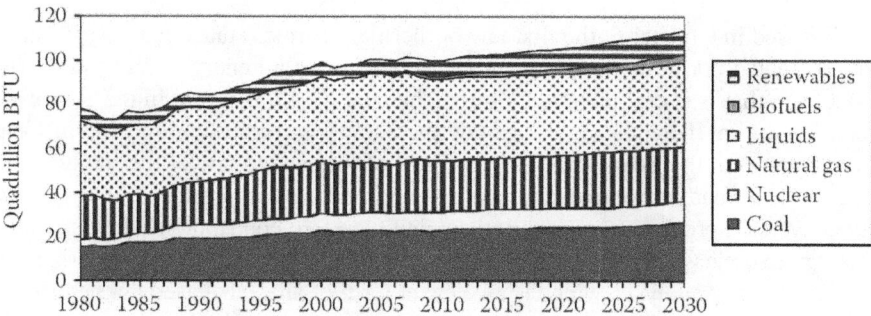

FIGURE 9.4 Primary energy use by fuel in the United States. (With permission from U.S. Department of Energy, Energy Information Administration. Annual Energy Review 1991, DOE/EIA-0384(91), Washington, DC, June 1992.)

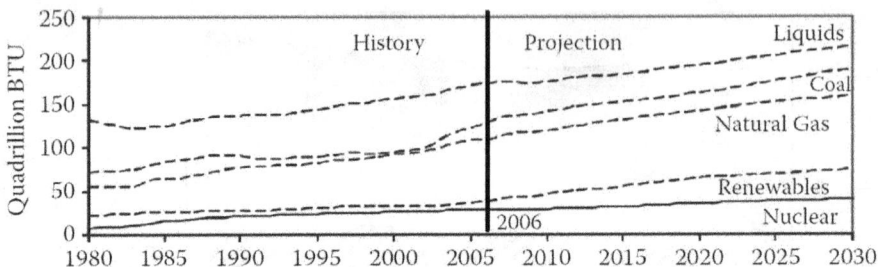

FIGURE 9.5 World energy use by fuel. (Data up to 2006, projections after 2006.) (With permission from U.S. Department of Energy, Energy Information Administration. Annual Energy Review 1991, DOE/EIA-0384(91), Washington, DC, June 1992.)

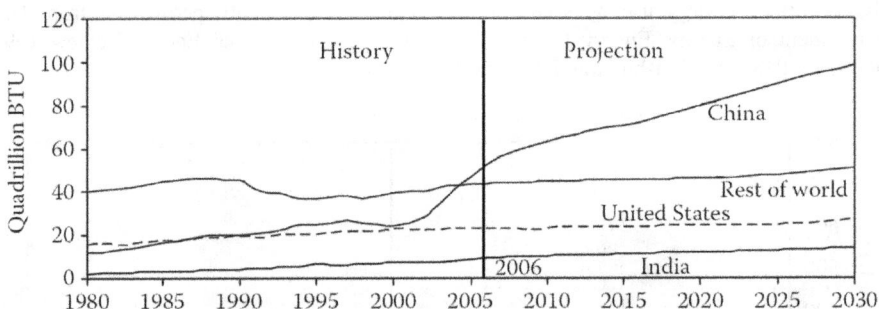

FIGURE 9.6 Coal consumption by selected world region. (Data up to 2006, projections after 2006.)

that coal will continue to play a large role, and perhaps an increasing role. Figure 9.5 shows that the use of coal will continue to grow, but the utilization will not be uniform throughout the world, but is projected to be concentrated in China (Figure 9.6).

9.3   GLOBAL EXERGY FLOWS

As discussed in Chapter 6, the first law of thermodynamics states that energy can be neither created nor destroyed, but it does not imply that all energy will be converted into a form that is useful (electricity or mechanical work). Inherent in the conversion processes are inefficiencies that disallow the complete conversion into useful forms of energy. It is at this point that the usefulness of the second law of thermodynamics is clear; the second law of thermodynamics, appropriately paraphrased, states that all *real* conversion processes are irreversible, and therefore contribute to the dissipation of energy to useless forms of energy. As described in earlier chapters, lost work, availability, or exergy analysis, which is an elegant form of analysis based on the first two laws of thermodynamics, can provide valuable insight into the location, the origin, and the nature of energy conversion inefficiencies and how they can be alleviated. Making a process more "energy-efficient" is attractive since it can allow for more useful energy production, that is, conversion into electricity or mechanical work.

GCEP   Global exergy flux, reservoirs, and destruction

34,000 Atmospheric Reflection   62,500 Extra-solar radiation

162,000 Solar radiation

Atmospheric absorption

0.06 Wind energy
60 Waves   Wave   0 energy
31,000   870 Wind

Ocean thermal gradient
Absorption   100   OTEC   0

86,000 Surface incident   41,000 Evaporation   300   7.2 Hydro-
90   5.4 electricity
43,000 Surface heating   Clouds   Rivers

0.36
5000 Surface reflection

Scattering

3.7 Tides   90 Photosynthesis   1.2 Traditional biomass   0.016 Solar energy
0.15 Commercial biofuels
3.5 Ocean tides   30 ZJ Plants   0.04 Carbon burial   3,100 ZJ Lithium
0.0005 Tidal energy   1000 ZJ Uranium   1 Nuclear fuel
270 ZJ Coal   3.6 Coal
0.2 Solid earth tides   110 ZJ Oil   5.0 Oil   300 ZJ Thorium   360,000 ZJ Seawater uranium
Methane hydrate   200 ZJ   3.2 Gas   1E10 ZJ Deuterium
Geothermal energy   0.03   50 ZJ Gas

32 Crustal thermal energy 1.5E7 ZJ

KEY   Thermal Kinetic   Natural exergy destruction   Energy accumulation [ZJ] (≈10²¹ J)
nuclear chemical   Human use for energy services   Energy flux [TW] (≈10¹² W)
radiation gravitational

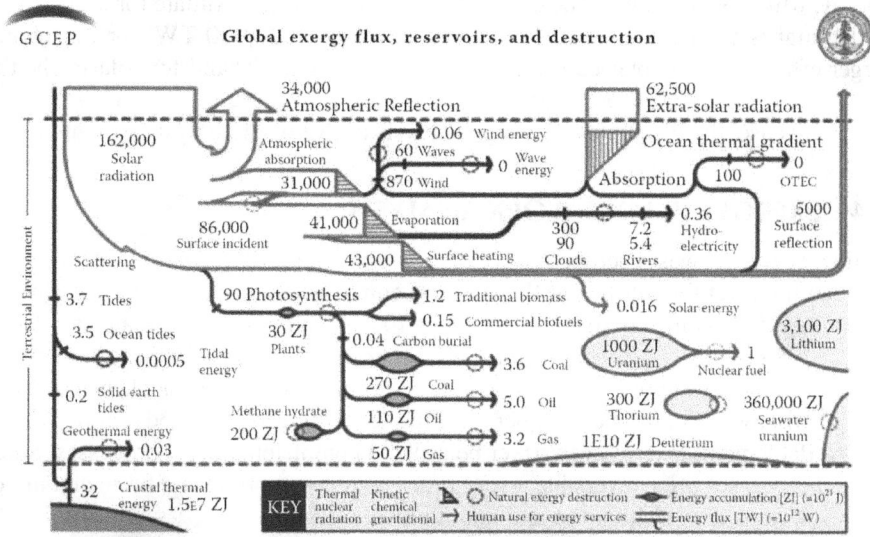

Energy is the useful portion of energy that allows us to do work and perform energy services. We gather energy from energy-carrying substances in the natural world we call energy resources. While energy is conserved, the energetic portion can be destroyed when it undergoes an energy conversion. This diagram summarizes the energy reservoirs and flows in our sphere of influence including their interconnections, conversions, and eventual natural or anthropogenic destruction. Because the choice of energy resource and the method of resource utilization have environmental consequences, knowing the full range of energy options available to our growing world population and economy may assist in efforts to decouple energy use from environmental damage.

Prepared by Wes Hesmann and A.J. Simon
Global Climate and Energy project at Stanford University (http://gcep.stanford.edu)   1:10 DCIP 2000-2006

FIGURE 9.7 Global exergy flux, reservoirs, and destruction. (With permission from Global Climate and Energy project at Stanford University. http://gcep.stanford.edu.)

Bearing the aforementioned in mind, together with the previous chapters, it is interesting to study Figure 9.7, which represents the global exergy flux, reservoirs, and destruction as calculated under the umbrella of the Global Climate and Energy Project at Stanford University (GCEP). Let us first examine what the total energy use was in 2006 and projected use will be in 2030, based on Figure 9.5 [4]. If we add all categories, we end up with 473 quadrillion BTU, or 0.5 ZJ (1 ZJ = 10^{21} J) in 2006 and 678 quadrillion BTU, or 0.7 ZJ in 2030. For simplicity, let us assume that this gets used evenly throughout a year, to estimate the energy requirement in Watts. In 2006, this means we take 0.5 ZJ, and divide it by $365 \times 24 \times 60 \times 60$ s to obtain 15.8 TW. Similarly, for 2030, this yields 22.7 TW.

Two observations can be made based on this simple calculation. If one examines Figure 9.7, one readily observes that resources such as coal, oil, and gas can last in the decades to century range, whereas nuclear options can last far longer. Second, solar energy seems to provide about 5000 TW of solar radiation that gets radiated back into space. For argument's sake, let us assume that the efficiency of solar cells is about 10%, and only 30% of the earth's surface area is land. This means that only $0.1 \times 0.3 \times 5000 = 150$ TW is potentially useful. However, it is inconceivable that all the earth's land area will be utilized for solar energy, so perhaps only 1% can be used. This then yields only 1.5 TW, which is fairly small compared to what the energy demand truly is.

This brief calculation is not meant to judge the viability of solar energy, but simply to provide discussion about what the appropriate methods should be to harvest the energy, which is incident on the earth. If one uses the larger estimate for exergy flux, which makes up the solar surface incident component (86,000 TW), or can obtain larger efficiencies in solar cells and use more of the earth's land for solar cells, the picture becomes much more optimistic.

In any case, the principles of exergy analysis or lost work analysis are important.

9.4   EXERGY OR LOST WORK ANALYSIS

In line with what was discussed in Chapter 6 with regard to the quality of the Joule, one can interpret Orwell [6], "All Joules are equal, but some Joules are more equal than others." This means that 1 J of heat at 1000 K is more useful than, say, 1 J of heat at 298 K. This is a direct consequence of the work available in these amounts of heat, as stated in Chapters 6 and 7, where precise definitions of physical and chemical exergy are given. A direct consequence of the second law of thermodynamics is that the available work (exergy) can never be utilized completely in real processes. Since all real processes are irreversible, every process step will produce a finite amount of lost work, thus diminishing the amount of useful work.

Lost work analysis is useful in analyzing energy conversion processes since it can pinpoint the process step where most work is lost or dissipated. This can provide guidelines for the improvement of the process as a whole. In this chapter, we will examine the generation of electricity from coal and natural gas, using simple combustion and cogeneration. In Section 9.5, we briefly survey commonly used power generation technologies and discuss the thermodynamic efficiency of combustion in Section 9.6. We then move on to Section 9.7, where we analyze power generation using gas turbines in combination with combustion chambers. A simple steam-based power plant is examined in Section 9.8, and in Section 9.9 a combined cycle and a combined cycle cogeneration power plant are discussed. We conclude this chapter with Section 9.10.

9.5   ELECTRIC POWER GENERATION

Power is defined as the rate at which work is performed [2]. The direct transformation of wind energy to mechanical work is convenient if a milling process is the user of this energy. In this day and age, however, transformation to electrical energy seems to be more useful, as many modern instruments depend on electricity as their energy source.[2] We must bear in mind that wind energy power generation is dependent on the availability of wind, and, as such, continuous production is not assured. Electric power is commonly generated using one of the following technologies/energy sources [2,3,7,8]:

1. Steam plants
2. Gas turbines
3. Combined cycle
4. Nuclear power
5. Hydropower
6. Wind power

7. Solar energy
8. Geothermal power

9.5.1 STEAM PLANTS

In a steam plant, steam is generated by the combustion of a fossil fuel, which releases the necessary heat. Water is pumped into the boiler, and the heat of combustion from the furnace forms wet steam. Directly passing wet steam through the turbine could damage the turbine blades, as condensation in the final stages would inevitably create small droplets. To avoid this, superheated steam is created by heating the wet stream even further, which can be passed through the turbine. After the turbine, the steam is condensed to water again and reused. A simple schematic of a steam plant is given in Figure 9.8 [9,10].

Steam plants usually operate on the Rankine cycle [9,10], as shown in Figure 9.9. This figure shows the ideal Rankine cycle in the sense that the pump and turbine operate isentropically, that is, reversibly or without entropy production. In practice, these will operate with entropy production.

9.5.2 GAS TURBINES

Gas turbines have enjoyed a resurgence in popularity in recent years, mainly due to substantial improvements in efficiency [2,3]. Many implementations are possible for gas turbines, but in its basic configuration, atmospheric air is drawn into a rotary

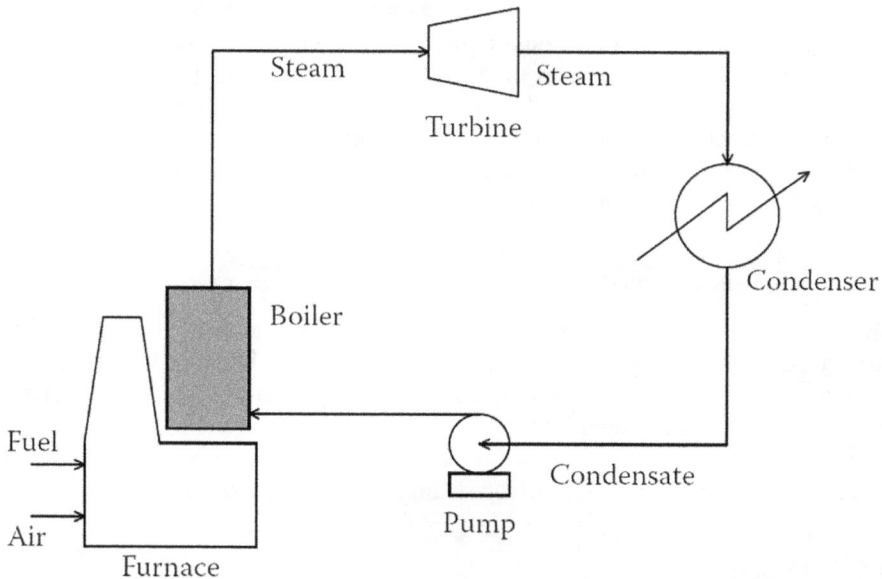

FIGURE 9.8 Schematic of a simple steam cycle in a steam plant. (With permission from Bisio, A. and Boots, S. (eds), *Energy Technology and the Environment*, Vol. 2, Wiley Encyclopedia Series in Environmental Science, New York, 1995.)

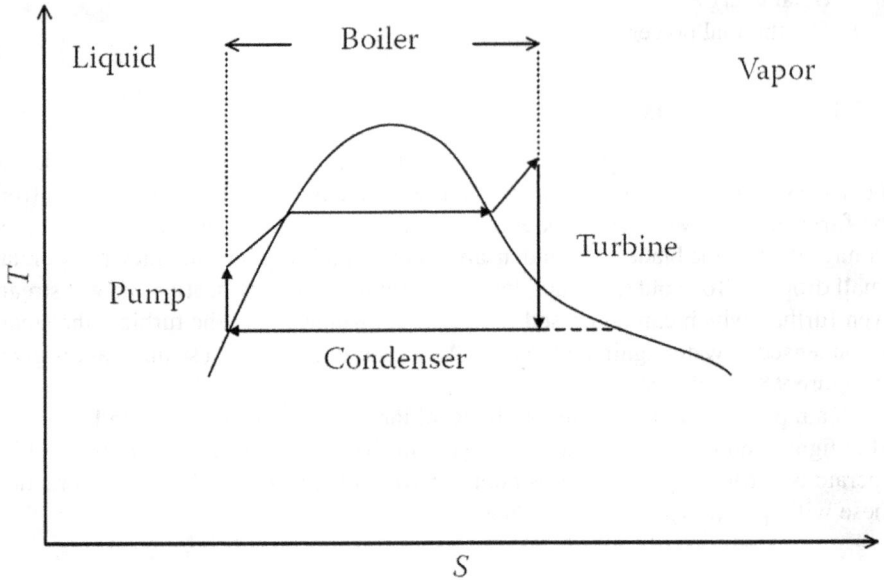

FIGURE 9.9 *T-S* diagram with ideal Rankine cycle.

compressor (Figure 9.10). The compressed air enters the combustor, where it mixes with fuel and combustion takes place. The hot gases subsequently enter a turbine where mechanical work is extracted. The gases then exit at atmospheric pressure.

The gas turbine process operates on the Brayton cycle, as illustrated in Figure 9.11 [2,8,9]. The compression reduces the volume and increases the pressure. The combustion takes place isobarically. The turbine results in a reduction of the pressure and an increase in volume of the combustion gases. The combustion gases then leave the turbine at atmospheric pressure. Since there is a flow of material into and out of the system, the system is not closed but constitutes an open system.

9.5.3 Combined Cycle

The combustion gases that leave the gas turbine still have a great deal of heat that can be utilized [2,3]. The combined cycle plant does not discard the hot combustion gases directly to the environment, but uses a heat recovery steam generator and drives a separate steam turbine (Figure 9.12).

It is understood that the steam exiting the steam turbine is condensed and returned to the boiler in a fashion similar to the steam cycle, as discussed earlier (Figure 9.8).

9.5.4 Nuclear Power

Instead of using fossil fuels to generate heat by combustion, which is an exothermic chemical reaction, a nuclear reaction can be used to generate the necessary

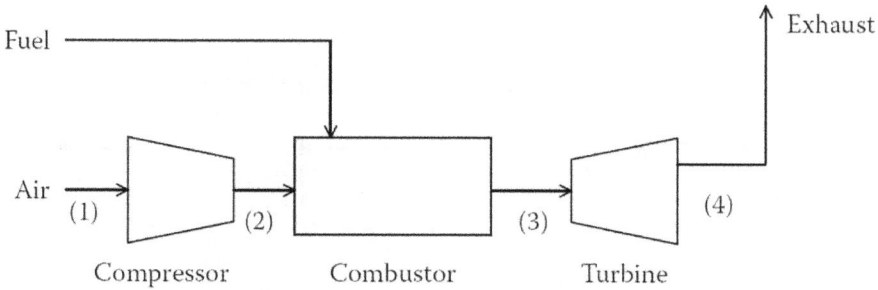

FIGURE 9.10 Schematic of a simple, singe-shaft gas turbine process (1) inlet, (2) compressed state, precombustion, (3) post-combustion, pre-compressor, (4) post-turbine to exhaust.

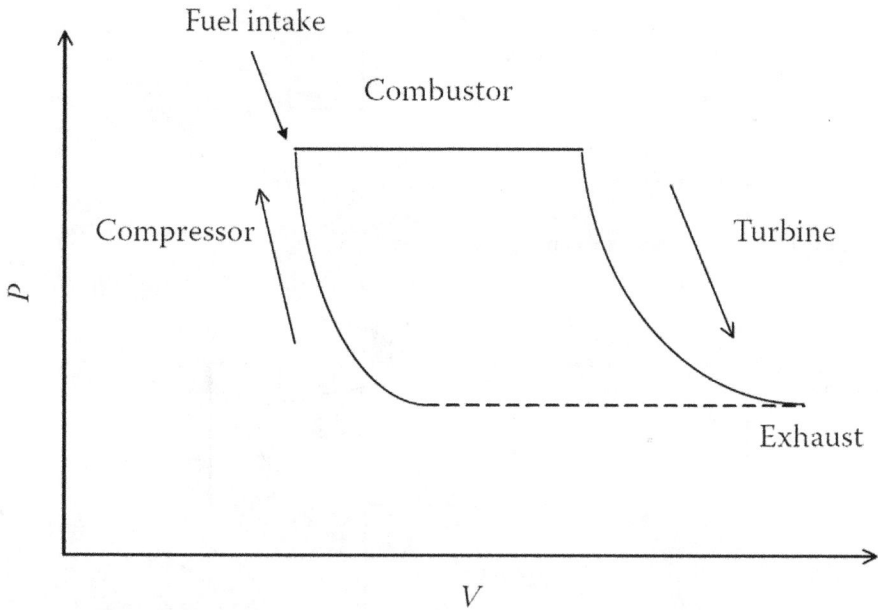

FIGURE 9.11 *P-V* diagram with Brayton cycle.

heat. In nuclear power plants, the heat is generated by nuclear fission. The heat, in turn, is used to generate steam. Broadly speaking, the pressurized water reactor and the boiling-water reactor are used today in the United States [2,3]. In the pressurized water reactor (Figure 9.13), water is pumped through the reactor by a reactor coolant pump. This is the last step in the closed loop, typically referred to as the primary loop. In the secondary loop, a feed water pump circulates water through a heat exchanger where the primary and secondary loops exchange heat. The water in the secondary loop is turned to steam here and feeds a turbine, where electricity is generated. In the boiling-water reactor, there is only one loop,

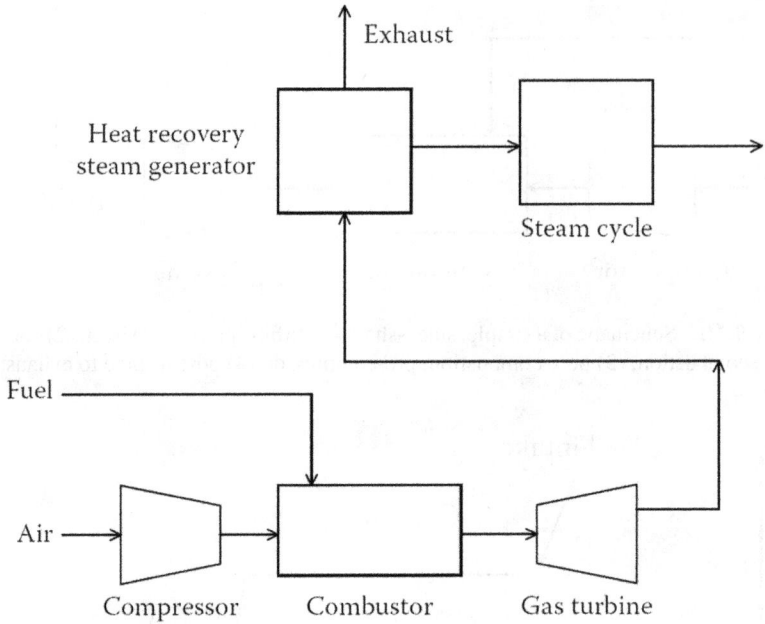

FIGURE 9.12   Schematic of a combined cycle plant.

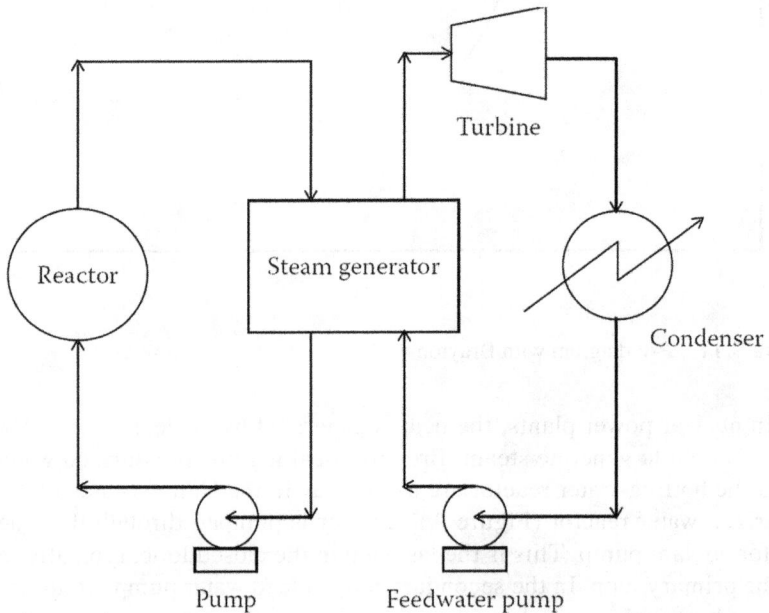

FIGURE 9.13   Schematic of a pressurized water reactor. (With permission from Bisio, A. and Boots, S. (eds), *Energy Technology and the Environment*, Vol. 2, Wiley Encyclopedia Series in Environmental Science, New York, 1995.)

and as a result, the overall efficiencies are higher at the added expense that the turbine becomes radioactive. We will not analyze the efficiency of nuclear reactors. The interested reader is referred to Refs. [11,12], where an analysis is presented. The efficiency of the process was found to be around 50%, where main losses were due to irreversible heat transfer from the nuclear heat source to the cooling water.[3]

9.5.5 HYDROPOWER

Hydrogenerators produce electricity by converting the potential energy stored in water to kinetic energy when water is allowed to fall in height through a turbine. The turbine shaft is generally arranged vertically, as opposed to the horizontal shafts of the gas and steam turbines. The efficiency of hydroelectric power generation plants can be estimated to be [11] around 80%, which is extremely high. However, when examining the hydroelectric power station, one must realize that in certain cases the formation of artificial lakes and the subsequent flooding of large areas are necessary. This will not show up in a thermodynamic efficiency analysis of the power generation process as such. In the case of the power station at Niagara Falls, no artificial flooding is necessary, and the environmental impact is probably minimal. We will not discuss the environmental impact in this chapter and only examine the thermodynamic efficiency of the process, but will touch on the environmental impact in Chapters 12 and 13.

9.5.6 WIND POWER

The use of wind energy to generate mechanical work has been known for centuries and has been used extensively in milling processes. In recent years, the interest in renewable energy has stimulated development in wind-generated electricity. Wind is a resource that is definitely renewable. In the United States, California is the leader in using wind energy for its power requirements, and wind energy accounts for more than 1% of the state's electricity requirement. This pales in comparison to Denmark, where 19.7% of the country's electricity is from wind power[4], and goals have been set to 50% by 2050. The current high contribution of wind energy to the Danish national requirement indicates that wind energy is definitely a viable option. In other European countries, such as Germany and The Netherlands, wind power is becoming increasingly more important. It is interesting to note that the sun is the ultimate source of wind power, as the sun creates atmospheric driving forces such as temperature and pressure differences, which ultimately cause wind.

9.5.7 SOLAR POWER

At present, the production of power from solar energy is not a significant contributor in electricity generation. However, this technology has attracted attention as a potential alternative for future energy production. This is discussed in Chapter 13.

9.5.8 GEOTHERMAL ENERGY[5]

The presence of geothermal energy is proven by visual evidence from phenomena such as volcanic activity, geysers, hot springs, pools of boiling mud, and so forth. Over the centuries, man has used this naturally occurring heat source. The simplest example is probably that of bathing in hot baths for therapeutic benefits (balneology). Most of these springs in ancient days were of thermal origin and can still be used even today.

In Iceland, almost all buildings are supplied with domestic heat from geothermal sources. Few Icelandic houses less than 40 years old have chimneys. In short, there are many applications of geothermal heat, of which we have only mentioned a small number. Probably the most spectacular advance for geothermal energy was the production of electricity from naturally occurring steam using steam engines in 1904 in Italy. Almost half a century passed before any other country followed Italy's pioneering work in geothermal energy generation. Now, New Zealand, the United States (California), Italy, and other countries have installed geothermal energy capacity. The production of electric power from geothermal energy is a well-established activity.

9.6 COAL CONVERSION PROCESSES

Power generation plants such as the steam plant, the gas turbine plant, and combined cycle plants require the combustion of a fossil fuel. Now, combustion is a chemical reaction of fuel with an oxidant (usually oxygen), and it makes sense to examine the combustion process more closely and analyze its thermodynamic efficiency. This means that we will examine the furnace/combustor of Figures 9.8, 9.10, and 9.12. We will examine coal and gas combustion *at the level needed for thermodynamic analysis*, after discussing some commonly used coal combustion processes.

Coal combustion processes can be classified based on process type (see Table 9.1), even though classification based on the particle size, the flame type, the reactor flow type, or the mathematical model complexity is also possible [7].

9.6.1 FIXED OR MOVING BEDS

The combustion of coal in a fixed bed (e.g., stokers) is the oldest and most common method of coal combustion. In recent decades, however, the fixed beds have lost some of their popularity due to the increased use of fluidized bed and suspended bed combustors [2,7].

9.6.2 SUSPENDED BEDS

In a suspended bed or entrained flow reactor technology, the coal is crushed, dried, and then pulverized to fine powder in a crusher and mill. As Table 9.1 shows, the coal particles used in entrained flow reactors are very small. The pulverized coal is transported with air to the furnace (primary air), and secondary air is heated and

TABLE 9.1
Classification of Coal Combustion by Process Type.

Process Type	Fixed or Moving Bed	Fluidized Bed	Suspended Bed
Particle size, μm Operating temperature, K	10,000–50,000 <2,000	1,500–6,000 1,000–1,400	1–100 1,900–2,000
Advantages	Established technology, low grinding, simple	Low SO_x and NO_x emissions, less slagging	High efficiency, large-scale possibilities, high capacity
Disadvantages	Emissions, especially particulate, less efficient than other methods	New technology	High NO_x, fly ash, pulverizing expensive
Commercial operations	Stokers	Industrial boilers	Pulverizing coal furnaces and boilers

Sources: Kroschwitz, J.I. and Howe-Grant, M. (eds.), *Encyclopedia of Chemical Technology*, Vol. 20, Interscience Publishers, New York, 1991; Bisio, A. and Boots, S. (eds.), *Energy Technology and the Environment*, Vol. 1, Wiley Encyclopedia Series in Environmental Science, New York, 1995.

fed into the combustor to ensure complete combustion. The residence time of the coal in the furnace is typically around 1–2 s, which usually suffices for complete combustion. However, not all coal burns completely, and fly ash will be generated (see Table 9.1).

9.6.3 FLUIDIZED BEDS

The combustion of coal in fluidized beds is becoming increasingly common. Atmospheric fluidized bed combustion (afbc) technology has been used commercially for the last two decades, whereas pressurized fluidized bed combustion (pfbc) is not yet as widespread. A simple schematic of a fluidized bed is given in Figure 9.14.[6]

9.6.4 THERMODYNAMIC ANALYSIS OF COAL COMBUSTION

We assume moist bituminous coal with 57.7% carbon, 4.1% hydrogen, 11.2% oxygen, 0.7% nitrogen, 1.3% sulfur, 10% water, and 15% ash on a mass basis. Based on this composition, the available work or standard chemical exergy value of this type of coal is 23,583 kJ/kg, as computed by Szargut et al. [11] along the methods of Chapter 7, and its experimentally determined higher heating value (HHV) is 21,860 kJ/kg.[7] Irreversible combustion and heat transfer occur simultaneously in a combustion chamber. For the analysis, we first consider adiabatic combustion followed by heat transfer. The combustion products will be CO_2, H_2O, oxides of nitrogen and sulfur, and nitrogen, assuming that air is used for combustion. Now, the useful product is not the combustion products but the heat, which is generated by the exothermic chemical reaction (combustion), and they "contain" the heat.

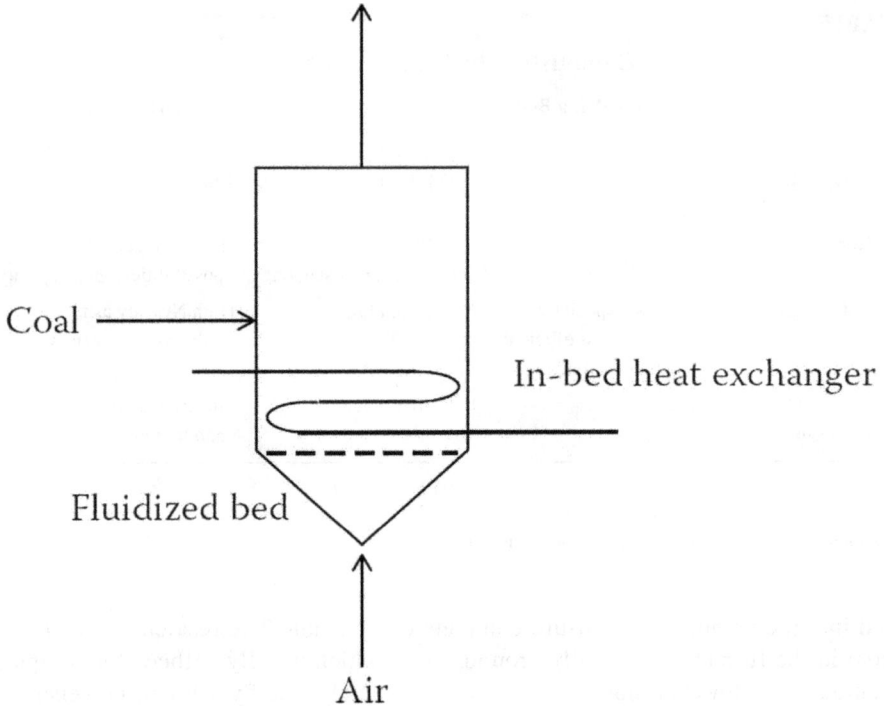

FIGURE 9.14   Schematic of a fluidized bed.

Consider 1 kg/s of coal that is combusted with an adequate amount of air (approximately zero exergy contribution). The rate at which exergy flows into the system is therefore 23,583 kW. The combustion releases heat, namely, at a rate of 21,860 kW at a temperature T. Since we have created a heat source at temperature T, it is straightforward to compute the work potential (exergy) of this heat source. All we need to do is multiply the heat release rate (21,860 kW) by the Carnot factor $1 - (T_0/T)$. This means that if the combustion takes place at temperature $T = 1200$ K for a fluidized bed reactor (Table 9.1), the efficiency of the combustion alone is $\eta_{combustion}$ = $(21,860/23,583) [1 - (T_0/T)] = 0.93 [1 - (T_0/T)] = 0.93 [1 - (298.15/1200)] = 0.7!$ This means that already 30% of the maximum work has been lost! We summarize this simplified analysis in Figure 9.15.

9.6.5   DISCUSSION

The heat is available at 1200 K, but there will be temperature differences in the heat exchanger, so more available work will be lost in the heat exchange process. What can we learn from this example? If we examine the Carnot factor, the answer seems to be clear. If we increase the operating temperature of the combustor, we can

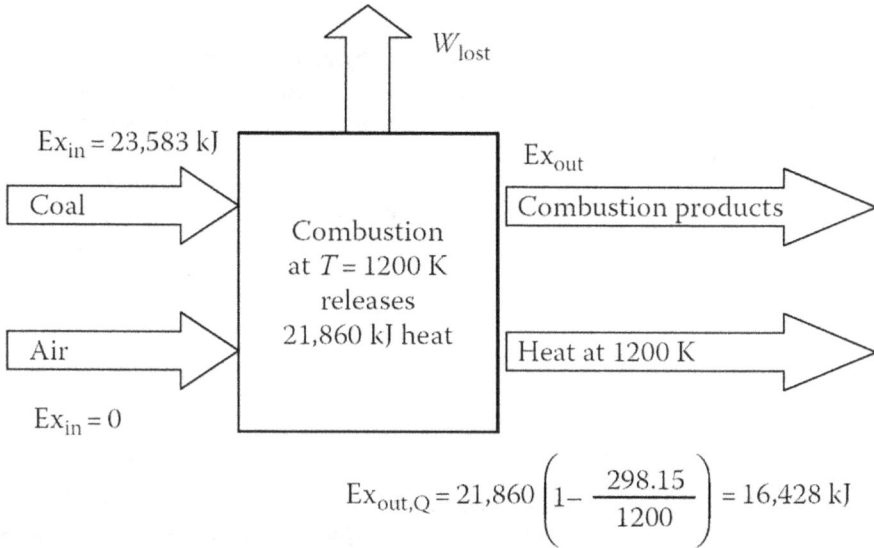

$$Ex_{out,Q} = 21,860 \left(1 - \frac{298.15}{1200}\right) = 16,428 \text{ kJ}$$

$$\eta_{combustion} = Ex_{out,Q}/Ex_{in} = 0.7$$

FIGURE 9.15 Flow of exergy in fluidized bed combustion.

increase the efficiency and lose less work in the process. For example, if we had cho-sen an operating temperature of 2000 K, as could be possible in the suspended bed, we would have obtained an efficiency of 0.79, which is quite considerable. However, any gain in efficiency could be offset by the increase in work necessary to pulver-ize the coal! For the sake of simplicity, we have not included these in this analy-sis. From the point of view of efficiency of combustion, the higher the combustion temperature, the better. This would of course mean that steam could be generated at higher temperatures or pressures, which is a direct technological consequence provided suitable materials are available.

We caution the reader that applying Carnot's analysis is based on the assump-tions that the heat is available at temperature T and that the heat reservoir is infinite. This means that if we use the adiabatic flame temperature for T, we will end up with a maximum attainable efficiency, since the exchange of heat will inevitably lead to a reduction in the temperature of the reservoir. From our analy-sis, it is not clear whether we used an adiabatic flame temperature [11]. Note that the adiabatic temperature is the highest temperature that can be reached by the system if all the heat generated is used to elevate the flame temperature. However, we can safely state that *at least* 30% of the maximum work potential has been lost. We will return to this subtle point at a later stage, when we examine the combus-tion of natural gas.

9.6.6   COAL GASIFICATION

A great deal of the coal demand today is driven by coal gasification plants. There is a great deal of similarities between coal combustion processes and coal gasification plants at the level that we will consider in this book. At the heart of a gasification process is the gasifier, which converts hydrocarbon feedstock into gaseous components by applying heat under pressure in the presence of steam (Figure 9.16).

Coal gasification is essentially the transfer of the exergy from solid coal to an excellent chemical gaseous fuel and base chemical. Suppose gasification is shown by

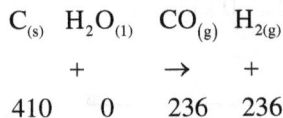

$$
\begin{array}{ccccc}
C_{(s)} & H_2O_{(1)} & & CO_{(g)} & H_{2(g)} \\
& + & \rightarrow & & + \\
410 & 0 & & 236 & 236
\end{array}
$$

where C represents coal and the corresponding exergy values are shown in kJ/mol below the symbols. It is clear that the right-hand side of the equation has more exergy than the left-hand side, which means that there has to be another source of exergy, which is obtained by combusting some coal. It is in this that a gasifier differs from a combustor. The amount of air or oxygen available inside the gasifier is carefully controlled so that only a relatively small portion of the fuel burns completely. This partial oxidation process provides the heat, or better said, the exergy. Most of the carbon-containing feedstock is chemically broken apart by the gasifier's heat and pressure, setting into motion chemical reactions that produce syngas. The composition

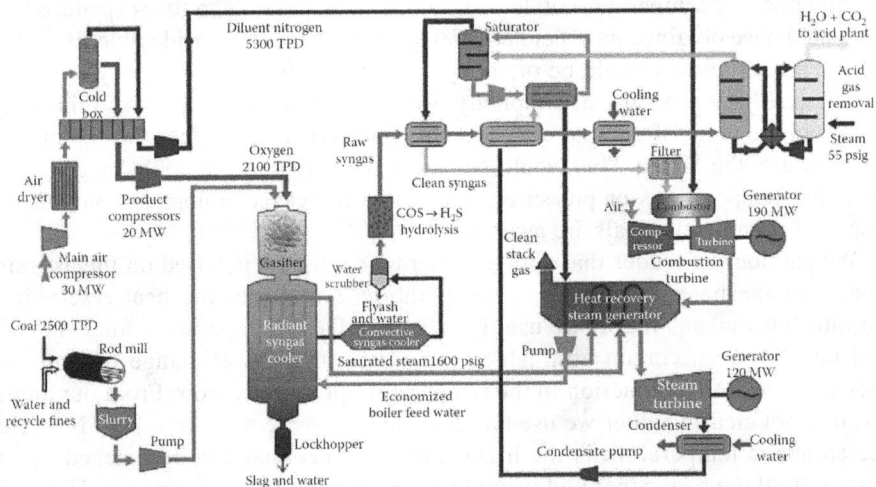

FIGURE 9.16   Schematic of coal gasification. (With permission from www.fossil.energy.gov/programs/powersystems/gasification/howgasificationworks.html; http://en.wikipedia.org/wiki/Integrated_gasification_combined_cycle.)

of the syngas is primarily hydrogen and carbon monoxide, but it varies based on the feedstock and process conditions.

Components that cannot gasify, such as mineral components in the fuel, leave the gasifier either as an inert glass-like slag or in a form useful to marketable solid products. A small fraction of the mineral matter is blown out of the gasifier as fly ash and requires removal downstream.

Sulfur impurities in the feedstock are converted to hydrogen sulfide and carbonyl sulfide, from which sulfur can be easily extracted, typically as elemental sulfur or sulfuric acid, both valuable by-products. Nitrogen oxides, potential pollutants, are not formed in the oxygen-deficient (reducing) environment of the gasifier; instead, ammonia is created by nitrogen-hydrogen reactions. The ammonia can be easily washed out of the gas stream.

In integrated gasification combined-cycle (IGCC) systems, the syngas is cleaned of its hydrogen sulfide, ammonia, and particulate matter and is burned as fuel in a combustion chamber of a gas turbine (much like natural gas is burned in a turbine). The combustion turbine drives an electricity generator. Exhaust heat from the combustion turbine is recovered and used to boil water, creating steam for a steam turbine-generator.

The use of these two types of turbines—a combustion turbine and a steam turbine—in combination, known as a "combined cycle," is one reason why gasification-based power systems can achieve high power generation efficiencies. Currently, commercially available gasification-based systems can operate at around 40% efficiencies; in the future, some IGCC systems may be able to achieve efficiencies approaching 60% with the deployment of advanced high-pressure solid oxide fuel cells, which use hydrogen as the fuel. (A conventional coal-based boiler plant, by contrast, employs only a steam turbine-generator and is typically limited to 33%–40% efficiencies.)

All or part of the syngas can also be used in other ways:

- As chemical "building blocks" to produce a broad range of higher-value liquid or gaseous fuels and chemicals using processes well established in today's chemical industry such as water-gas shift and Fischer-Tropsch chemistry.
- As a fuel producer for highly efficient fuel cells by hydrogen or perhaps in the future, hydrogen turbines and fuel cell-turbine hybrid systems.
- As a source of hydrogen that can be separated from the gas stream and used as a fuel or as a feedstock for refineries (which use the hydrogen to upgrade petroleum products).

Another advantage of gasification-based energy systems is that when oxygen is used in the gasifier (rather than air), the CO_2 produced by the process is in a concentrated gas stream, making it easier and less expensive to separate and capture. Once the CO_2 is captured, it can be sequestered. Carbon sequestration will be discussed later. It is clear that gasifiers can play a crucial role in the so-called coal-to-liquids technology.

9.7 THERMODYNAMIC ANALYSIS OF GAS COMBUSTION

We will now consider the combustion of 1 mol per second of natural gas (we assume 100% methane).

9.7.1 EXERGY IN

The heating value of methane (see Chapter 6) is almost equal to its exergy value [11,14]. For the sake of simplicity in the ensuing analysis, we will set this exergy value of the gas to be equal to the energy value or, equivalently, the value of the heat of reaction. In Table 9.2, we have tabulated the exergy of methane at a number of different conditions.

As the table shows, the exergy is not a strong function of temperature or pressure. For the illustrative purposes of our analysis, we use the value of 831.6 kJ/mol.

9.7.2 AIR REQUIREMENTS

We assume that the oxygen from the air (assumed to be 20 mol% oxygen and 80% nitrogen for simplicity) and the methane react as follows:

TABLE 9.2
Standard Exergy Value of Methane at Different Conditions.

Substance	Ex, kJ/mol
CH_4 (1 bar, 298.15 K)	831.6
CH_4 (100 bar, 298.15 K)	842.6
CH_4 (100 bar, 373.15 K)	842.9

Source: Smoot, L.D., *Fossil Fuel Combustion: A Science Source Book*, Bartok, W. and Sarofim, A.F. (eds.), John Wiley & Sons, Inc., New York, 1991.

TABLE 9.3
Product Distribution in Combustion Gas.

Species	Number of Moles per Second	Gas Mole Fraction, y_i
H_2O	2	$\dfrac{2}{5\lambda+1}$
CO_2	1	$\dfrac{1}{5\lambda+1}$
N_2	4λ	$\dfrac{4\lambda}{5\lambda+1}$
O_2	$\lambda-2$	$\dfrac{\lambda-2}{5\lambda+1}$

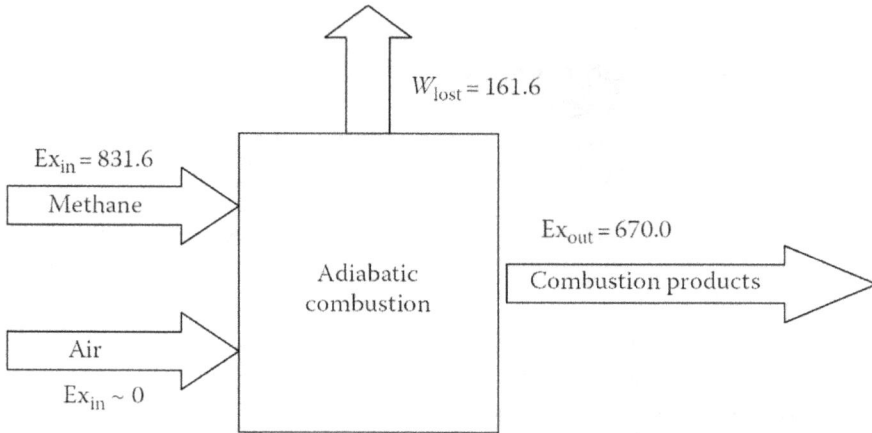

FIGURE 9.17   Flow of exergy in kW for adiabatic gas combustion.

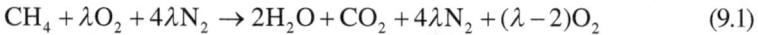

$$CH_4 + \lambda O_2 + 4\lambda N_2 \rightarrow 2H_2O + CO_2 + 4\lambda N_2 + (\lambda - 2)O_2 \qquad (9.1)$$

which expresses essentially complete oxidation. With $\lambda > 2$, we make use of the excess air to ensure complete combustion, and we will again assume a feed rate of 1 mol methane per second. Table 9.3 shows the product distribution in the gas flowing out of the combustor. We can calculate λ if we set the temperature of combustion as we will show later.

The exergy of the ambient air is nearly zero, will be small for the compressed air flowing into the combustor compared to the exergy of the fuel, and can be neglected (when chemical transformations take place, the bulk of the exergy is usually due to the chemical component of the exergy).

In adiabatic combustion, the chemical energy of the natural gas is converted into the same amount of thermal energy stored in the effluent stream. For the effluent stream, however, the exergy will no longer be equal to the energy content, since some work would have been lost. (See Figure 9.17. The numbers are calculated in Section 9.7.3.)

We can compute λ since we made the assumption that the combustion is adiabatic. At this point, we invoke the first law for flowing systems, which reduces to

$$\Delta \dot{H} = 0 \qquad (9.2)$$

because no work is performed in the combustor, and no heat exchange takes place (adiabatic). This means that the *sum* of the rate of enthalpy change of reaction and the enthalpy change of the combustion gases equals zero. This equation can be written as

$$-\Delta_r \dot{H} = \dot{n} \int_{T_2}^{T_3} C_p(T)dT = \dot{n} C_p(T_m)(T_3 - T_2) \qquad (9.3)$$

TABLE 9.4

Heat Capacity of Combustion Products at 860.65 K.

Component	C_p (860.65 K) J/mol K
H_2O	39.4
CO_2	51.5
O_2	33.6
N_2	31.55

where $\dot{n}$ is the number of moles[8] in the combustion stream. We approximate the molar heat capacity of the mixture $C_p(T_m)$ by evaluating C_p at a mean temperature. For convenience and mathematical simplicity we use the arithmetic mean, though this is formally incorrect, but it simplifies our calculations greatly without adding unnecessary complications to the computations. For a better and more accurate method of calculating this integral, we refer to standard textbooks on thermodynamics. The temperatures T_3 and T_2 denote the outlet and inlet temperatures of the combustor, respectively. Here, we assume an inlet temperature of $T_2 = T_1 = 298.15$ K and an outlet temperature of 1423.15 K, which is a common inlet temperature for the gas turbine [9]. We therefore compute the heat capacity at a mean temperature of $T_m = (1/2)\,(298.15 + 1423.15) = 860.65$ K.

Table 9.4 shows the value of the heat capacity at 860.65 K for the constituents of the effluent stream.

This yields $C_p = \Sigma\; y_i C_{p,i} = (63.1 + 159.8\lambda)/(5\lambda + 1)$. If we insert this result into Equation 9.3, and substitute the heat of reaction and the values of T_3 and T_2, we obtain

$$(1423.15 - 298.15) \times \frac{63.1 + 159.8\lambda}{5\lambda + 1} \times (5\lambda + 1) = 831.6 \times 10^3 \qquad (9.4)$$

which yields $\lambda = 4.23$. We can compute the corresponding value of $C_p = 33.4$ kJ/mol K.

9.7.3 EXERGY OUT

We can now compute the exergy of the gaseous effluent, by using the definition

$$\dot{EX} = \dot{EX}_{ph} + \dot{Ex}_{ch}^0 + \dot{EX}_{mix} \qquad (9.5)$$

where the chemical exergy of the stream can be computed as (using Table 9.5)

$$\dot{Ex}_{ch}^0 = 2\dot{Ex}_{ch}^0\left(H_2O\right) + 16.92\dot{Ex}_{ch}^0\left(N_2\right) + \dot{Ex}_{ch}^0\left(CO_2\right) + 2.23\dot{Ex}_{ch}^0\left(O_2\right)$$
$$\dot{Ex}_{ch}^0 = 2 \times 9.5 + 16.92 \times 0.72 + 19.9 + 2.23 \times 3.97 = 59.9 \text{kW} \qquad (9.6)$$

TABLE 9.5
Standard Chemical Exergy of Selected Components.

Component	Ex^0_{ch} kJ/mol
H_2O (l)	0.9
CO_2	19.9
O_2	3.97
N_2	0.72

---

and the mixing contribution as

$$\dot{Ex}_{mix} = \dot{n}RT_0 \sum_i y_i \ln y_i$$
$$= 22.15RT_0[0.09\ln 0.09 + 0.05\ln 0.05 + 0.76\ln 0.76 + 0.10\ln 0.10] \qquad (9.7)$$
$$= -20.4\text{kW}$$

The computation of the chemical exergy of water requires some explanation. It is not equal to zero, since the water in the steam is not in the *liquid* phase, but in the *vapor* phase. Therefore, we can compute it from

$$Ex^{0g}_{ch}(H_2O) - Ex^{01}_{ch}(H_2O) = RT_0 \ln\left(P_0 / P^{sat}_{H_2O}\right)$$
$$= 8.314 \times 298.15 \times \ln\left(\frac{1.013 \times 10^5}{3.166 \cdot 10^3}\right) = 8.6\text{kJ / mol}$$

Since $Ex^{01}_{ch}(H_2O) = 0.9$ kJ/mol, this yields $Ex^{0g}_{ch}(H_2O) = 9.5\text{kJ / mol}$. The physical exergy rate can be expressed as

$$Ex_{ph} = \dot{n}\left(\Delta H - T_0 \Delta S\right) \qquad (9.8)$$

where ΔH and ΔS are the enthalpy and entropy increase of changing a mole of combustion gas from T_0, P_0 to T, P. From Cengel and Boles [9], we take $P = 14$ bar and we had already taken $T = 1150°C = 1423.15$ K. The quantities ΔH and ΔS can be calculated by integrating standard thermodynamic relations and the assumption that the gas behaves ideally with constant C_p:

$$\Delta S = \int_{T_0,P_0}^{T,P} dS = C_p \int_{\ln T_0}^{\ln T} d\ln T - R \int_{\ln P_0}^{\ln P} d\ln P \qquad (9.9)$$

$$\Delta H = \int_{T_0,P_0}^{T,P} dH = C_p \int_{T_0}^{T} dT \qquad (9.10)$$

where we use the value of $C_p = 33.4$ kJ/mol K, computed earlier. Ex_{ph} can now be computed:

$$\Delta H = C_p \int_{T_0}^{T} dT = C_p \left(T - T_0\right) = 33.4(1423.15 - 298.15)$$
$$= 37.5 \text{kJ} / \text{mol} \tag{9.11}$$

$$\Delta S = C_p \int_{\ln T_0}^{\ln T} d\ln T - R \int_{\ln P_0}^{\ln P} d\ln P$$
$$= C_p \ln \frac{T}{T_0} - R \ln \frac{P}{P_0} = 33.4 \ln \frac{1423.15}{298.15} - 8.314 \ln \frac{14}{1} \tag{9.12}$$
$$= 30.3 \text{J} / \text{mol.K}$$

which yields the physical exergy rate

$$Ex_{ph} = 22.15\left(37.5 - 298.15 \times 30.3 \times 10^{-3}\right) = 630.5 \text{kW} \tag{9.13}$$

9.7.4 EFFICIENCY

The total value of the exergy flowing out is $630.5 + 59.9 - 20.4 = 670$ kW. The amount of exergy flowing in is equal to 831.6 kW, so the efficiency of this combustion is 0.81 (see Figure 9.17). Now, if we use the method described for the coal combustion, we can also compute the thermodynamic efficiency of the combustion:

$$\eta_{\text{combustion}} = \left(1 - \frac{T_0}{T}\right) = 1 - \frac{298.15}{1423.15} = 0.79 \tag{9.14}$$

which is close to 0.81 and can be considered to be the same, considering the approximations we made. Now, this is no coincidence, because we have simply computed the thermodynamic efficiency in two different ways. The analysis shows that if a higher combustion temperature can be used, the process will be more efficient. This can be accomplished by using less air, so the hot stream is less diluted. The example highlights that (complete) combustion, or direct spontaneous reaction, is a process that is limited in its efficiency. We point out that in fuel cells, or indirect reactions, Carnot limitations are not present, *but* other irreversibilities may reduce the efficiency. A discussion of fuel cells is not covered in this book, but we believe that this technology will become increasingly more popular and common. The tools developed in this book will most certainly be applicable.

9.7.5 DISCUSSION

At this point, it is useful to point out that simply computing the efficiency based on the Carnot factor is an exercise that should be performed with care. The Carnot factor, C, is given by $C = 1 - (T_0/T)$. As shown in the example, we can compute the group $-C\Delta_r H/Ex_{in}$ to get the efficiency! Strictly speaking, this is not always true, but why? The answer is very subtle. Carnot's analysis holds only for infinite heat reservoirs, and if the heat (which can be viewed as the useful product) is transferred, the temperature of the reservoir (the product mixture) also changes! So the correct way of computing the efficiency based on the heat is to take into account the fact that the temperature T is not constant, but is variable and will decrease to T_0:

$$\text{Ex} = \int_T^{T_0} d\dot{\text{Ex}}_Q = \int_T^{T_0} \left(1 - \frac{T_0}{T}\right) d\dot{Q} = \int_{T_0}^T n'C_p \left(1 - \frac{T_0}{T}\right) dT \tag{9.15}$$

Note that we have used $d\dot{Q} = -\dot{n}C_p dT$ to ensure that when heat is transferred out of the "reservoir," the temperature of the reservoir decreases. A closer inspection of this equation reveals that it can be simplified as follows:

$$\text{Ex} = \int_{T_0}^T \dot{n}C_p \left(1 - \frac{T_0}{T}\right) dT = \int_{T_0}^T n\dot{C}C_p dT - T_0 \int_{T_0}^T \frac{\dot{n}C_p}{T} dT$$
$$= \dot{n}\left[\Delta H - T_0 \Delta S\right] \tag{9.16}$$

This is the definition of the physical exergy of the effluent stream! The computation of the terms will yield the physical component of the stream, and the combination with the chemical and mixing components will allow for the computation of the efficiency. The question now remains: Why did the computation of the efficiency based on the Carnot factor give the correct number? The answer is that since the temperature of the effluent gases is *fixed*, it mimics an infinite heat reservoir, and therefore $\dot{n}\left[\Delta H - T_0 \Delta S\right]$ simplifies to $\dot{n}\Delta H\left[1 - T_0(\Delta S / \Delta H)\right] = \dot{n}\Delta H\left[1 - (T_0 / T)\right]$, since $\Delta G = \Delta H - T\Delta S = 0$ at equilibrium.

9.8 STEAM POWER PLANT

Steam power plants are commonplace in the power-generating industry (see Figures 9.8 and 9.18), where the heat is generated in the furnace.

State-of-the-art first law efficiencies of boilers are typically between 85% and 90%, that is, the fraction of fuel energy value (heat of combustion) captured in steam. It is not surprising that the efficiency is not 100%, since the flue gas that exits the boiler is much hotter than the temperature of the environment and therefore takes with it some of the energy ("heat goes up the chimney"). Typical conditions in a steam cycle are as given in Table 9.6 [3,10]. In this table, we have not included the exergies of the fuel, since we will perform our analysis based on 1 kJ/s fuel feed rate. Now, assume the first law efficiency of the boiler is 87%. This means that 1 kJ/s of fuel is fed into the boiler, and 0.87 kJ/s is absorbed in the enthalpic value of steam. This means that the steam rate equals the following (using values of enthalpy from the steam tables):

$$\dot{m}_{steam} = \frac{0.87}{h_1 - h_4} = \frac{0.87}{3391.6 - 203.4} = 2.73 \times 10^{-4} \text{ kg} / \text{s} \tag{9.17}$$

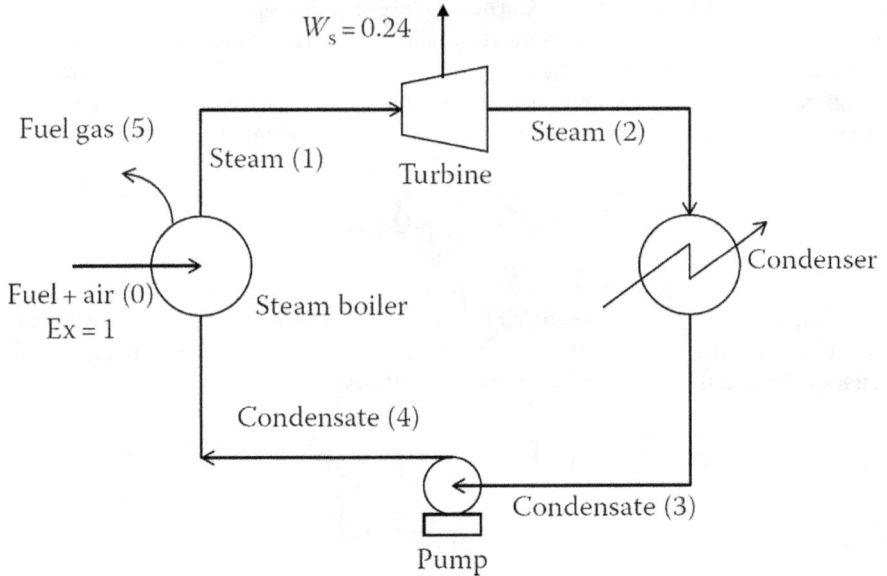

FIGURE 9.18　Computation of the thermodynamic efficiency in a steam cycle (1) inlet, (2) pre-combustion, (3) post-combustion, pre-compressor, (4) post-turbine to exhaust, (5) gas inlet.

TABLE 9.6
Typical Process Conditions in Steam Cycle.

Point	Thermodynamic State	T, P	Exergy (kJ/kg)
(0)	Gas + solid/gas + liquid in natural gas + air	—	—
(1)	Superheated vapor	500°C, 8600kPa	1402.8
(2)	Wet vapor, quality = 0.9378	45°C, 10 kPa	149.5
(3)	Saturated liquid	45°C, 10 kPa	11.8
(4)	Subcooled liquid	45°C, 8600 kPa	12.8
(5)	Gas[a]	93°C, 101 kPa	—

[a] We have given the computation of the *rate* of exergy lost through this stream in the text.

The steam production allows for the computation of the exergy flow at various points (Table 9.7). For example, the exergy flow at point (1) is simply $2.73 \times 10^{-4} \times 1402.8 = 0.38$ kJ/s. The exergy value of the flue gas stream (5) is computed by using the *first law* efficiency and realizing that the flue gas stream is a heat stream being discarded:

$$(1-0.87)\left(1-\frac{298.15}{366.15}\right) = 0.043\text{kJ}/\text{s} \tag{9.18}$$

TABLE 9.7
Exergy Streams for Steam Cycle.

Point	Exergy (kJ/s)
(0)	1.0
(1)	0.38
(2)	0.04
(3)	3.2214×10^{-3}
(4)	7.6×10^{-4}
(5)	0.043

Table 9.7 shows the exergy streams (based on 1 kJ/s exergy input by fuel), which are calculated by multiplying the exergy values, shown in Table 9.6, by the steam mass flow rate.

Now, the maximum amount of work the turbine can extract from the steam is given by the difference of the exergy streams flowing in and out of the turbine (which represent the work potential of the steam), which is $0.38 - 0.04 = 0.34$ kJ/s. Assuming a turbine efficiency of 75%, we readily see that 0.25 kJ/s electricity is generated. Therefore, the exergetic efficiency of the steam power plant is 25% if the electricity input of the pump is not taken into account. The pump duty can be calculated (per unit mass steam) by using the enthalpy of water: $W_{pump} = h_4 - h_3 = 203.4 - 191.8 = 11.6$ kJ/kg$_{steam}$ or, equivalently, 3.17 J/s for an ideal pump, by multiplying by the steam rate. Using an efficiency of 75%, the "real" pump duty equals 4.2 J/s. The new exergetic efficiency is 25%. We note that the thermodynamic efficiency (*second law*) of the boiler-furnace is 38%, which is in contrast to the *first law* efficiency of 87% (see Figure 9.18).

9.9 GAS TURBINES, COMBINED CYCLES, AND COGENERATION

If a fossil fuel is combusted in a steam plant, the combustion proceeds to release thermal energy and uses this energy to supply the heat used to generate steam. The analysis given earlier can be adapted to any type of (fossil) fuel. In contrast, in gas turbines, the combustion takes place for a different reason, namely, to produce high-velocity gases that already produce work before they are sent to a steam cycle.

9.9.1 GAS TURBINES

Gas turbines are frequently used since the inlet temperatures can be much higher than the maximal steam temperatures, and currently inlet temperatures around 1400 K are commonplace. However, in order to use a gas turbine, expansion to a lower pressure has to occur. This is the reason that both the fuel and the air have to be elevated in pressure at room temperature. A typically used pressure is 2.1 MPa, and after combustion, the pressure typically drops to about 2.0 MPa. The expansion to atmospheric pressure occurs at high temperatures, and because of the elevated temperatures, the volume of the gas that will undergo the expansion is much larger than the volume of gas initially compressed, thus leading to a net production of mechanical energy (consider Figure 9.10). Air enters the compressor at ambient conditions and is compressed to some higher pressure. No heat is added, but the temperature of the air rises due to compression, which means that the temperature and pressure of the compressor discharge are higher than at the inlet.

Upon leaving the compressor, the compressed air enters the combustor, where fuel is injected and combustion takes place. The combustion process takes place at essentially constant pressure. In the turbine section, part of the energy of the hot gases is converted into work. This conversion takes place in two steps. The hot gases are expanded, a portion of the thermal energy is converted into kinetic energy, and the kinetic energy into an increase in the enthalpy that can be converted into work. Some of the work developed by the turbine is used to drive the compressor. Typically, more than 50% of the work is used to drive the compressor, which generally includes compressor inefficiencies.

9.9.2 THERMODYNAMIC ANALYSIS OF GAS TURBINES

We can make an estimate of the efficiency of a power plant using a gas turbine (see Figure 9.10). We assume that the combustor inlet temperature of the gases is $T_2 = 298.15$ K and turbine inlet temperature of $T_3 = 1423.15$ K and use the data from the methane combustion example. We further assume that the turbine has a total efficiency of 0.75. We will first compute the turbine work if the expansion is isentropic to 1 bar. Consider Equation 9.9, where we set $\Delta S = 0$. We can derive the following [10]:

$$\frac{T_4}{T_3} = \left(\frac{P_4}{P_3} \right)^{R/C_p} \tag{9.19}$$

We can solve this equation iteratively (C_p is a function of the arithmetic mean temperature) to obtain $T_4 = 747.5$ K, which is the isentropic exit temperature for which $C_p = 33.4$ J/mol K as was shown earlier. The isentropic work rate is given by

$$\dot{W}_s = -\dot{n}\Delta H = -\dot{n}C_p\left(T_4 - T_3\right) = 22.1 \times 34.07(1423.15 - 747.5) \tag{9.20}$$
$$= 508.992 \text{kW}$$

If we assume a turbine efficiency[9] of 0.75, the "real" work becomes 508.9 × 0.75 = 381.675 kJ, and the exit temperature can be computed to be 892.1 K. Now in general, around 50% of the work generated by the turbine is used to drive the compressor, which means that the net generation of electricity is only 0.5 × 381.657 = 190.8 kJ. This means that the total system has a thermodynamic efficiency of 190.8/831.6 = 0.23! Now the hot gas stream allowed to exit the system at 892.1 K still has exergy. We can compute the total exergy rate by the method described earlier:

$$\dot{Ex} = \dot{Ex}_{ph} + \dot{Ex}_{ch}^{0} + \dot{Ex}_{mix} \qquad (9.21)$$

Since the composition did not change, we do not need to recompute Ex_{ch} and Ex_{mix}. We need to recalculate Ex_{ph} since the temperature is different. We compute C_p at 595.13 K, which is the arithmetic mean of 892.1 and 298.15 K, to obtain 33.104 J/mol K. In this fashion, we obtain $\Delta H = 19{,}662.12$ J/mol and $\Delta S = 36.28$ J/mol K, which yields the rate $\dot{Ex}_{ph} = \dot{n}(\Delta H - T_0 \Delta S) = 195.47$, which in turn gives the rate $\dot{Ex} = 225.1$ kW (see Figure 9.19). In terms of the initial input rate of exergy (831.6 kW), this is 27.1%, which means that 0.271 is going to waste! If this hot gaseous stream could be used somewhere, the efficiency could be increased. We have dealt with this in Section 9.8 on the steam power plant.

9.9.3 Combined Cycles, Cogeneration, and Cascading

The gases leaving the turbine are still sufficiently hot to be useful (i.e., contain exergy). For instance, gases exiting the turbine typically have temperatures of around 800 K [9], which means that this heat has a quality (see Section 6.3) of $(1 - 300/800) = 0.625$, which is fairly significant. For this reason, practical power generation plants usually have heat recovery systems that either preheat the gases flowing into the combustor or generate steam. This steam can then be used to drive a steam cycle and, despite losing much of its quality, can still be used in district heating. This is the principle of *cascading*. Power plants that use a gas

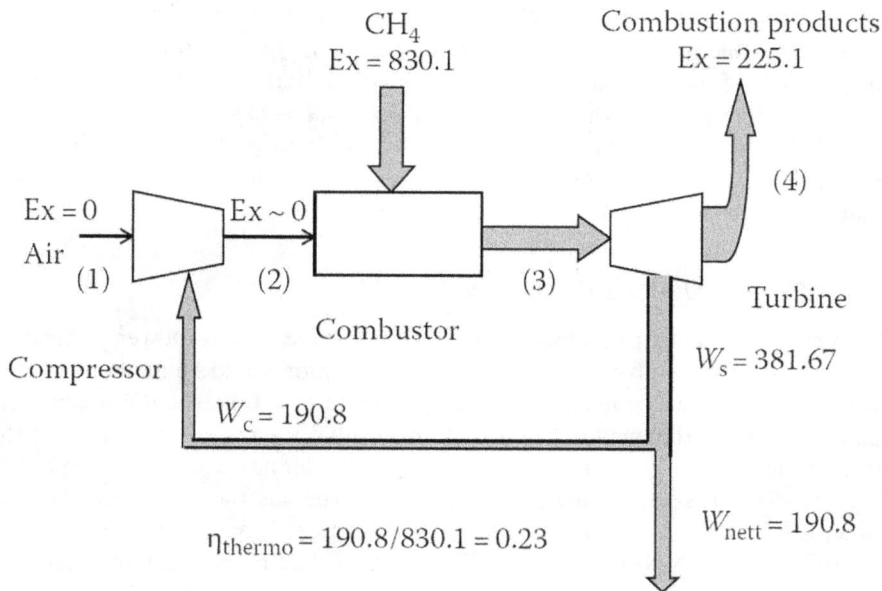

FIGURE 9.19 Thermodynamic efficiency in a gas cycle (1) inlet, (2) compressed state, pre-combustion, (3) post-combustion, pre-compressor, (4) post-turbine to exhaust.

turbine and a steam cycle are referred to as *combined cycle* plants, since they use two cycles for the generation of power (see Figure 9.12). In addition to using two cycles, a power plant can also generate heat, which is a useful product.[10] In the power industry, the simultaneous generation of heat and power is referred to as *cogeneration*.

9.9.4 EXAMPLE

In the gas turbine example, there was waste heat, which was discarded to the environment. Suppose we use the waste heat to drive a steam cycle. The steam cycle described earlier generates steam at 500°C (= 773.15 K) from liquid water, supplied at 45°C (318.15 K). To do this, $h_1 - h_4 = 3391.6 - 203.4 = 3187.6$ kJ/kg$_{steam}$ are necessary. At our disposal is a gas stream (22.1 mol/s) at 892.1 K. We will assume that we cool the gas stream from 892.1 to 373.15 K (we choose these temperatures arbitrarily for illustrative purposes only). We compute the average C_p of the gas stream at 632.65 K, to obtain 35.36 J/mol K. The heat transferred is therefore equal to 22.1 × 35.36 (892.1 − 373.15) = 405.5 kJ/s, which means that the work the steam turbine (see Section 9.8) can generate is 405.5 × 0.25 = 101.375 kJ/s. So the combined cycle generates 101.375 + 190.8 = 292.175 kJ/s. The net efficiency of the combined cycle is therefore 292.175/831.6 = 0.35. Now in practice, combined cycle plants have efficiencies of around 0.5. The reason we did not recover this value is that we simply pasted two cycles together, while in practice the conditions are chosen carefully to optimize power output. However, the concept is clear: Combined cycles have higher efficiencies. If we now turn this "combined cycle" plant into a cogeneration combined cycle plant, we use the heat as a useful product. The heat is available at 373.15 K. We consider only the physical component of the exergy $\dot{Ex}_{ph} = \dot{n}(\Delta H - T_0 \Delta S)$. We can compute ΔH, ΔS using the standard methods described earlier to obtain $\Delta H = 2977.5$ J/mol, $\Delta S = 8.91$ J/mol K, which gives the rate $\dot{Ex}_{ph} = 7.1$ kJ/s. The new efficiency is now 0.373, which is an improvement on the original figure of 0.35. It is interesting to consider the gain if this waste heat (for a shower for example) was directly generated from natural gas!

9.10 CONCLUDING REMARKS

We presented the thermodynamic analyses of some simple power generation technologies. The analysis hinted that the combustion could be made more efficient in certain cases by using higher temperatures. The boiler in the steam cycle had a fairly low efficiency (0.38), which meant that the overall efficiency of the steam cycle was low as well. The reason that the boiler had a low thermodynamic efficiency is that steam is generated at 500°C, whereas the heat is available at higher temperatures; in other words, the heat transfer is across a large temperature difference, which reduces the efficiency, as was made clear in Chapter 5 (Figure 5.3).

Other noteworthy points are that electricity requirements of the pump in the steam cycle (incompressible fluid) are less important than the requirements of the compressor in the gas turbine cycle (compressible fluid) and that inefficiencies in the latter can reduce the overall thermodynamic efficiency a great deal.

The combined cycle and the combined cycle cogeneration plant showed that waste heat could be put to good use, thus generating less lost work. The process conditions used in the analysis of the combined cycle plant were chosen for illustrative purposes and were arbitrary. In practice, the conditions are generally chosen to increase power output (see, e.g., Chapter 5).

We wish to alert the reader that in the preceding analyses, the results were essentially independent of the type of fuel used. From an efficiency point of view, this may be true, but from a sustainability point of view, it is not. In general, gas is a much cleaner burning fuel than coal and requires less pre- and post-treatment. Even though the standard power generation plants can be made more efficient using thermodynamic analysis (lost work, availability, or exergy analysis), we note that power generation based on fossil fuels is not sustainable since the combustion of these fuels leads to increased CO_2 levels and the depletion of finite resources. Alternatives such as hydroelectric and solar power energy should be considered if man is to live in a truly sustainable society. The necessary tools and insights will be developed in Part IV.

NOTES

1 The steam turbine provides mechanical energy, which can be transformed into electrical energy by means of a generator. The advantage of this is that it allows for transportation of the electricity at high voltages.
2 A notable exception is of course the internal combustion engine in automobiles (alluded to earlier) where locomotion is the primary objective and the chemical exergy of fuel is partly converted to mechanical work.
3 The assumption is typically made that the nuclear fuel can be treated as a heat source at an infinite temperature [12].
4 Announced in mid-2000.
5 Section 9.5.8 was adapted from [13].
6 In practice, the gases exiting the fluidized bed reactor contain a certain amount of ash and have to be cleaned. Also, the combustion products of coal are sometimes corrosive, which means that in addition to air being fed into the reactor, various other chemicals are added to ensure "clean" combustion products that will not corrode turbine blades or violate environmental standards. Coal combustion is a very active field of research, and many exciting developments are occurring there. In this analysis, we make certain assumptions that illustrate the *thermodynamic* concepts as clearly as possible. Therefore, we do not examine the effect of hydrodynamics, heat, and mass transfer, which are very important in the combustion of the coal particle and the distribution of combustion products. We do not expect that this will have a significant impact on the analysis.
7 The higher heating value gives the gross heat content for the fuel, including the heat of vaporization, and is commonly used in the United States. In Europe, the lower heating value is often used, which does not include the heat of condensation [2].
8 In our example $n = 5\lambda + 1$.
9 An increase in the efficiency of the turbine is extremely important for the overall result, and a great deal of research is directed to improving this. Discussion of this, however, is beyond the scope of this book.

10 Approximately 46% of all energy use in residential buildings goes to space heating and 15% goes to water heating, whereas for commercial buildings these figures are 31% and 4%, respectively. This means that low-quality heat can definitely be used, instead of using electricity or fossil fuels for heating purposes.

REFERENCES

1. McDougal, A. *Fuel Cells*, John Wiley & Sons: New York, 1976.
2. Kroschwitz, J.I.; Howe-Grant, M. (eds.). *Encyclopedia of Chemical Technology*, Vol. 20, Interscience Publishers: New York, 1991.
3. Bisio, A.; Boots, S. (eds.). *Energy Technology and the Environment*, Vol. 2, Wiley Encyclopedia Series in Environmental Science: New York, 1995.
4. Energy Information Administration. *Report No.: DOE/EIA-0226 (2009/05) Data for February 2009 report released*, EIA: Washington, DC, May 15, 2009.
5. Energy Information Administration, Office of Integrated Analysis and Forecasting, U.S. Department of Energy, International Energy Outlook 2001, DOE/ELA-0484, 2001. http://tonto.eia.doe.gov/ftproot/forecasting/04842001.pdf
6. Orwell, G. *Animal Farm*, Harcourt, Brace, & Company: New York, 1946.
7. Bisio, A.; Boots, S. (eds.). *Energy Technology and the Environment*, Vol. 1, Wiley Encyclopedia Series in Environmental Science: New York, 1995.
8. Bisio, A.; Boots, S. (eds.). *Energy Technology and the Environment*, Vol. 3, Wiley Encyclopedia Series in Environmental Science: New York, 1995.
9. Çengel, Y.A.; Boles, M.A. *Thermodynamics—An Engineering Approach*, 2nd edn., McGraw-Hill: New York, 1994.
10. Smith, J.M.; Van Ness, H.C.; Abbott, M.M.; Van Ness, H. *Introduction to Chemical Engineering Thermodynamics*, 6th edn., McGraw-Hill International Editions: New York, 2000.
11. Szargut, J.; Morris, D.R.; Steward, F.R. *Exergy Analysis of Thermal, Chemical, and Metallurgical Processes*, Hemisphere Publishing Corporation: New York, 1988.
12. Pruschek, R. Exergy availability of nuclear fuels. *Brensst Wärme Kraft* 1970, *22*, 429.
13. Armstead, H.C.H. *Geothermal Energy: Its Past, Present and Future Contributions to the Energy Needs of Man*, E. & F.N. Spon: London, 1983.
14. Smoot, L.D. *Fossil Fuel Combustion: A Science Source Book*, Bartok, W.; Sarofim, A.F. (eds.), John Wiley & Sons, Inc.: New York, 1991.

10 Separations

In this chapter, we examine the thermodynamic efficiency of the propanepropylene separation process by distillation. The tools necessary for this analysis are developed using the first and second laws of thermodynamics. Sources of thermodynamic inefficiency are pinpointed and, finally, some options are discussed to improve the efficiency of the separation.

10.1 INTRODUCTION

Operations that deal with the separation of mixtures into pure components have long been part of industry. For example, distillation has played a very important role in both the chemical and petrochemical industries. Crystallization has been used for centuries in refining sugar in the food industry, and it is also frequently encountered in the pharmaceutical industry. Stricter demands on product quality have forced industries to devise complicated separation units and, as a result, separation processes, now more than ever, play a central role in industry. To get an idea of the importance of separation processes, it makes sense to study some of the statistics [1]. Distillation consumes over $6 billion of U.S. energy annually, which amounts to 3% of the national total. For chemical plants, one-third of the typical capital investment is meant for separation units. For petroleum refineries and biochemical factories, this is typically 70%. Since separations are important consumers of energy, it makes sense to make efforts to reduce this energy consumption. Reduced energy consumption will, in general, lead to less cost and less burden on the environment, resulting in a more sustainable society. As an example, we will consider the distillation of propane and propylene [2], the raw material for polypropylene. The model used will contain some simplifying assumptions to illustrate the concepts clearly.

10.2 PROPANE, PROPYLENE, AND THEIR SEPARATION

Propane is a colorless, easily liquefied, gaseous hydrocarbon, the third member of the paraffin series following methane and ethane. The chemical formula for propane is $CH_3CH_2CH_3$. It is separated in large quantities from natural gas, light crude oil, and oil-refinery gases, and is commercially available as liquefied propane or as a major constituent of liquefied petroleum gas (LPG).

As with ethane and other paraffin hydrocarbons, propane is an important raw material for the ethylene petrochemical industry. The decomposition of propane in hot tubes to form ethylene also yields another important product, propylene. The oxidation of propane to compounds such as acetaldehyde is also of commercial interest.

Propylene is used principally in organic synthesis to produce the following materials: acetone, isopropylbenzene, isopropyl alcohol, isopropyl halides, and propylene

DOI: 10.1201/9781003304388-13

oxide. Propylene is also being polymerized to form polypropylene. A colorless, flammable, gaseous hydrocarbon, propylene, also called propene, has the chemical formula ($CH_2 = CHCH_3$) and is obtained from petroleum; large quantities of propylene are used in the manufacture of resins, fibers, elastomers, and numerous other chemical products. Although propylene is an important raw material in the chemical industry, it is produced almost exclusively as a by-product in steam cracking and in catalytic cracking [3].

In general, a C_3 stream is obtained that contains propane, propylene, propadiene, and propyne, and these are separated in a C_3 distillation column, also referred to as the C_3 *splitter*. Propane-propylene separation and, as a rule, olefin-paraffin separation, are energy-intensive, and some estimates are that 1.27×10^{17} J are used for olefin-paraffin separation on an annual basis [4] while roughly 3% is used by paraffin-olefin distillation columns [5]. This provides an incentive to examine the propane-propene separation, which is an example of paraffin-olefin separation.

Propane and propylene have similar atmospheric boiling points (propane: $-42.1°C$, propylene: $-47.70°C$) and, as a result, the separation of these compounds requires highly complicated units. Distillation is by far the most commonly used separation process in the chemical industry today. The variants in use are:

1. Single-column process
2. Double-column process
3. Heat pump process

10.2.1 SINGLE-COLUMN PROCESS

This process requires a large number of trays (150–200), resulting in units of about 100 m. The reflux can be condensed with cooling water (column pressure 16–19 bar) or in air coolers (column pressure 21–26 bar).

10.2.2 DOUBLE-COLUMN PROCESS

For the large throughputs that are common today, the double-column process is preferred over the single-column process, since it does require smaller columns with smaller column diameters, which makes the transportation of these units easier. A simplified schematic of the double-column process is given in Figure 10.1. Only the reflux from the second column is condensed with cooling water. The pressure of the first column is sufficiently high (ca. 25 bar) that the overhead vapors (ca. 59°C) can be condensed in the reboiler of the second column and serve as the heat carrier. Heating the first column with steam is not necessary, and warm water can be used for heating purposes. Both columns provide approximately half the propene product. Since the reboiler for the second column also serves as the condenser of the first column, the first column does not require any cooling water. As a result, the cooling water requirements are about half that of the single-column process.

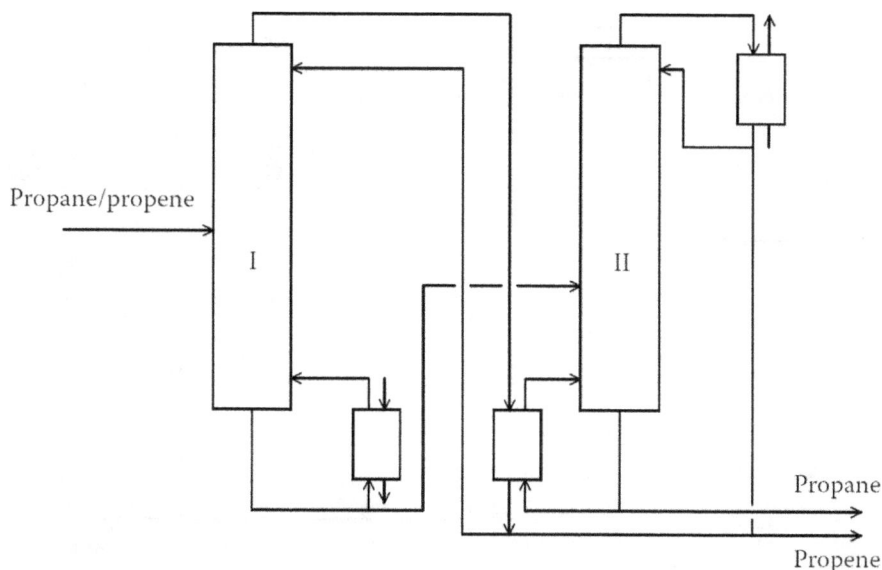

FIGURE 10.1 Schematic of double-column process. (Adapted from *Ullmann's Encyclopedia of Industrial Chemistry*, Vol. A22, VCH Publishers, Inc., New York, 1993.)

10.2.3 HEAT PUMP PROCESS

In the aforementioned process, the heat for the reboiler is usually available as waste heat from the steam cracker, for example, and is essentially cost-free. If this heat is not available, a heat pump can be used. The heat pump can upgrade the heat, at an exergetic cost, to the desired temperature level. If the separation is viewed in isolation, this means that the heat rejected by the condenser at relatively low temperature, can be upgraded to be the higher temperature heat input for the reboiler. A schematic of the heat pump process is given in Figure 10.2. The overhead vapors are heated slightly in the reflux subcooler, which enables these vapors to be compressed and cooled in the condenser-reboiler.

In the example discussed in this chapter (Sections 10.3–10.6), we restrict ourselves to the single-column process for the simple reason that the analysis is straightforward and illustrates the concepts best. In addition, data for this analysis are readily available [6].

10.3 BASICS

10.3.1 FLASH DISTILLATION

As stated earlier, distillation is a widely used separation technique for liquid mixtures or solutions. The formation of these mixtures is straightforward, and is usually spontaneous, but the separation of a mixture into its separate constituents requires energy. One of the simplest distillation operations[1] is flash distillation. In this process, part

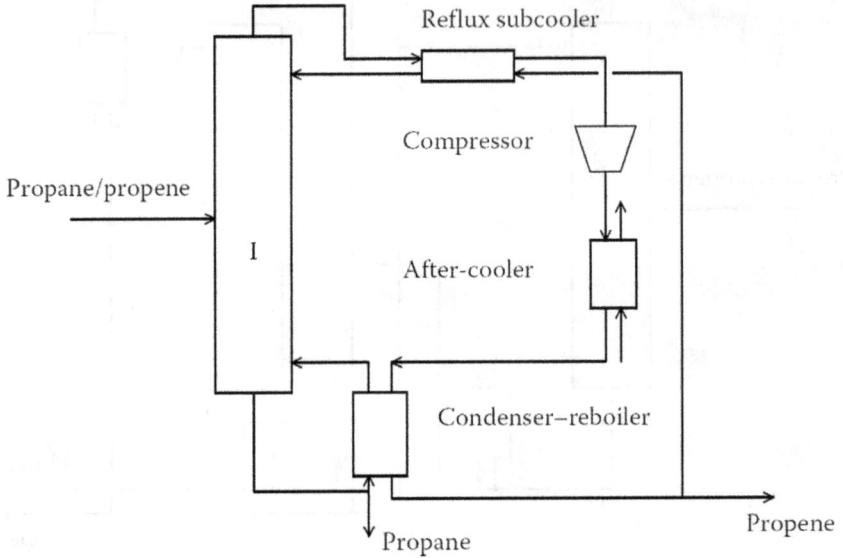

FIGURE 10.2 Schematic of heat pump process. (Adapted from *Ullmann's Encyclopedia of Industrial Chemistry*, Vol. A22, VCH Publishers, Inc., New York, 1993.)

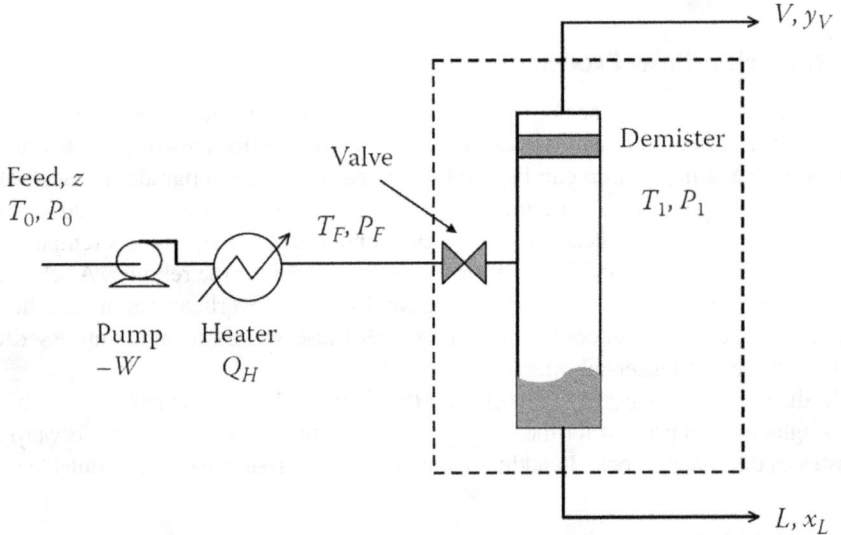

FIGURE 10.3 Schematic representation of equipment necessary for flash distillation.

of the feed stream vaporizes in a flash chamber, and the vapor-liquid mixture, which is at equilibrium, is separated. The vapor is rich in the more volatile component, but complete separation is usually not achieved. A simple schematic showing the necessary equipment for flash distillation is given in Figure 10.3. We will illustrate the concepts by using a simple case of the flash distillation of a binary mixture.

The liquid feed, at temperature T_0 and pressure P_0, and mole fraction z, enters the system. It is elevated to a higher pressure and temperature by a pump and heat exchanger, both requiring input of energy. The binary liquid mixture, now at T_F and P_F, proceeds to a valve, which reduces the pressure and usually the temperature, and the mixture separates into a vapor and a liquid phase of composition y_V and x_L, respectively, which are separated physically in the flash vessel. The resulting vapor effluent is rich in the volatile component, and the liquid stream is lean in the same.

Figure 10.4 shows how the separation can be visualized in a phase diagram, in this case a Txy-diagram at P_1. The feed, with composition z, is initially at T_F and P_F in the liquid phase. At $P < P_F$, the Txy-diagram is shifted to lower temperatures. The pressure is then reduced (the "flash"), which results in the feed's changing pressure and temperature to T_1 and P_1. The trajectory of this change in the phase space is given by the dotted line, since the starting condition (T_F, P_F) is not in the Txy-diagram at P_1. The equilibrium state at T_1 and P_1 is that of a vapor-liquid mixture, and the feed phase

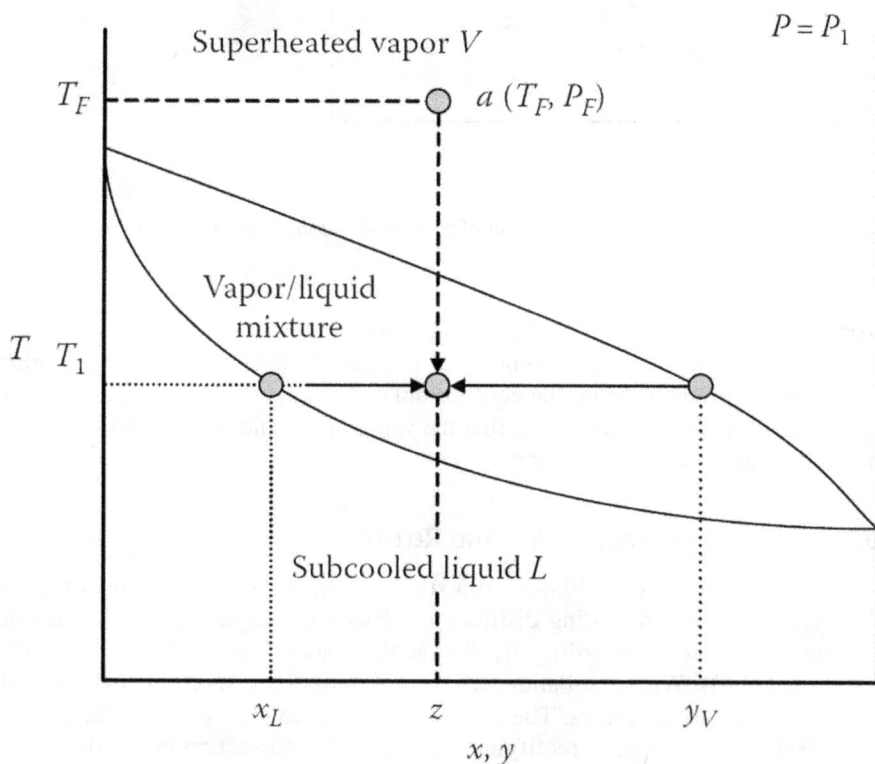

FIGURE 10.4   Txy-diagram at $P = P_1$ of an adiabatic flash (distillation).

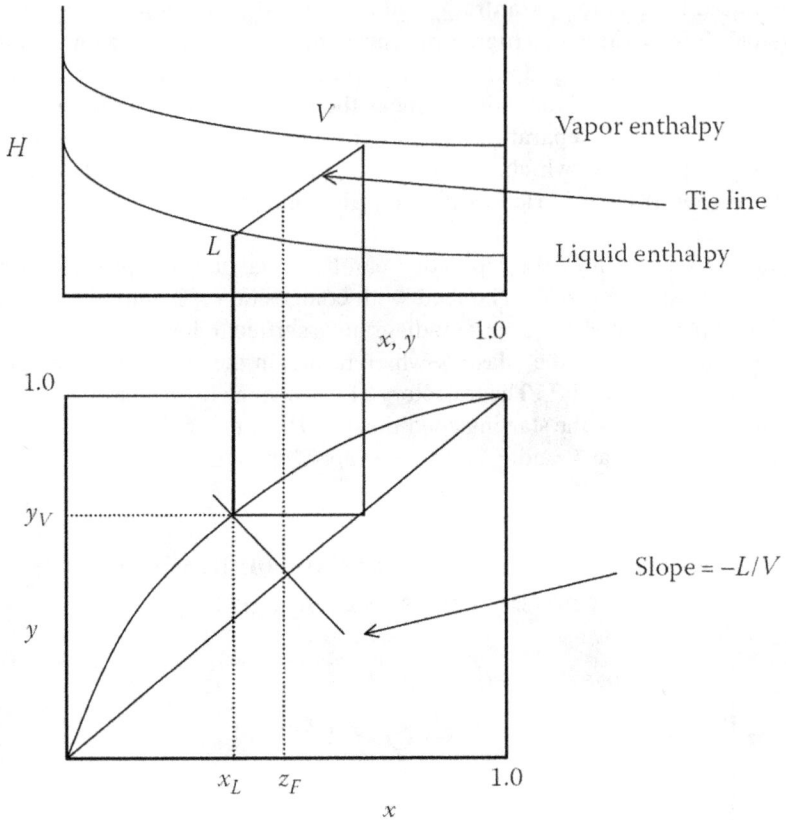

FIGURE 10.5   Schematic representation of an enthalpy-composition (Hxy) diagram and yx projection for vaporization.

separates into a vapor and a liquid at composition y_V and x_L, respectively, if equilibrium is reached. The line connecting L and V is called the tie-line. This equilibrium is represented as one point on the equilibrium curve $y(x)$, as given schematically in Figure 10.5. Since we are assuming that the vapor and liquid are at equilibrium, the line connecting L and V is a tie-line.

10.3.2   Multistage Distillation and Reflux

We can describe binary distillation in a similar way. We will briefly review some of the main concepts regarding distillation relevant to our discussion. For further information and details regarding distillation, the reader is referred to books on distillation (e.g., [7]). With distillation, the feed is introduced more or less centrally into a vertical cascade of trays. The section above the feed is typically referred to as the absorption, enriching, or rectifying section, while the section below the feed is called the stripping or exhausting section (see Figure 10.6). The column has N trays or stages, where the numbering typically starts at the top of the column and ends at

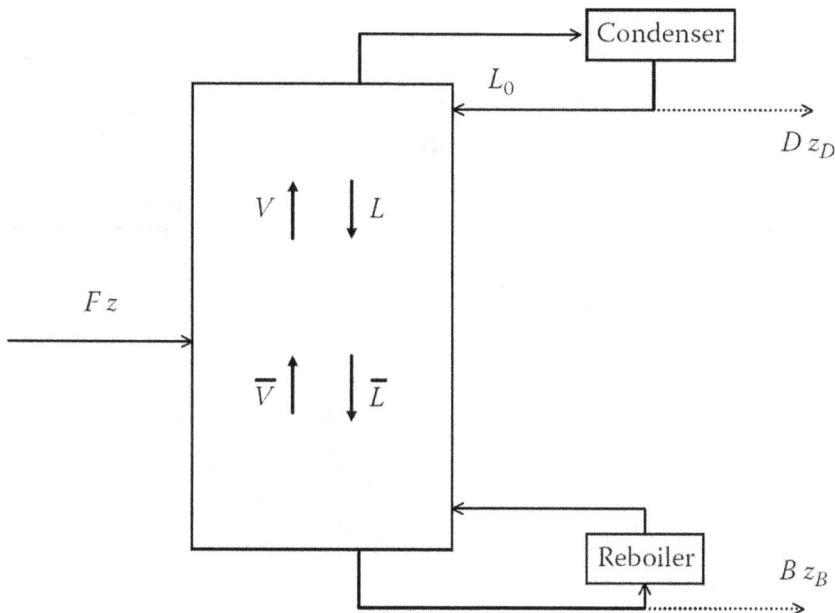

FIGURE 10.6   Schematic representation of a distillation column.

the bottom tray. A schematic representation of a tray j is given in Figure 10.7. It is important to note that for an ideal tray, the vapor flow V_{j+1} and liquid flow L_{j-1} are not in equilibrium, but the tray equilibrates them to flow V_j and L_j. However, for a real tray, with a tray efficiency less than 100% (e.g., 75%), this means that equilibrium is not reached, and that more stages N_{real} are needed than the ideal case (N_{ideal}). The ideal number of stages can be computed using a McCabe-Thiele diagram, as is shown in Figure 10.8. Here, q is the liquid fraction of the feed, the operating lines are used to determine the number of ideal stages N_{ideal}. The simplifying assumption was made that specific heat changes are negligible compared to latent heat changes, the heat of vaporization per mole is constant, and the column is operated adiabatically. The reflux r is given by $r = L_0/D = L_0/(V_1 - L_0)$. We can rewrite this as

$$\frac{L_0}{V_1} = \frac{r}{r+1} \tag{10.1}$$

In case of the propane-propylene separation, the $y = x$ line and the yx-equilibrium line are very close, as shown in Figure 10.9. This means that a large number of trays is necessary, which is the case in practice. Another consequence of the close proximity of the $y = x$ line and yx-equilibrium line is that the slope of the top operating line, L/V, will be close to unity. This implies that the reflux ratio will be high and the condenser will have to "liquefy" large amounts of vapor, leading to a large cooling duty.

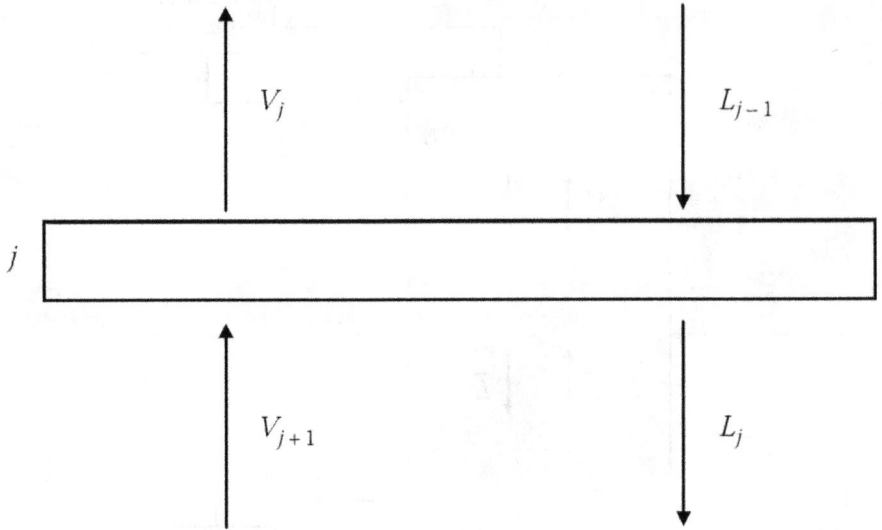

FIGURE 10.7   Schematic representation of tray j.

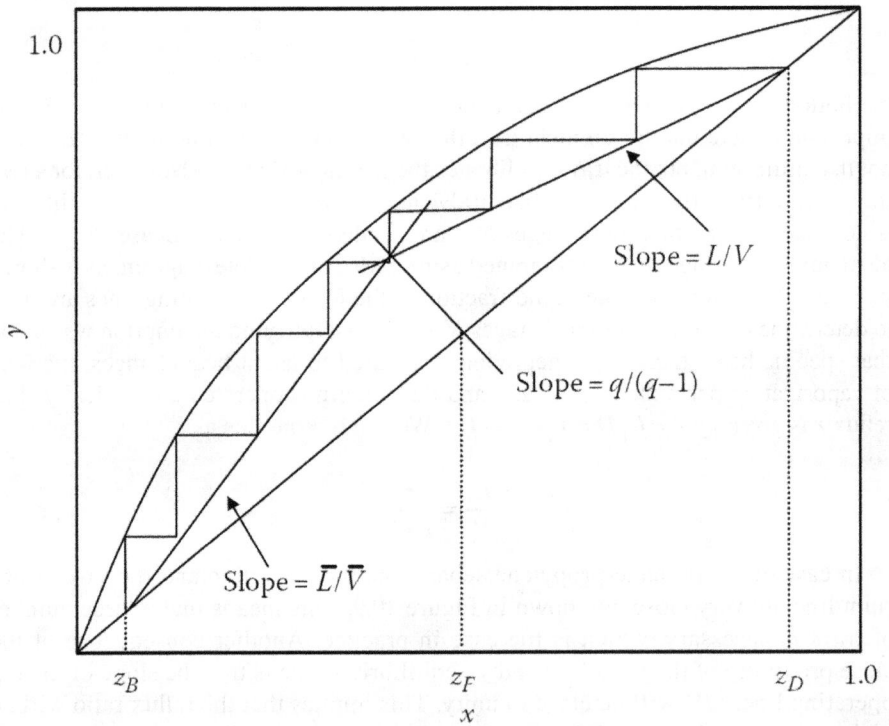

FIGURE 10.8   McCabe-Thiele diagram to determine the number of stages.

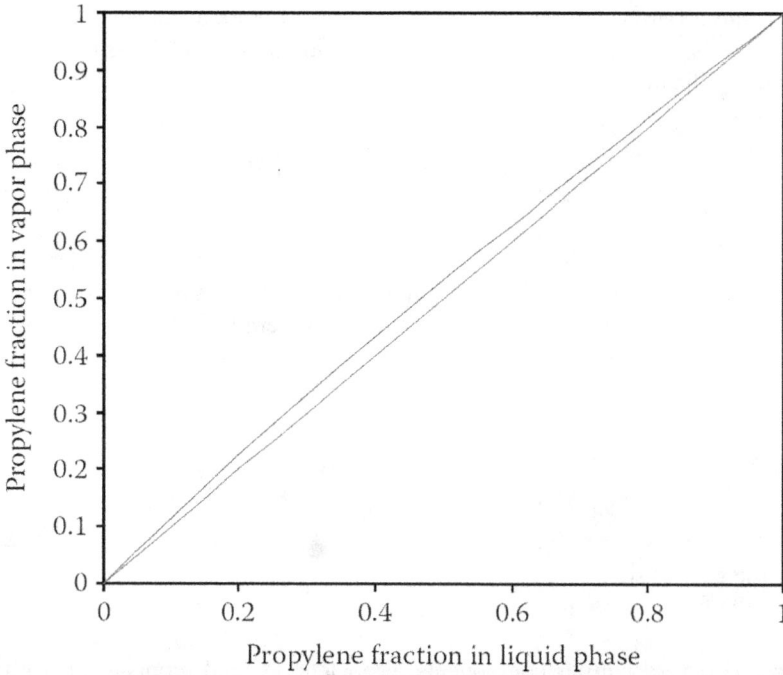

FIGURE 10.9 The *yx* diagram for propane-propylene at 16 bar.

10.4 THE IDEAL COLUMN: THERMODYNAMIC ANALYSIS

As stated earlier, the separation of mixtures into their constituents requires energy. In distillation, this energy is supplied as heat. To this end, it is useful to recall that Carnot (see Chapter 3 for details) showed that the maximum amount of work that can be extracted from a heat source Q at $T > T_0$ with respect to the surroundings at T_0 is given by

$$W_{out}^{max} = Q\left(1 - \frac{T_0}{T}\right) \tag{10.2}$$

It is straightforward to show that if heat is available at temperature levels T_h and T_l, the maximum amount of work that can be extracted from the former temperature level is $W_{out}^{max} = Q\left[1 - (T_0)/(T_h)\right]$, whereas $W_{out}^{max} = Q\left[1 - (T_0)/(T_l)\right]$ is the maximum amount of work for the latter. From this, it can be shown that if heat flows spontaneously from a temperature level T_h to a temperature level T_l (with $T_h > T_l > T_0$), the amount *of available* work that has been lost is given by

$$W_{lost} = QT_0\left(\frac{1}{T_l} - \frac{1}{T_h}\right) \tag{10.3}$$

which is the difference of the maximum amounts of work available at the two temperature levels. If, on the other hand, heat at a temperature level T_l has to be upgraded to a level T_h, the minimum amount of work required is

$$W_{in}^{min} = QT_0 \left(\frac{1}{T_l} - \frac{1}{T_h} \right)$$
(10.4)

Another useful equation is the Clausius-Clapeyron equation. It states that, provided the ideal gas law holds and the enthalpy of vaporization, $\Delta_v H$, is independent of T (which is a reasonable assumption for a small temperature range), the slope of the vapor pressure curve is given by

$$\frac{d \ln p^{sat}}{d(1/T)} = -\frac{\Delta_v H}{R}$$
(10.5)

where

R is the gas constant
p^{sat} is the vapor pressure

Consider two very similar components, propylene [1] and propane [3], with close boiling points. The ideal relative volatility is now defined as

$$\alpha_{12}^{ideal} \equiv \frac{p_1^{sat}}{p_2^{sat}}$$
(10.6)

From Equations 10.5 and 10.6, it follows that over the temperature range of distillation

$$\ln \alpha_{12}^{ideal} = -\frac{\Delta_v H}{R} \left(\frac{1}{T_{bottom}} - \frac{1}{T_{top}} \right)$$
(10.7)

where T_{bottom} and T_{top} denote the temperatures at the bottom and top of the column, respectively. At this point, it is useful to note that the assumptions we have made are that (1) the mixture is *ideal*, which is an approximation of the behavior of the real mixture, and (2) the enthalpy of vaporization is the same for both components. In reality, the relative volatility α_{12} will differ slightly from the ideal value. A simplified scheme for the separation of an equimolar mixture of propylene and propane is given in Figure 10.10. The feed is a liquid mixture introduced at that point where the liquid has the same composition and temperature.

Per mole of feed F, the distillate D amounts to 1/2 mol and so does the bottom product B, since the feed is *equimolar*. With a reflux ratio $r = L/D$, in which L is the number of moles of liquid reintroduced at the top of the column, $L = rD = \frac{1}{2}r$ moles above the feedpoint and $(\frac{1}{2}r + 1)$ below the feedpoint. The vapor flow $V = \frac{1}{2}(r + 1)$ throughout the column. The heat introduced at the bottom of the column is therefore

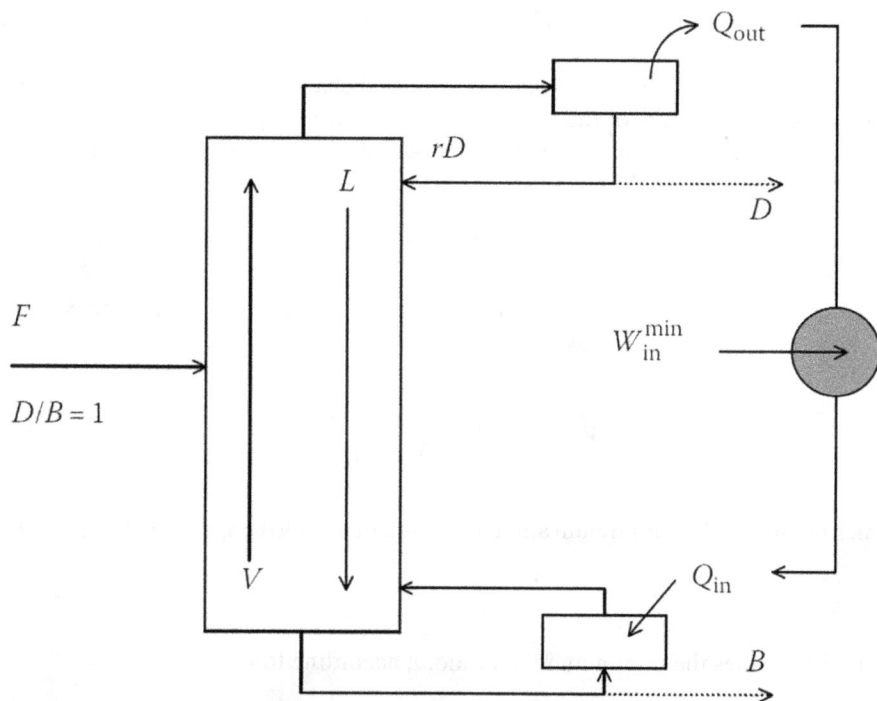

FIGURE 10.10 Simplified schematic to separate propane-propylene.

$$Q_{in} = \frac{1}{2}(1+r)\Delta_v H \tag{10.8}$$

We assume that the heat of vaporization is roughly the same for both components and that the temperature dependence is negligible over the range of the column. From this, it follows that

$$Q_{in} = Q_{out} \tag{10.9}$$

in which Q_{out} is the cooling duty of the condenser. Now, it is interesting to note that the overall separation does not require any energy! The number of Joules entering the column equals the number of Joules leaving the column, which is a direct consequence of the first law of thermodynamics. However, the "quality" of these heat streams or, equivalently, the available work of these heat streams, is not the same, due to the Carnot factor, the term in brackets in Equation 10.2, which is due to the second law of thermodynamics. In an ideal column, that is, a column operating under *reversible* conditions, the heat is stripped of its quality and "pays" for the separation of the liquid mixture into its constituents in the liquid state. The minimum work required to separate the liquid mixture into its constituents is given next (see Figure 10.10):

$$W_{\text{sep}}^{\text{ideal}} = -RT_0 \sum_i x_i \ln x_i \tag{10.10}$$

where the assumption is made that the mixture behaves in an ideal fashion and is close to the temperature of the surroundings, which, for the propanepropylene mixture, is a fair assumption. If we further insert the assumption that the mixture is equimolar, Equation 10.10 reduces to

$$W_{\text{sep}}^{\text{ideal}} = RT_0 \ln 2 \tag{10.11}$$

which is the minimum amount of work that needs to be introduced into the column.
  According to Equation 10.4,

$$W_{\text{in}}^{\text{min}} = Q_{\text{in}}^{\text{min}} T_0 \left(\frac{1}{T_{\text{top}}} - \frac{1}{T_{\text{bottom}}} \right) \tag{10.12}$$

which has to equal the minimum amount of work that has to be spent on the separation:

$$W_{\text{in}}^{\text{min}} = W_{\text{sep}}^{\text{ideal}} \tag{10.13}$$

This also defines the minimum reflux ratio, r, according to

$$Q_{\text{in}}^{\text{min}} = \frac{1}{2}\left(r^{\text{min}} + 1\right) \Delta_v H \tag{10.14}$$

The minimum amount of heat with an exergy value that corresponds to the minimum amount of work required to separate the mixture prescribes the minimum value of the reflux ratio. This value is obtained by combining Equations 10.7, 10.11, 10.12, and 10.14 to yield

$$r^{\text{min}} = \frac{2 \ln 2}{\ln \alpha_{12}^{\text{ideal}}} - 1 \tag{10.15}$$

We stress that this equation dictates the minimum reflux ratio based purely on *thermodynamic* arguments. As $\alpha_{12}^{\text{ideal}} = 1.11$ in our case, the value of $r^{\text{min}} = 12.28$. In general, the mixture will not be equimolar, and if the products are not pure but satisfy a less strict specification, the value of r^{min} will be smaller. Now, a column operated under these conditions will have an efficiency of 100% since it is using the minimum amount of work necessary to separate the components.

10.5   THE REAL COLUMN

The previous analysis begs the following question. *What will the efficiency be of a "real" propane-propylene distillation column?* To answer this question, we must realize that heat cannot be transferred into the column without a finite temperature difference. In a "real" column with less stringent product quality constraints, assume that heat is supplied at $T_R = 377$ K (see Figure 10.11), the bottom temperature of the column is $T_b = 331$ K, the temperature at the top of the column is $T_t = 320$ K and this heat is transferred to the surroundings at $T_0 = 298$ K [6], as shown in Figure 10.11. The minimum heat required for separation is, according to Equation 10.14,

$$Q_{in}^{min} = \frac{1}{2}\left(r^{min} + 1\right)\Delta_v H \qquad (10.16)$$

with $r^{min} = 9.64$ from the data in [6]. The separation inside the column does not take place according to thermodynamic ideal processes, and the real heat input is larger:

$$Q_{in}^{real} = \frac{1}{2}\left(r^{real} + 1\right)\Delta_v H \qquad (10.17)$$

where $r^{real} = 15.9$.

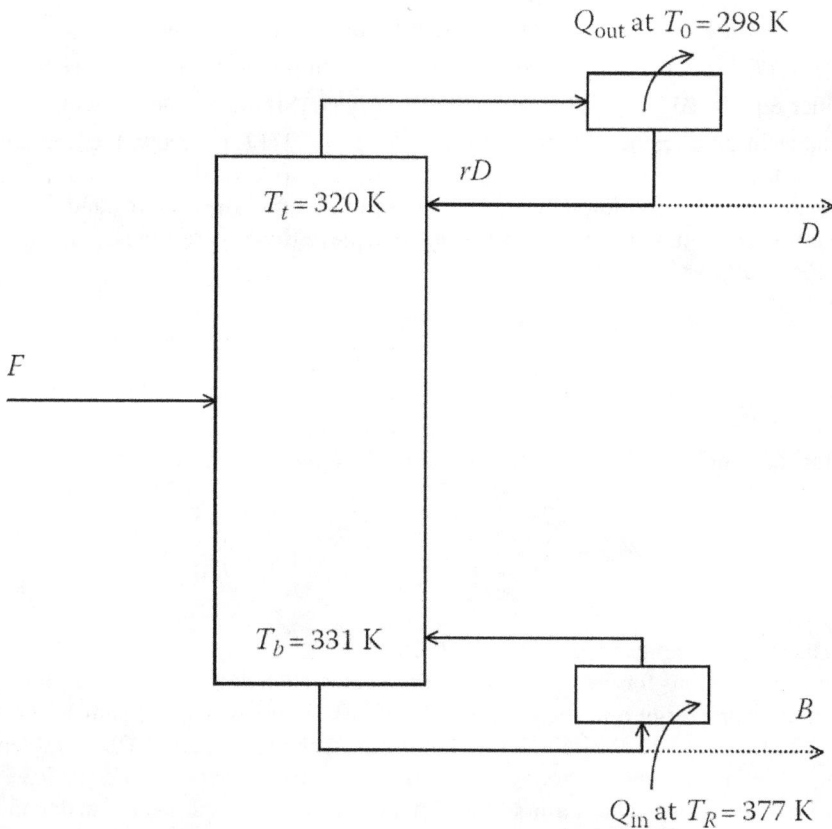

FIGURE 10.11 Important temperatures for the distillation column.

At the bottom of the column, the heat has to be transferred over a temperature difference of $377 - 331 = 46$ K, and the resulting lost work can easily be calculated with Equation 10.3. Then the heat flows from 331 to 320 K inside the column and is used to perform the separation. Finally, the heat is discarded at the top of the column at 320 K to the surroundings at 298 K.

The overall thermodynamic efficiency of the column can be computed as follows:

$$\eta_{\text{overall}} = \frac{Q_{\text{in}}^{\min} \Delta(1/T)_{\text{column}} T_0}{Q_{\text{in}}^{\text{real}} \left[\Delta(1/T)_{\text{bottom}} + \Delta(1/T)_{\text{column}} + \Delta(1/T)_{\text{top}} \right] T_0} \tag{10.18}$$

which yields $\eta_{\text{overall}} = 0.093$! This means that the thermodynamic efficiency of the distillation column is only 9.3%. Closer scrutiny of Equation 10.18 reveals that the main sources of inefficiency are the temperature-driving forces in the condenser and the reboiler (the ratio $Q_{\text{in}}^{\min}/Q_{\text{inn}}^{\text{raal}}$ equals 0.63 and reflects work losses *inside* the column). We can visualize the sources of inefficiency by tracking the fate of heat that enters the separation process at the reboiler. The heat, $Q_{\text{in}}^{\text{real}}$, is supplied at T_R, which means that the available work (the exergy) of this heat is $Q_{\text{in}}^{\text{real}} \left[1 - (T_0)/(T_R) \right]$. Since the heat enters the column at T_b, the available work of the heat entering the column is $Q_{\text{in}}^{\text{real}} \left[1 - (T_0)/(T_b) \right]$. So the amount of work lost in the heat transfer process in the reboiler equals $W_{\text{lost}} = Q_{\text{in}}^{\text{real}} T_0 \left[(1)/(T_b) - (1)/(T_R) \right]$. Similarly, the amount of work available in heat leaving the column is $Q_{\text{in}}^{\text{real}} \left[1 - (T_0)/(T_t) \right]$. Now, the amount of work lost in the column *is not* the difference of the work available in the heat flowing in and out of the column. The heat is used to separate the propane and propene. The minimum amount of work necessary to achieve this separation is, according to Equation 10.12,

$$W_{\text{in}}^{\min} = Q_{\text{in}}^{\min} T_0 \left(\frac{1}{T_t} - \frac{1}{T_b} \right)$$

So, the amount of work lost in the column is given by

$$W_{\text{lost}} = Q_{\text{in}}^{\text{real}} T_0 \left(\frac{1}{T_t} - \frac{1}{T_b} \right) - Q_{\text{in}}^{\min} T_0 \left(\frac{1}{T_t} - \frac{1}{T_b} \right)$$

We can summarize these findings in Table 10.1.

The calculations for Table 10.1 have been scaled such that they relate to the input of one unit of exergy into the column. Figure 10.12 shows that the sum of the lost work (lost exergy) in the reboiler, column, and condenser equals 0.907 in scaled form, which means the thermodynamic efficiency is $1 - 0.907 = 0.093$, or 9.3%, as we obtained earlier. The main sources of inefficiency are the heat transfer in the reboiler and the condenser, which contribute to $0.525 + 0.328 = 0.853$ units of lost

TABLE 10.1
Overview of Exergy in Distillation Column.

	Exergy In	Exergy Out	Exergy Lost
Reboiler	$Q_{in}^{real}\left(1-\dfrac{T_0}{T_R}\right)$	$Q_{in}^{real}\left(1-\dfrac{T_0}{T_b}\right)$	$Q_{in}^{real}T_0\left(\dfrac{1}{T_b}-\dfrac{1}{T_R}\right)$
Column	$Q_{in}^{real}\left(1-\dfrac{T_0}{T_b}\right)$	$Q_{in}^{real}\left(1-\dfrac{T_0}{T_t}\right)$	$\left(Q_{in}^{real}-Q_{in}^{min}\right)T_0\left(\dfrac{1}{T_t}-\dfrac{1}{T_b}\right)$
Condenser	$Q_{in}^{real}\left(1-\dfrac{T_0}{T_t}\right)$	$Q_{in}^{real}\left(1-\dfrac{T_0}{T_0}\right)$	$Q_{in}^{real}T_0\left(\dfrac{1}{T_0}-\dfrac{1}{T_t}\right)$

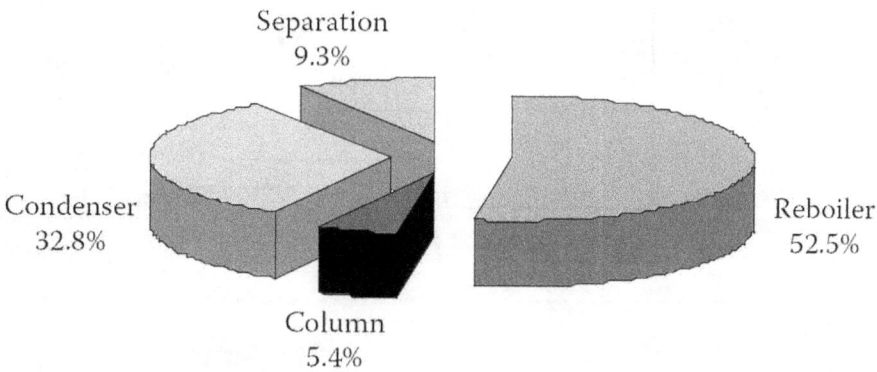

FIGURE 10.12 Ultimate fate of one unit of exergy in distillation column: 9.3% is used for the separation, 32.8% is lost in the condenser, 52.5% is lost in the reboiler, and 5.4% is lost in the column.

work (see Figure 10.12). More than 85% is thus lost outside the column. Inside the column, the work lost is about half the work spent on separation.

10.6   EXERGY ANALYSIS WITH A FLOW SHEET PROGRAM

To obtain the thermodynamic efficiency it is not necessary to perform the analysis as shown earlier. We can easily estimate the thermodynamic efficiency by setting up an available work or exergy balance (see Figure 10.13). We begin by defining our control volume. In this case, we choose the volume to include the reboiler, the column, and the condenser. The propane-propene feed enters the control volume and, with it, exergy enters as well (Ex_F). Exergy also enters the system at the reboiler in the form of heat (Ex_{Q1}). Exergy flows out of the system with the distillate (Ex_D), the bottoms product (Ex_B), *and* as "heat" rejected at the condenser (Ex_{Q2}).

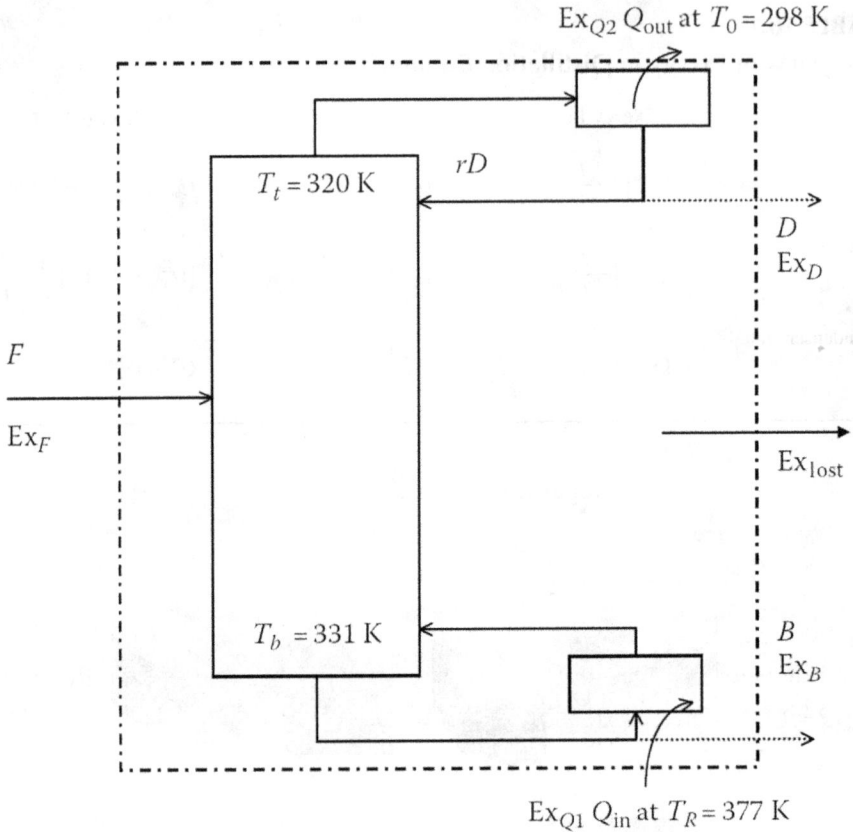

FIGURE 10.13   Schematic representation of the exergy flow in a separation process.

In the hypothetical case that the process is reversible in the thermodynamic sense, the flow of exergy into the system will equal the flow of exergy out of the system. But since real processes always have a degree of irreversibility, some exergy is lost (Ex_{lost}). We can now set up the exergy balance

$$Ex_F + Ex_{Q1} = Ex_B + Ex_{Q2} + Ex_D + Ex_{lost} \qquad (10.19)$$

Because chemical transformation does not occur, we do not need to include the chemical exergy in the exergy flows. We can now write expressions for the terms in the exergy balance based on 1 mol feed entering the system for both the reversible and real separation processes.

Note that for both the reversible and irreversible, real, processes, we are once again invoking the assumption that the mixing behavior is ideal, thus that there is zero enthalpy of mixing, and the entropy of mixing is that for an ideal mixture. In flow-sheeting programs, it is very simple to extract the heating and cooling duties and the values of the enthalpies and entropies of the streams entering a process unit. It is then straightforward to compute the exergy of these streams for the real process case.

Note that the Ex_{Q1} and Ex_{Q2} terms differ for the reversible and irreversible cases. The reversible case has the minimum heat input Q_{min}^{in} at temperature level T_b since heat is transferred reversibly without a temperature gradient. Similarly, it is rejected at T_r. It is in these two terms that the reversible and irreversible exergy balances differ. Now, first consider the reversible exergy balance, where Ex_{lost} is, by definition, equal to zero. From Equation 10.19, for the reversible case, we obtain

$$
\begin{aligned}
&\frac{1}{2}\left(\left(h_{C_=} + h_{C_3}\right) - T_0\left(s_{C_=} + s_{C_3}\right)\right) - \frac{1}{2}\left(\left(h_{C_=} + h_{C_3}\right) - T_0\left(s_{C_=} + s_{C_3}\right)\right)\Bigg|_0 \\
&-RT_0\ln 2 + Q_{min}^{in}\left(1 - \frac{T_0}{T_b}\right) = \frac{1}{2}\left(h_{C_3} - T_0 s_{C_3}\right) - \frac{1}{2}\left(h_{C_3} - T_0 s_{C_3}\right)\Bigg|_0 \\
&+ \frac{1}{2}\left(h_{C_=} - T_0 s_{C_=}\right) - \frac{1}{2}\left(h_{C_=} - T_0 s_{C_=}\right)\Bigg|_0 + Q_{min}^{in}\left(1 - \frac{T_0}{T_t}\right)
\end{aligned}
\tag{10.20}
$$

which is the exergy balance for the reversible separation process. Similarly, we obtain the following balance for the real process:

$$
\begin{aligned}
&\frac{1}{2}\left(\left(h_{C_=} + h_{C_3}\right) - T_0\left(s_{C_=} + s_{C_3}\right)\right) - \frac{1}{2}\left(\left(h_{C_=} + h_{C_3}\right) - T_0\left(s_{C_=} + s_{C_3}\right)\right)\Bigg|_0 \\
&-RT_0\ln 2 + Q_{real}^{in}\left(1 - \frac{T_0}{T_R}\right) = \frac{1}{2}\left(h_{C_3} - T_0 S_{C_3}\right) - \frac{1}{2}\left(h_{C_3} - T_0 S_{C_3}\right)\Bigg|_0 \\
&+ \frac{1}{2}\left(h_{C_=} - T_0 S_{C_=}\right) - \frac{1}{2}\left(h_{C_=} - T_0 s_{C_=}\right)\Bigg|_0 + Ex_{lost}
\end{aligned}
\tag{10.21}
$$

Note that the $Ex_{Q2} = 0$ for the real case, since heat is rejected at T_0. Upon simplification, Equation 10.21 yields, after combination with Equation 10.20,

$$
Ex_{lost} = -Q_{min}^{in} T_0\left(\frac{1}{T_0} - \frac{1}{T_R}\right) - Q_{real}^{in} T_0\left(\frac{1}{T_t} - \frac{1}{T_b}\right)
\tag{10.22}
$$

Ex_{lost} can now be calculated in scaled units (i.e., per unit exergy input):

$$
\frac{Ex_{lost}}{Q_{real}^{in} T_0\left((1/T_0) - (1/T_R)\right)} = 1 - \frac{Q_{min}^{in} T_0\left((1/T_t) - (1/T_b)\right)}{Q_{real}^{in} T_0\left((1/T_0) - (1/T_R)\right)}
\tag{10.23}
$$

which upon substitution of the data yields 0.907, or equivalently, a thermodynamic efficiency of $0.093 = 9.3\%$, which is in agreement with the earlier value. In this case, we also could have obtained an estimate of the lost work (lost exergy) by substituting values from a flow-sheeting program into the real exergy balance, Equation 10.21. The only quantities necessary for this computation are (1) the heating duty of the reboiler, and (2) the enthalpies and entropies of the entering and exiting stream at the process conditions and the base state, which are all readily available! We stress that since the control volume was chosen to include the distillation column, the reboiler, and the condenser, the thermodynamic calculation yields a net thermodynamic efficiency and an estimate for the total lost work. If information regarding the separate units (i.e., the reboiler, the column, and the condenser) is desired, the control volume should be chosen differently. For example, by choosing the control volume to only include the reboiler, the efficiency of the heat transfer process can be computed. We already have computed the fraction of work used for the separation and lost in the reboiler, column, and condenser (Figure 10.12), so it is not necessary to compute these. As discussed earlier, the main sources of the losses are the reboiler and the condenser.

10.7   REMEDIES

Both the reboiler and the condenser lose considerable amounts of work due to the heat transfer over large temperature differences (Figure 10.12). A simple way of improving these heat transfer processes is, therefore, to reduce the temperature-driving forces.

10.7.1   MAKING USE OF WASTE HEAT

A noteworthy point is, however, that the heat for the reboiler is supplied at 377 K, which is often available as waste heat in a chemical plant. Waste heat is usually rejected (hence the term "waste"), so heat integration could improve the process efficiency and reduce total exergy losses for the entire chemical plant (see Figure 10.14). Now, what Figure 10.14 shows is that heat integration can sometimes reduce the exergy losses of a chemical plant as well as reduce the need for utilities such as steam. In the particular case of the propane-propene single-column distillation, waste heat is frequently used to satisfy the heating duty of the reboiler.

10.7.2   MEMBRANES

Another way to improve the single-column process is to use a membrane to split the feed into two different feed streams (see Figure 10.15). Since the membrane does part of the separation, the column has to do less work to complete the separation, which means less reboiler duty and less stages [8]. Figure 10.16 shows a "cartoon" of the y-x projection of the system, including the operating lines. The membrane splits the feed in two streams, which enter the column at different feed locations. The operating lines are lowered in the region around the feed, resulting in a reduction of the reflux

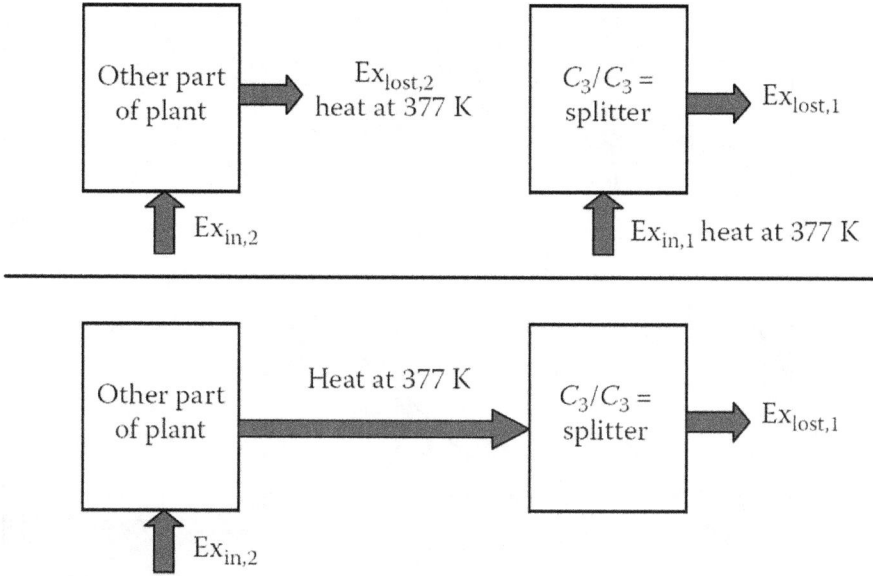

FIGURE 10.14 Schematic representation of the effect of heat integration on lost exergy (lost work).

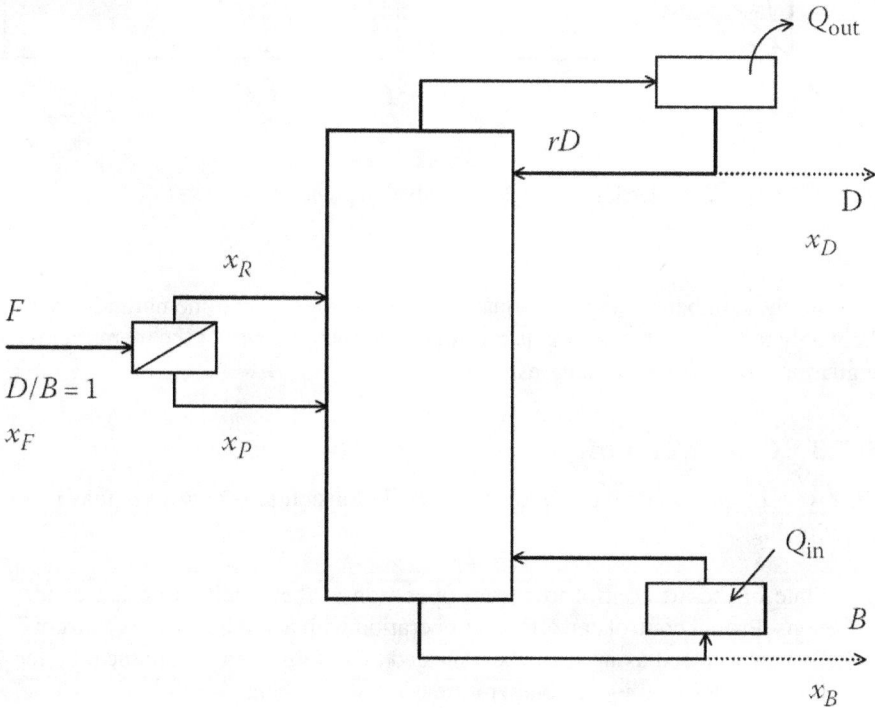

FIGURE 10.15 Hybrid distillation of propane-propene.

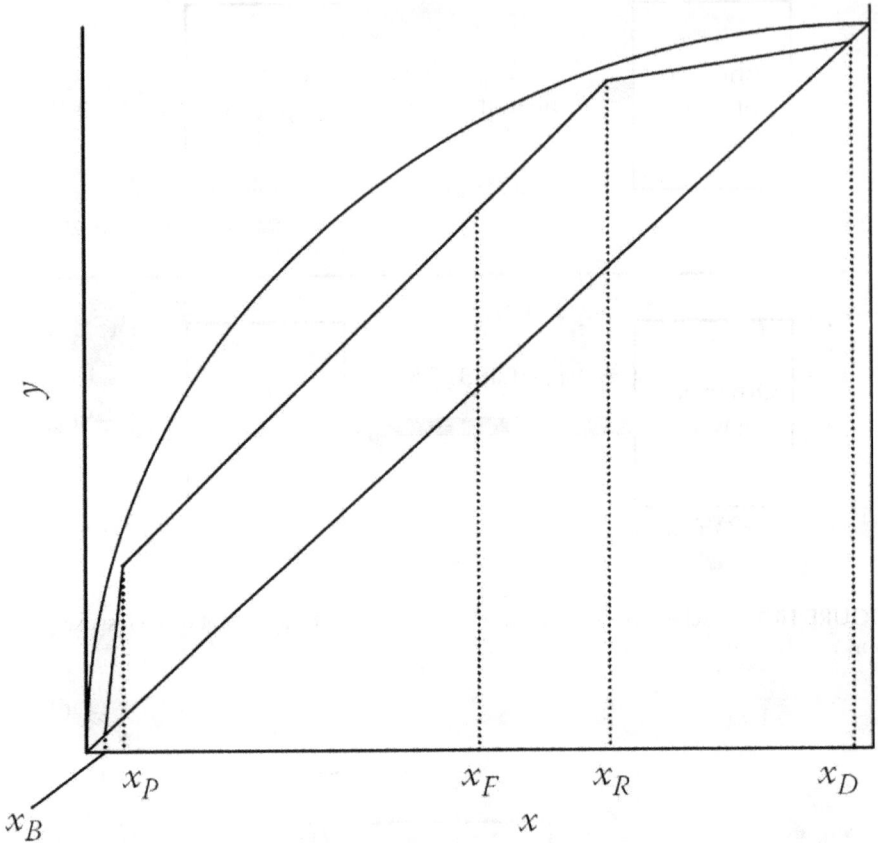

FIGURE 10.16   *yx* projection for a hybrid distillation process.

ratio and the number of stages. It must be noted, however, that membranes are most likely only to be useful if fouling is minimal, and the integrity of the membranes can be guaranteed for long run-lengths.

10.7.3   OTHER METHODS

Other ways of improving the efficiency of distillation columns *in general* may include the following:[2]

1. State-of-the-art control to ensure operation at the specified design conditions. Proper control can allow for operation with a smaller "safety margin."
2. Preheat the feed using waste heat instead of adding extra heat to the reboiler that operates at a higher temperature level, or use heat pumps.[3]
3. Reduce the pressure drop in the column by using packings instead of trays, if possible.

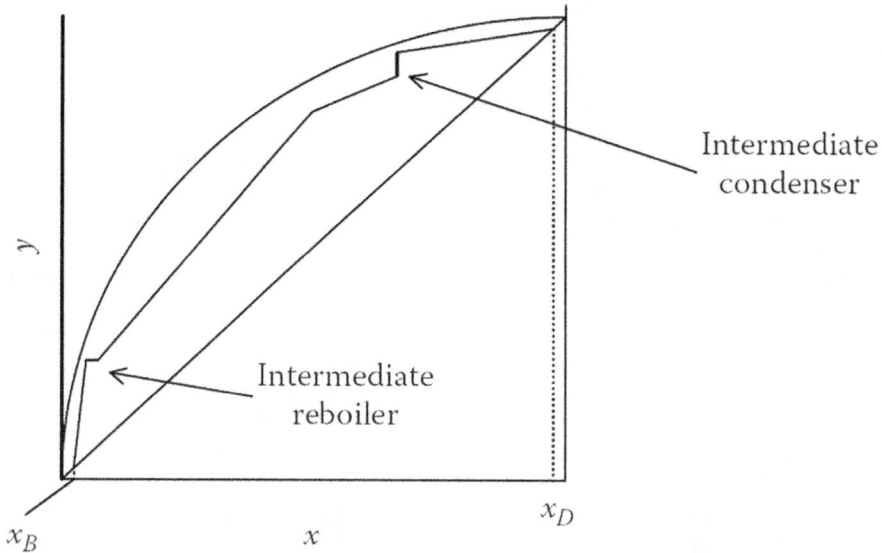

FIGURE 10.17 *yx* projection for a distillation process with intermediate condenser and reboiler.

4. Use intermediate condensers and reboilers (if possible). This pushes the operating lines closer to the equilibrium line (Figure 10.17), thus reducing inherent column inefficiencies. The use of waste heat may also be possible now without upgrading to the reboiler temperature. A potential disadvantage may be that extra trays are needed.
5. Separate to lower purity, and continue separation with other separation techniques.
6. Operate the rectifying section at a higher temperature, thus higher pressure, and use the heat of condensation to supply heat to the reboiler.
7. For certain applications, such as the separation of ternary mixtures, divided wall columns may be of interest, [9] which have shown to reduce the energy requirements and increase exergy efficiency.

10.8   CONCLUDING REMARKS

The analysis presented in this chapter is an example of how the principles of thermodynamics can be applied to establish efficiencies in separation units. We have shown how exergy analysis or, equivalently, lost work or availability analysis can be used to pinpoint inefficiencies in a distillation column, which in this case were the temperature-driving forces in the condenser and the reboiler. The data necessary for this analysis can easily be obtained from commonly used flow sheeters, and minimal extra effort is required to compute thermodynamic (exergetic) efficiencies of various process steps. The use of hybrid distillation has the potential to reduce column inefficiencies and reduce the number of trays. We note that for smaller propane-propene separation facilities (less than 5000 bbl/day [10]), novel technologies such as adsorption and reactive distillation can be used.

NOTES

1 For a detailed discussion of distillation, we refer to textbooks on the subject (see, e.g., [7]).
2 The options we enumerate for improving efficiency of a column are by no means exhaustive or unique. We simply state them here to alert the reader of *possible* improvement options. We note that "cookbook" techniques to improve efficiencies are foredoomed to obsolescence since technology advances and they do not capture "out-of-the-box" solutions, which can shift paradigms.
3 Improvements in the efficiency of propane-propylene splitters have been reported in actual industrial settings leading to net energy, or better said, exergy savings in the order of 15%, depending on the conditions.

REFERENCES

1. Li, N.N.; Strathmann, H. (eds.). *Separation Technology*, United Engineering Foundation: New York, 1987.
2. *Ullmann's Encyclopedia of Industrial Chemistry*, VCH Publishers, Inc.: New York, 1993; Vol. A22.
3. Humphreys, J.L.; Seibert, A.F.; Koort, R.A. Separations technologies advances and priorities. *U.S. Department of Energy Report 12920–1*, U.S. Department of Energy, Washington, DC, 1991.
4. Gokhale, V.; Hurowitz, S.; Riggs, J.B. A comparison of advanced distillation control techniques for a propane/propene splitter. *Industrial & Engineering Chemistry Research* 1995, *34*, 4413–4419.
5. de Swaan Arons, J.; van der Kooi, H.J. *The Thermodynamic Analysis of Distillation, Some Parallels with Living Systems* (Course for industry organized by Prof. Ohe), Science University of Tokyo: Tokyo, 1999.
6. Seader, J.D. *Thermodynamic Efficiency of Chemical Processes*, The MIT Press: Cambridge, MA, 1982.
7. Seader, J.D.; Henley, E.J. *Separation Process Principles*, John Wiley and Sons: New York, 1997.
8. Pressly, T.G.; Ng, K.M. A break-even analysis of distillation-membrane hybrids. *The AIChE Journal* 1998, *44*, 93–105.
9. Serra, M.; Espuna, A.; Puigjaner, L. Control and optimization of the divided wall column. *Chemical Engineering Science* 1999, *38*, 549.
10. Eldridge, R.B. Olefin/paraffin separation technology: A review. *Industrial & Engineering Chemistry Research* 1993, *32*, 2208–2212.

11 Chemical Conversion

This chapter examines two industrial polymer processes. Thermodynamic analysis is used as a tool to locate, in a matter-of-fact fashion, process inefficiencies. Some sample process improvement options are discussed.

11.1 INTRODUCTION

Chemical industry owes its existence to society's need for various products. Chemical industry can be viewed as a system that converts various raw materials into useful and waste products using energy, and sometimes producing energy (Figure 11.1). At this point, we will introduce our working definition of sustainability. Angela Markel [1] defined sustainability as the use of resources no faster than they are regenerated and releasing pollutants to no greater extent than natural resources can assimilate them. This definition will be refined later in Chapter 13 but will suffice for the moment.

Within the framework of sustainability and durable technology, it is easy to generalize this concept to include chemical industry. Figure 11.1, albeit rather naive, provides a simple framework to start the discussion of how a chemical process can be made more sustainable. The reduction of energy inputs can lead to a more sustainable chemical industry, as does the reduction of waste streams. Exergy analysis can help provide insights into how the process can be made more energy-efficient, but does not, in general, provide an answer as to how waste streams can be reduced. Catalysts with higher specificity and processing of the "waste" products are a better answer to that query.

This chapter gives the example of an exergy analysis of two low-density polyethylene (LDPE) production processes, namely, the high-pressure (HP) and the gas-phase processes. Section 11.2 provides a general overview of some existing polyethylene processes. Section 11.3 develops some of the tools necessary to estimate exergies of the polymers. The actual result of the exergy analysis for the HP process is given in Section 11.4, and some process improvement options are discussed in Section 11.5. Section 11.6 presents the exergy analysis of the gas-phase process followed by process improvement options in Section 11.7. We conclude with a brief summary in Section 11.8.

11.2 POLYETHYLENE PROCESSES: A BRIEF OVERVIEW

Polyethylene[1] (PE) is the largest synthetic commodity polymer in terms of annual production and is widely used throughout the world in a variety of applications. Based on the density, PE is classified as LDPE at 0.910–0.930g/cm³, high-density polyethylene (HDPE) at 0.931–0.970 g/cm³, and linear low-density polyethylene (LLDPE) based on the polymer chain microstructure. At present, processes that produce PE use the following raw materials:

DOI: 10.1201/9781003304388-14

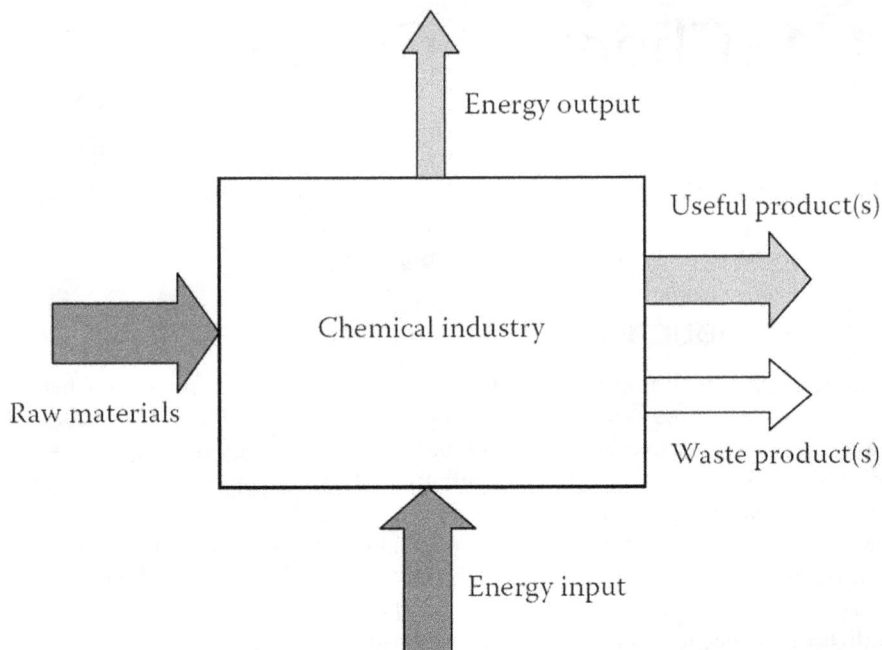

FIGURE 11.1   Abstract representation of chemical industry.

Ethylene (for homopolymers)
Ethylene and α-olefins (for copolymers)

The main reaction to obtain homopolymers[2] from ethylene can be depicted as follows:

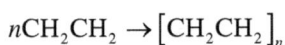

$$n\text{CH}_2\text{CH}_2 \rightarrow \left[\text{CH}_2\text{CH}_2\right]_n$$

The transformation can be achieved in a number of ways. PE is commercially produced exclusively by means of continuous processes. The oldest process, due to Imperial Chemical Industries (ICI), is the so-called conventional HP process. In the 1950s, Prof. Ziegler developed special catalysts that made it possible to polymerize ethylene at lower pressures. Since then, many new processes have been developed. On the basis of polymerization mechanisms and reactor configurations, the PE processes can be classified broadly into five categories, as shown in Table 11.1.

Among these polymerization processes, the gas-phase process is the most recent addition to the list. Since its emergence, the process has challenged many other existing technologies for market share. In addition to differences between the gas-phase process and the HP-free radical process shown in Table 11.1, more differences can be noted, which are shown in Table 11.2. Because the HP process is a well-established technology, and the gas-phase process is a relatively new process with certain advantages, we have chosen these two processes as examples for the application of our thermodynamic analysis.

TABLE 11.1
Polymerization Processes and Reactor Operating Conditions.

	Conventional HP	HP Bulk Process	Solution Process	Slurry Process	Gas-Phase Process
Reactor type	Tubular or autoclave	Autoclave	CSTR	Loop or CSTR	Fluidized or stirred bed
Pressure, bar	1200–1300	600–800	–100	30–35	30–35
Temperature, °C	130–350	200–300	140–200	85–110	80–100
Mechanism	Free radical	Free radical/ coordination	Coordination	Coordination	Coordination
Location of polymerization	Monomer phase	Monomer phase	Solvent	Solid surface	Solid
Product grades	LDPE	LDPE/HDPE[a]	HDPE/LDPE	HDPE	LDPE/HDPE

Source: Kroschwitz, J.I. (editor-in-chief), Encyclopedia of Polymer Science and Engineering, Wiley, New York, 1990.

[a] High-pressure plants have been modified to produce HDPE.

TABLE 11.2
Comparison of High-Pressure Free Radical Process and Low-Pressure Gas-Phase Process.

High-Pressure Process	Low-Pressure Process
High energy requirement	Low energy requirement
High capital cost	Low capital cost
Limited to low-density PE	Can produce both high- and low-density PE
No comonomer required (homopolymer) for LDPE production	Up to 10%–15% comonomer required (in case of copolymer)
Catalyst is less sensitive to impurities in feedstock	Catalyst is very sensitive to impurities
Low raw material cost	Higher raw material cost
Can produce ethylene and vinyl acetate copolymer with long chain branching	Currently a limited range of comonomers, mainly linear polymer chains with short chain branching

11.2.1 POLYETHYLENE HIGH-PRESSURE TUBULAR PROCESS

The HP tubular process is, as stated earlier, a continuous process and has been in use for decades in the chemical process industries. A simplified version of a generic HP tubular process is given in Figure 11.2 [1].

A primary compressor increases the pressure of the entering ethylene gas (and propylene gas, which is added as a molecular weight control agent) from between 5 and 15 bar to about 250 bar. The secondary compressor further increases the gas pressure from 250 bar to the desired reactor pressure (approximately 2500 bar). An initiator

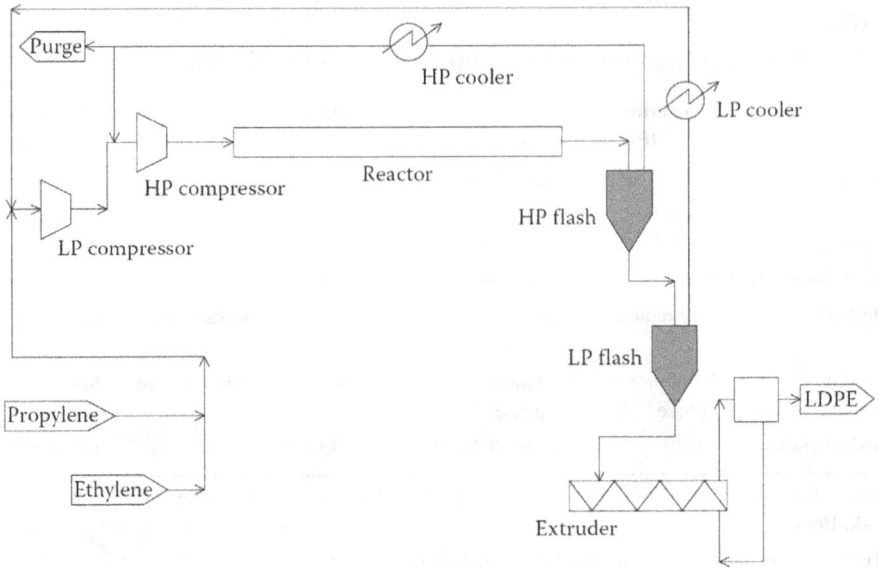

FIGURE 11.2   Simplified flow diagram of HP LDPE process.

is added to the gas as it enters the reactor. The reactor is operated to ensure a per-pass conversion of 15%–35% and is a wall-cooled reactor where the cooling water can be used to produce steam. The reaction mixture then enters the HP separator (~250 bar), where the mixture is flashed to produce two distinct phases: a PE-rich melt phase and an ethylene-rich gas phase. The separated gas then enters the recycle loop. The ethylene gas is cooled before entering the secondary compressor. The PE enters the low-pressure separator. This low-pressure separator, also referred to as a hopper, performs the final degassing step. The separated ethylene gas is cooled and some components are removed. This step takes place in the low-pressure return gas loop. The hopper also functions as a storage buffer before the PE enters the extruder.

The extruder granulates the PE. In addition to this function, various quality enhancers can be added to obtain certain polymer properties, and the PE is further degassed. The granulation is achieved by pressing the hot PE melt through a perforated plate, directly cooling with water and cutting the solidified PE. The water-PE mixture is then led to a separator in which the water is removed from the PE. The PE is then dried, bagged, and ready for transport.

11.2.2   POLYETHYLENE GAS-PHASE PROCESS

Gas-phase polymerization processes, as mentioned earlier, are also used. In this section, we will discuss a plant geared toward the production of LLDPE. A simplified version of the flow diagram is given in Figure 11.3 [1].

The monomer ethylene is first led through a series of dryers (not shown), mixed with recycle gas, hydrogen, and butylene, and enters the fluidized bed reactor (R1)

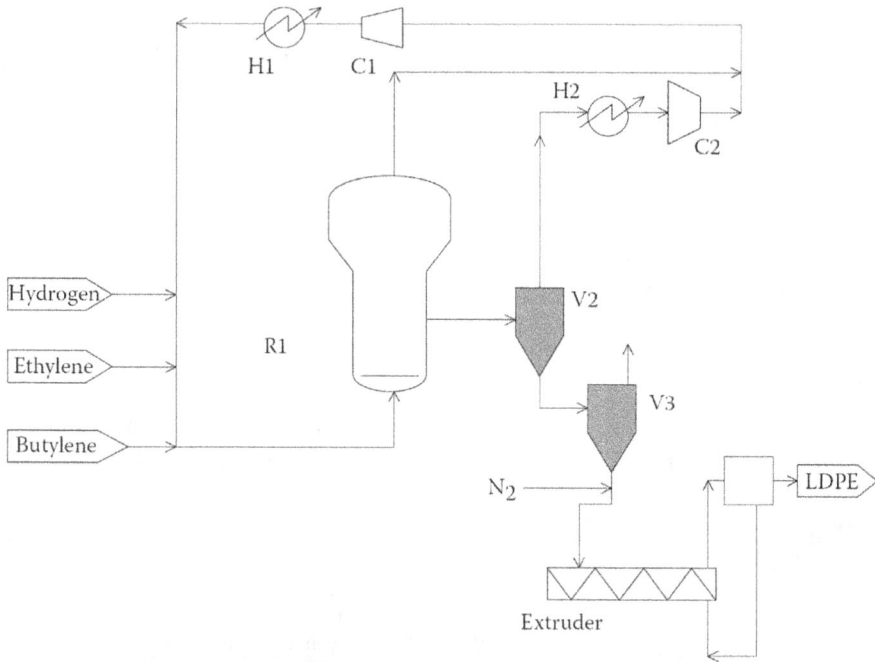

FIGURE 11.3 Simplified flow diagram of a gas-phase polymerization process.

at the bottom. A catalyst is added from numerous locations on the side of the reactor (not shown). The reactor operates at approximately 85°C and 20 bar (the temperature depends on the desired grade). The heat of reaction is removed as sensible heat of the gas stream and by the evaporation of the partially liquid reaction mixture. The reactor gas exits the top of the reactor.

The gaseous reactor exit stream is compressed (C1) and cooled (H1) and recycled to the bottom of the reactor. The product is removed from the bottom end of the reactor and is led to the polymer discharge vessel (V2). The polymer is discharged to the purge vessel (V3). Nitrogen is passed through the polymer in this vessel to remove any remaining gas in the interstices of the polymer particles, as well as to purge any of the monomers left in the solid. The polymer then proceeds to the finishing section. In this section, additives are added to the product in an extruder. The product is pelletized, bagged, and stored in a warehouse.

11.3   EXERGY ANALYSIS: PRELIMINARIES

In order to perform an exergy analysis, it is vital to have accurate information regarding the physical and chemical exergies of the various process streams. The first is accomplished by having an accurate flow sheet of the processes, which readily gives the enthalpies and entropies of the streams if proper care has been taken in selecting thermodynamic models. This allows for the computation of the physical exergy for liquid-vapor mixtures according to

$$\text{E}\dot{\text{x}}_{\text{ph}} = \Delta_{0 \to \text{actual}} \left[\dot{L} \left(\sum_{i=1}^{n} x_i h_i^l - T_0 \sum_{i=1}^{n} x_i s_i^l \right) + \dot{V} \left(\sum_{i=1}^{n} y_i h_i^v - T_0 \sum_{i=1}^{n} y_i s_i^v \right) \right] \qquad (11.1)$$

where

$\dot{L}$ = liquid molar flow rate
$\dot{V}$ = vapor molar flow rate
x_i = liquid mole fraction
y_i = vapor mole fraction

In general, a flow-sheeting program can give the total entropy and enthalpy of the stream, in which case the physical *and* mixing exergies can be computed as

$$\text{E}\dot{\text{x}}_{\text{ph}} = \left[\dot{L} \left(H^l - T_0 S^l \right) + \dot{V} \left(H^v - T_0 S^v \right) \right] - \dot{L} \left(\sum_{i=1}^{n} x_i h_i^l - T_0 \sum_{i=1}^{n} x_i S_i^l \right)_{p_0, T_0}$$
$$- \dot{V} \left(\sum_{i=1}^{n} y_i h_i^v - T_0 \sum_{i=1}^{n} y_i S_i^v \right)_{p_0, T_0} \qquad (11.2)$$

The computation of the chemical exergy of the various streams is achieved by using documented values of the chemical exergy [2] of the chemical species or from estimation methods [2]. The chemical exergy values used for the calculations are given in Table 11.3 (here, the value of PE is calculated using group contribution methods and accounts for the crystallinity of PE [3]). Note that the values given are not the cumulative exergy values (see Chapter 7 for an explanation of this concept), and the efficiencies and losses computed in this case study are for this process only and do not take into account the losses incurred in producing ethylene from naphtha or some other

TABLE 11.3
Values of Chemical Exergy for Species.

Compound	Ex_{ch} (kJ/mol)
Ethylene	1361.1
Nitrogen	0.72
Ethane	1495.8
Hydrogen	236.1
Polyethylene[a]	4650×10^3
Butylene	2659.7
Butane	2805.8

Source: Szargut, J. et al., *Exergy Analysis of Thermal, Chemical and Metallurgical Processes,* Hemisphere Publishing Corporation, New York, 1988.

[a] Calculated using group contribution methods and accounting for the crystallinity of PE.

fossil fuel. The efficiencies of producing high purity ethylene from naphtha are approximately 40%–50%, depending on the technology used and raw material source [4,5].

Detailed Aspen (version 9.2) flow sheets were created to simulate (1) the production of 39% crystalline PE from solely ethylene as monomer for the LDPE HP process, and (2) LLDPE (92 wt% ethylene and 8 wt% 1-butylene).

11.4   RESULTS OF THE HP LDPE PROCESS EXERGY ANALYSIS

The results of the exergy analysis of the HP LDPE process are given in Figure 11.4 in the form of a Grassmann diagram, which shows the flow of exergy in the process.

The efficiency of the HP LDPE process is 91%. If the *steam bonus*[3] is taken into account, the process efficiency, based on exergy, is 91 + 1.1 + 0.05 = 92.2%. The main losses can be identified in the pie chart, which are given in Figure 11.5 and in Table 11.4. If one considers the cumulative efficiency of PE production via the high-pressure LDPE route from ethylene that was made from naphtha [3,4], one ends up with an efficiency of 0.922 × 0.41 = 38%.

As can be seen from the pie chart in Figure 11.5, the Grassmann diagram in Figure 11.4, and Table 11.4, the main losses can be attributed to the following sections:

1. The reactor
2. The HP cooler
3. The HP flasher
4. The Extrusion section

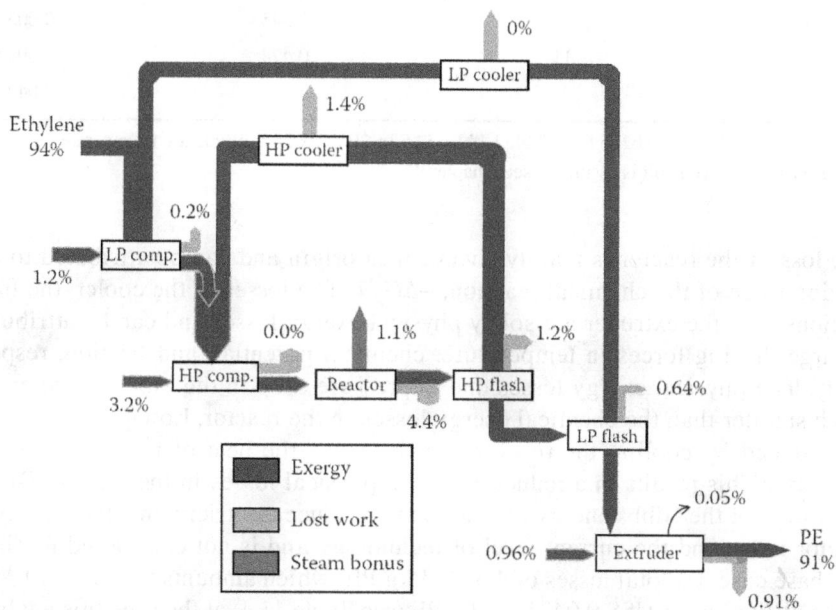

FIGURE 11.4   Grassmann diagram of the HP LDPE process. Available work, lost work, and potential steam bonus streams are given as a percentage of the total input.

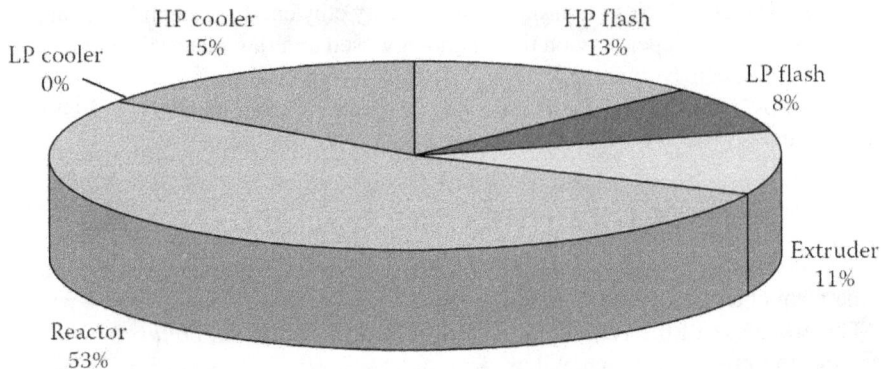

FIGURE 11.5   Pie chart of main exergy losses in the HP LDPE process.

TABLE 11.4
Overview of Main Losses in the HP LDPE Process.

Unit	Exergy Loss (%)	Exergy Loss (MJ/kg$_{PE}$)	Exergy Loss[a] ($/kg$_{PE}$)
Reactor	53	2.284	0.0225
HP cooler	15	0.647	0.0064
HP flash	13	0.560	0.0055
LP flash	8	0.345	0.0034
Extruder	11	0.474	0.0047
Total	100	4.31	0.0425

[a] Crude oil = 38.4 MJ/L = 6105.6 MJ/bbl, 1 bbl = $15 in 2002. A 25% efficiency to generate electricity (exergy) from the crude oil is assumed (see Chapter 9).

The loss in the reactor is mainly chemical in origin and can be attributed to the driving force of the chemical reaction, $-\Delta G_r/T$. The losses in the cooler, the flash sections, and the extruder are solely physical exergy losses and can be attributed to large driving forces in temperature, chemical potential, and friction, respectively. The physical exergy losses due to pressure drop in the tubular reactor are much smaller than the chemical exergy losses in the reactor. Low-pressure steam is produced by cooling the reactor, which means the heat of reaction is put to good use. This results in a reduction of the physical losses in the reactor. Direct utilization of the Gibbs energy of reaction to produce electricity in a fuel cell-type reactor is beyond the current level of technology and is not considered feasible. The base case has total losses of 4.31 MJ/kg PE, which amounts to 2.82×10^{-3} bbl crude oil/kg PE, or US$ 0.0425 per kg PE (see Table 11.4) at the time this analysis was done.

11.5 PROCESS IMPROVEMENT OPTIONS

Losses could potentially be reduced by considering, for example, the following options:

1. The losses in the HP flash can be reduced by using a turbine.
2. An alternative to extruders can be developed.

11.5.1 Lost Work Reduction by the Use of a Turbine

The theoretical[4] maximum amount of work ($W_{turbine}$) that can be obtained from the turbine is given by the following expression:

$$W_{turbine} = \frac{\dot{W}_s}{\dot{m}} = -\Delta H_s \cdot \eta_{turbine} \tag{11.3}$$

where $\dot{W}_s$, $\dot{m}$, and Δh_s denote the shaft work, the mass flow rate, and the isentropic enthalpy change, respectively. If the efficiency, $\eta_{turbine}$, is chosen to be 0.75, and the enthalpy difference is calculated for the pressure drop from the reactor exit pressure (2200 bar) to 1000 bar,[5] the following turbine duty can be computed:

$$W_{turbine} = 0.32 \, MJ \, / \, kg \, PE \tag{11.4}$$

The savings, therefore, are 0.32 MJ/kg PE. The original loss was (see Table 11.4) 0.56 MJ/kg PE for the HP flash, which means that 0.24 MJ/kg PE will still be lost. A noteworthy point, however, is that turbines do not exist yet that can operate at these kind of conditions.

11.5.2 Alternative to the Extruder

The extruder is used since it mixes, melts, and degases the polymer. However, as said in the previous paragraph, from an exergetic point of view, it is not an efficient apparatus, since it dissipates mechanical work into heat by frictional forces. A possible alternative to the extruder could therefore be a separate degasser, a static mixer, or a gear pump to push the polymer melt through the mixer and perforated plate (Figure 11.6).

The extruder is substituted by a deep flash vessel (which operates at 150 mbar), a gear pump, and a static mixer. The only exergy input is the energy requirement of the pump and the compressor, which removes the gas. The pump only needs to increase the pressure of the polymer such that it can pass through the static mixer and the granulating head (perforated plate). This results in the following expression for the exergy input:

$$\dot{Ex}_{in} = \frac{\dot{m}}{\eta_{mech}\rho}\left(\Delta P_{static\text{-}mixer} + \Delta P_{granulating\text{-}head}\right) + \dot{Ex}_{compressor} \tag{11.5}$$

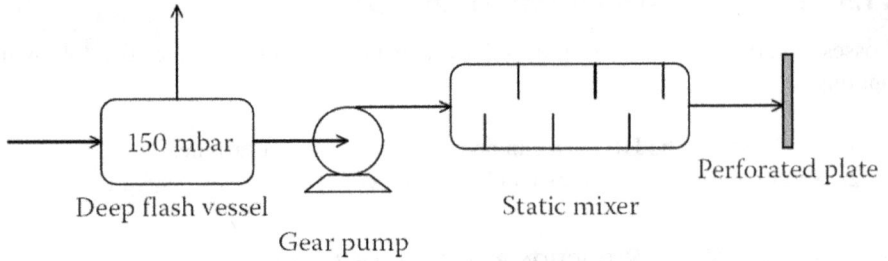

FIGURE 11.6   Alternative to the extruder.

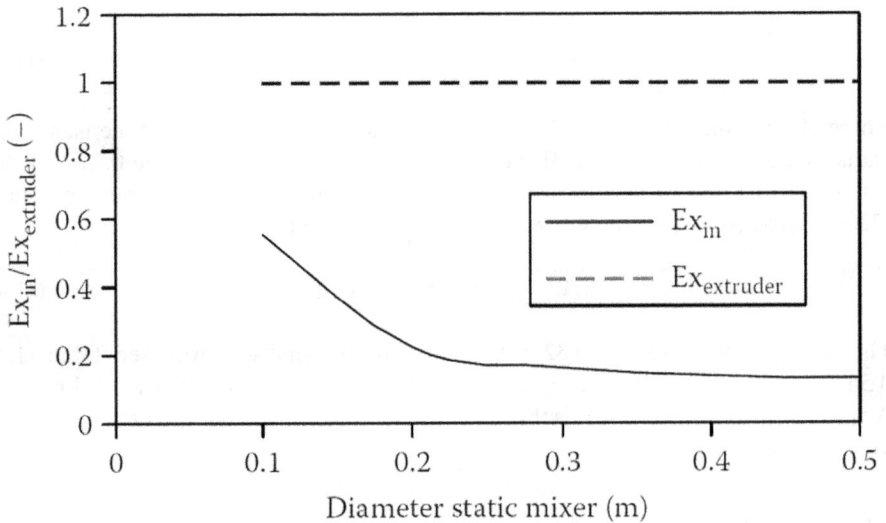

FIGURE 11.7   Scaled exergy input of the alternative extruder and regular extruder plotted against static mixer diameter.

where

   $\dot{m}$ is the mass flow rate of polymer passing through the mixer
   η_{mech} is the mechanical efficiency of the gear pump (50%)
   ρ is the density of the polymer

   By computing the pressure drop in the static mixer and the pressure drop in the granulating head and adding this to the exergy requirements of the compressor, it is possible to compute the exergy input of the alternative scheme, and compare this with the regular extruder (Figure 11.7).

   As the graph suggests, it is beneficial to use the proposed setup. It is, of course, unclear what the effect of the new setup will be on product quality, which stresses the use of laboratory experiments with the new setup. The initial exergy loss with the extruder was 0.47 MJ/kg PE (see Figure 11.5 and Table 11.4). The estimated exergy loss with the new setup will now be between $0.6 \times 0.47 = 0.285$ MJ/kg PE (for a diameter of 0.1 m) and $0.15 \times 0.47 = 0.071$ MJ/kg PE (for a diameter greater than 0.4 m) depending on the diameter of the static mixer.

11.5.3 PROCESS IMPROVEMENT OPTIONS: ESTIMATED SAVINGS

The total initial losses were 4.31 MJ/kg (see Figure 11.5 and Table 11.4). The estimated savings per improvement option are as follows:

1. Turbine to address HP flash losses: 0.32 MJ/kg PE savings, which gives a new loss of 0.24 MJ/kg compared to the original 0.56 MJ/kg PE. This translates into a savings of US$ 0.003 per kg PE.
2. Alternative to extruder: savings between 0.19 and 0.40 MJ/kg PE, which gives a loss between 0.07 and 0.28 MJ/kg PE compared to 0.47 MJ/kg PE.

Option 1 is not yet technologically viable but option 2 is. In the remainder of this chapter, we will continue to focus on the LDPE process from ethylene only.

11.6 RESULTS OF THE GAS-PHASE POLYMERIZATION PROCESS EXERGY ANALYSIS

The results of the exergy analysis of the gas-phase process are given next as a Grassmann diagram (see Figure 11.8). The efficiency of the gas-phase process is 91%. The main losses can be identified and are shown in the pie chart given in Figure 11.9

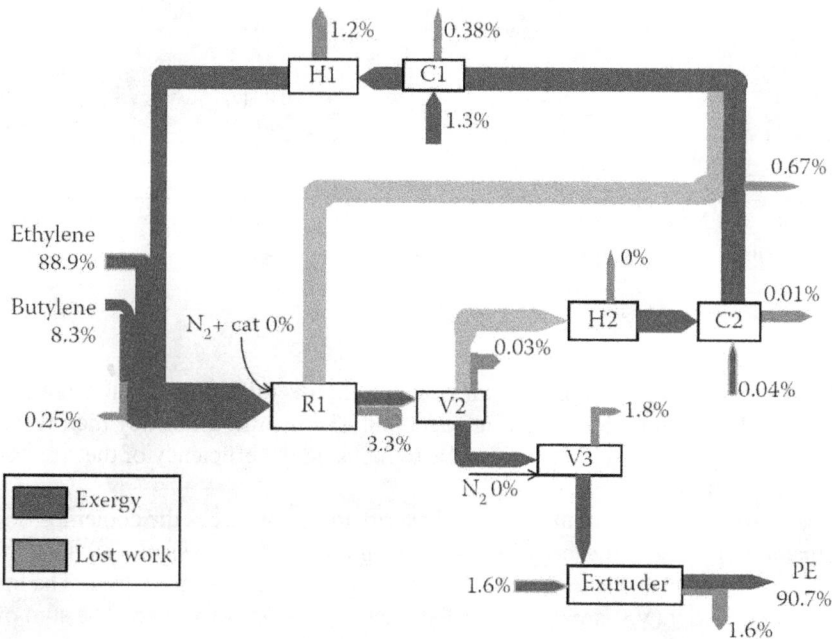

FIGURE 11.8 Grassmann diagram of the gas-phase process. Exergy and lost work streams are shown as a percentage of the total input.

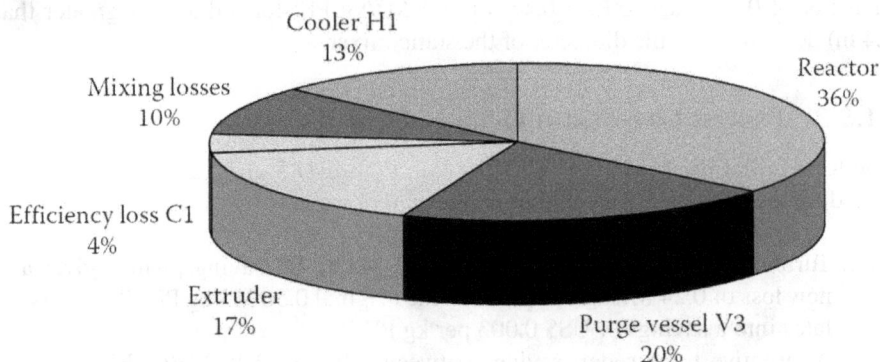

FIGURE 11.9   Pie chart of main exergy losses in the gas-phase process.

TABLE 11.5
Overview of Main Losses in the Gas-Phase PE Process.

Unit	Exergy Loss (%)	Exergy Loss (MJ/kg_{PE})	Exergy Loss $(\$/kg_{PE})^a$
Reactor	36	1.703	0.0166
Cooler H1	15	0.710	0.0069
Efficiency loss C1	4	0.189	0.0017
Mixing losses	10	0.473	0.0046
Extruder	17	0.804	0.0076
Surge vessel V3	20	0.946	0.0092
Total	100	4.73	0.0465

^{a} Crude oil = 38.4 MJ/L = 6105.6 MJ/bbl, 1 bbl = $15 in 2002. A 25% efficiency to generate electricity (exergy) from the crude oil is assumed (see Chapter 9).

and in Table 11.5. Here, too, one can consider the overall efficiency if the ethylene production from naphtha is taken into account. The overall efficiency then is 0.91 × 0.41 = 37%. Again, we will now continue to focus on the efficiency of the conversion from ethylene to PE.

The losses in the reactor are chemical exergy losses, whereas the cooler losses can be attributed to physical exergy losses. Mixing constitutes physical losses as do the losses in the extruder due to the dissipation of mechanical energy to heat. The losses in the purge vessel (V3) are due to the fact that the gas is incinerated. The sum of all losses equals 4.73 MJ/kg PE, or US$ 0.0465 per kg PE.

11.7   PROCESS IMPROVEMENT OPTIONS

Losses in the gas-phase polymerization process can be reduced, for example, by taking the following steps:

1. Coupling reactions to reduce the chemical exergy loss in the reactor
2. Reuse the gas that leaves the purge vessel instead of burning it
3. Use the heat discarded by cooler Hl to heat the polymer in the extruder after upgrading with a heat pump
4. Preheat the polymer to its melting point, thus reducing the exergy consumption of the extruder
5. Develop an alternative to extruders

Note that for the gas-phase polymerization process, the PE enters the extruder in *solid* form between 70°C and 80°C, which is approximately the upper limit for solids handling. For the safe and easy handling of solid PE, the handling temperature is usually chosen 40°C below the melting point, which results in 85°C. At higher temperatures, the morphology of PE becomes such that handling is difficult (sticky solids).

11.7.1   COUPLING REACTIONS AND CHEMICAL HEAT PUMP SYSTEM

The exergy loss in the reactor is formally defined as

$$\dot{\mathrm{Ex}}_{\mathrm{loss}} = \dot{m}\left(\mathrm{Ex}_{\mathrm{in}} - \mathrm{Ex}_{\mathrm{out}}\right) \qquad (11.6)$$

Since the chemical exergies in this case are much larger than the physical exergies, the exergy loss in this specific case is dominated by the difference in chemical exergy $(= -\Delta_r G^0)$. The reaction is accompanied by changes in entropy and enthalpy (heat of reaction). These losses can, in theory, be reduced by building the reactor in a way that it resembles a fuel cell, since fuel cells have the capacity to transform, in the reversible limit, the full Gibbs energy change into electricity. For polymer systems, this is unfortunately not a viable option. The only option that remains is to somehow use the heat liberated during reaction. The heat of reaction, which is currently not being utilized, could potentially be harnessed if a chemical heat pump were used. A chemical heat pump is simply a system that uses a chemical reaction to absorb heat at a certain temperature and uses the reverse reaction at a different temperature to release this heat. A list of proven chemical heat pump systems is given in Kyaw et al. [6]. Unfortunately, none of these systems is viable in a fluidized bed reactor since additional reactions in the reactor, known as direct coupling, may reduce product quality, poison the catalyst, or necessitate exergetically costly separation steps to remove PE from the species used in the chemical heat pump systems. Indirect coupling would require a heat exchanger in the reactor, which would change the hydrodynamics in the reactor, thus potentially affecting conversion.

At this stage, therefore, chemical heat pump systems are not considered viable for the PE gas-phase polymerization reactor [5].

11.7.2 EXERGY LOSS REDUCTION BY RECOVERING BUTYLENE AND ETHYLENE FROM PURGE GAS

This option is relatively simple to implement and would save a maximum of 0.946 MJ/kg PE if reused completely, which is a theoretical upper bound. The exergy of this purge stream is dominated by the chemical exergies of the constituents. If all butylene is recovered using a cooling cycle, the exergy savings will be 0.416 MJ/kg PE, and the losses would have been reduced to 0.530 MJ/kg, as the ethylene is not recovered. If the ethylene is recovered as well, the savings will be 0.946 MJ/kg PE.

Alternately, the purge stream can be sent to a power plant where (at 25% efficiency) electricity is produced, resulting in a loss reduction of $0.25 \times 0.946 = 0.2365$ MJ/kg. Hybrid options are also possible, where part of the steam is sent to a power plant to generate the electricity necessary to power the cooling of the reactor.

11.7.3 HEAT PUMP AND PREHEATING OF POLYMER

A potentially viable improvement to increase the exergetic efficiency is to feed solid PE to the extruder, and to recycle a part of the molten PE, which is then heated to a much higher temperature by utilizing the waste heat of the process upgraded by a heat pump system (see Chapter 10 for details). The hot molten PE quickly melts the solid feed upon entrance in the extruder, thus resulting in lower exergy requirements (Figure 11.10). The theoretical power consumption (per unit mass PE passing through the extruder) of an extruder is given by the following equation, since the extruder heats the polymer and transports it against a pressure gradient:

$$W_{\text{extruder}} = \int_{T_1}^{T_2} C_p \, dT + \frac{\Delta P}{\rho} \qquad (11.7)$$

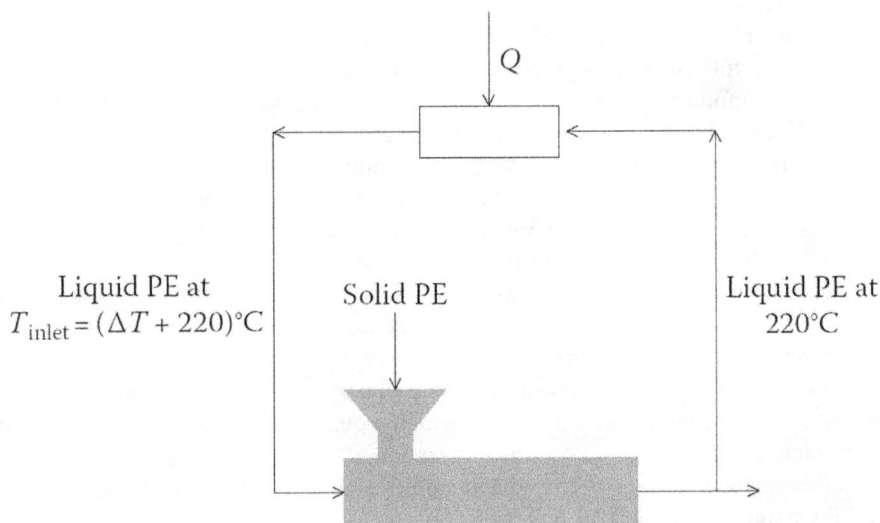

FIGURE 11.10 Extruder using recycled and upgraded heat.

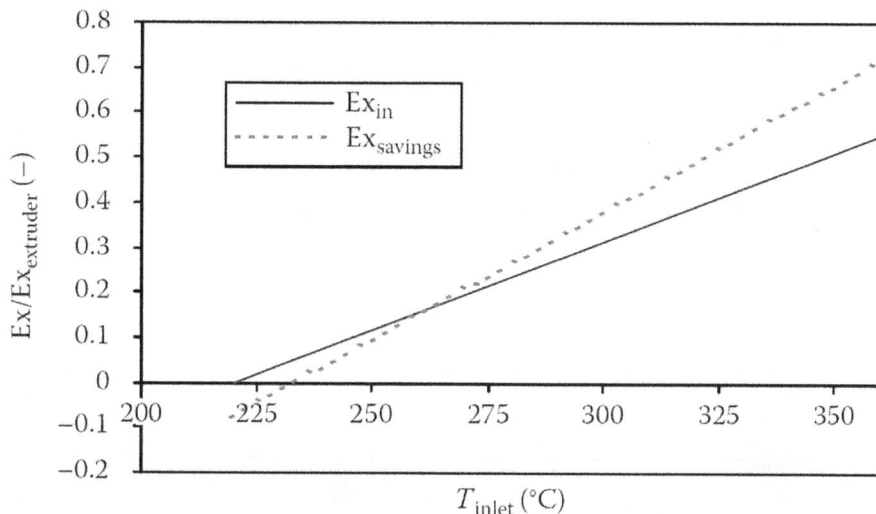

FIGURE 11.11   Exergy savings and input for recycle option (see Figure 11.13).

Now, with the recycle option, the total mass flow rate through the extruder will increase and, therefore, the power consumption *based on PE produced* (i.e., leaving the recycle loop) will increase. The graph in Figure 11.11 shows the exergy savings of the extruder and the exergy input to heat the PE in the heat exchanger as a function of the final temperature of the liquid PE after passing through the heat exchanger (T_{inlet}). In this graph, the mass flow rate of PE in the extruder has doubled, since an equal amount of PE leaves the recycle loop as stays inside it.

Since waste heat is available, this heat can be used to be upgraded with a heat pump. If compressor and pump efficiencies of 75% are assumed, and a simple two-stage vapor-compression heat pump is used, the total savings (extruder exergy savings and waste heat savings) can be calculated, as well as the exergy input for the heat pumps (see Figure 11.12). As at higher inlet temperature,[6] the total savings are larger than real exergy input for the extruder, so it is useful to consider this option.

11.7.4   An Alternative to the Extruder

If the setup shown in Figure 11.13 is chosen instead of an extruder, it can be calculated that exergetic gains are obtained as shown earlier.

For the gas-phase process, in which solid PE is fed into the extruder, the extruder can be substituted by a kneader (to melt the polymer by the dissipation of mechanical energy), a gear pump, a static mixer, and a granulator. By using an enthalpy-temperature correlation for PE, the kneader duty necessary to melt the polymer can be calculated. The calculation of the gear pump duty was shown earlier (Section 11.5.3). The graph in Figure 11.14 shows the (scaled) exergy requirements of the alternative extruder with respect to the original extruder.

From an efficiency point of view, the new setup is advantageous. As the pressure drop across the static mixer is approximately 25 MPa (for diameter 0.5 m; see Figure 11.15),

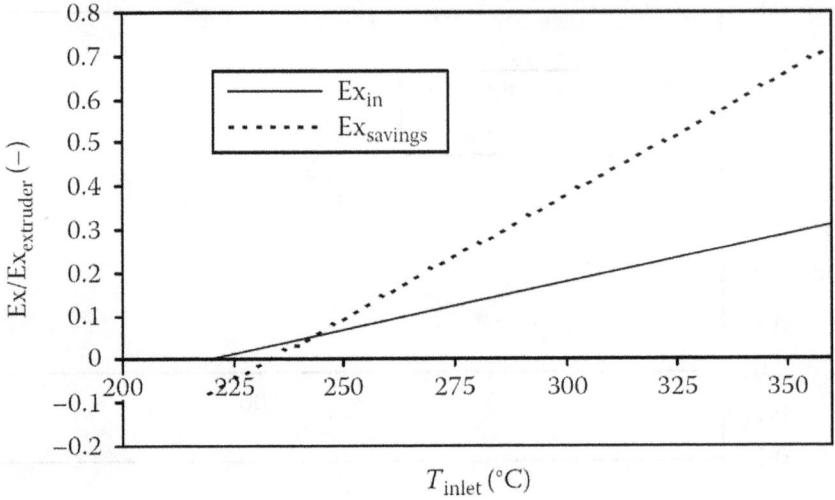

FIGURE 11.12   Total exergy savings (extruder and H1) and exergy input.

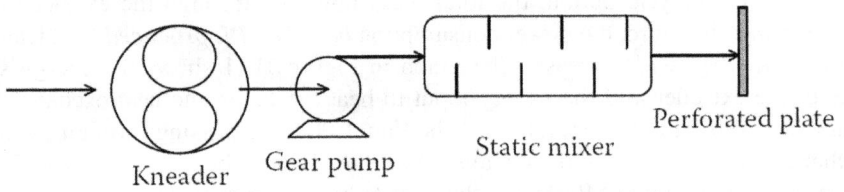

FIGURE 11.13   An alternative to the extruder.

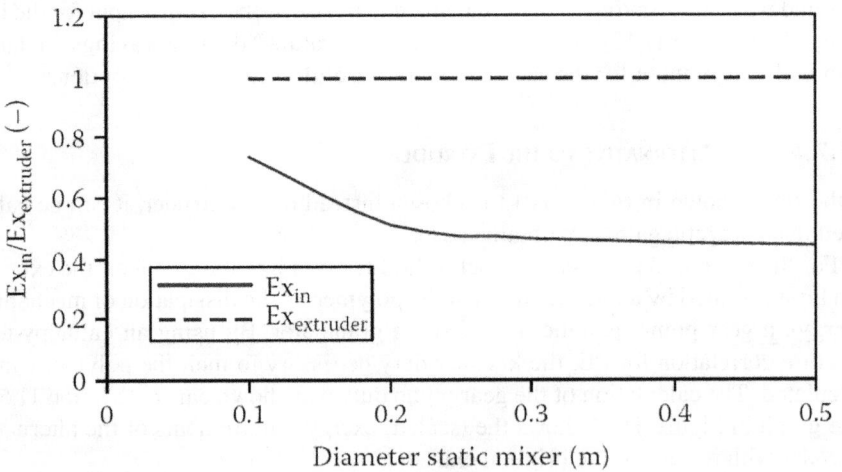

FIGURE 11.14   Exergy requirement of alternative and existing extruder as a function of the static mixer diameter.

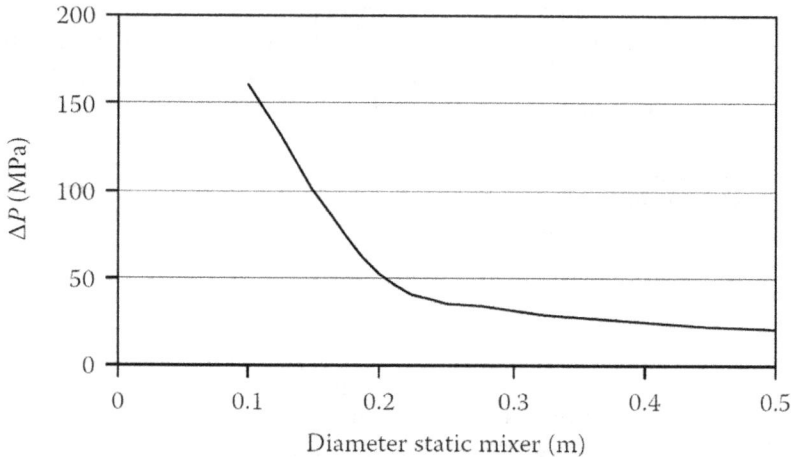

FIGURE 11.15 Pressure drop across static mixer for different diameters.

the proposed setup is feasible,[7] as gear pumps are known to deliver the pressure of 35–40 MPa. Another advantage of the proposed setup is that the polymer can be degassed further than is possible in the extruder. The present extruder loses $0.17 \times 4.73 = 0.8041$ MJ/kg PE. The savings will be between $(1 - 0.44) \times 0.17 \times 4.73 = 0.45$ MJ/kg PE and $(1 - 0.74) \times 0.17 \times 4.73 = 0.21$ MJ/kg PE. The values of 0.44 and 0.74 can be determined graphically from Figure 11.14 in combination with Figure 11.15.

11.7.5 PROCESS IMPROVEMENT OPTIONS: ESTIMATED SAVINGS

The original losses were 4.73 MJ/kg PE (see Figure 11.9 and Table 11.5). The estimated savings per improvement option are as follows:

1. Recover gases from purge stream: 0.946 MJ/kg PE, which reduces the losses to 0 MJ/kg PE. Recovery costs in terms of exergy are minimal and have not been shown.
2. Use the gases in the purge stream in a power plant: 0.24 MJ/kg PE, which reduces the loss to 0.71 MJ/kg PE.
3. Heat pump and preheating of polymer in the extruder: 0.2 MJ/kg PE, which reduces the losses from the original 0.8 to 0.6 MJ/kg PE.
4. Alternative to extruder: Between 0.21 and 0.45 MJ/kg PE, which reduces the losses from 0.8 MJ/kg PE to between 0.59 and 0.35 MJ/kg PE.

All these options are technologically viable.

11.8 CONCLUDING REMARKS

The objective of this chapter was to show how exergy analysis can help pinpoint steps where lost work is produced or, equivalently, exergy losses are greatest in chemical processes. Since the analysis was based on a flow sheet of a process, exergy analysis can be used as a tool to improve the efficiency of processes when they are still in the design stage. This can help move chemical processes closer to the ideal of a sustainable chemical process. In this specific case, the PE processes had an efficiency of around 90% and the cumulative efficiency was around 40%. The process improvement options are by no means definitive, and as technology evolves, better options will become available. The cases presented in this chapter do, however, indicate how *good chemical engineering* coupled with exergy analysis can make a difference in making processes more efficient and durable.

NOTES

1   As the title of this section already suggests, this is only a very brief survey of the commonly used PE production processes. A detailed description of the processes, the kinetics, and so on shall not be given, as this is not relevant within the framework of this case study.
2   The reaction to obtain copolymers is analogous.
3   Steam bonus: If the steam generated by the process is used elsewhere on-site or for, say, heating purposes, the steam can be treated as a valuable *bonus* product. The heat liberated in the reactor can be used to produce LP steam.
4   The figure calculated is theoretical, as turbines that can handle this pressure have not been built yet.
5   At lower pressures a two-phase mixture is obtained, which is not desirable for turbine blades.
6   It is undesirable to elevate the temperature of PE far above 300°C due to product quality concerns. The maximal gains are therefore 0.4 − 0.2 = 0.2 MJ/kg PE.
7   The change from extruder to gear pump/static mixer could potentially increase the gel count. As LDPE and LLDPE primarily find their uses in foils, laboratory scale tests are recommended to assess the impact of the proposed setup on the gel count. Extruders usually decrease the gel count by kneading the polymer. It is unclear whether the proposed setup can accomplish the same.

REFERENCES

1. Kroschwitz, J.I. (editor-in-chief). *Encyclopedia of Polymer Science and Engineering*, Wiley: New York, 1990.
2. Szargut, J.; Morris, D.R.; Steward, F.R. *Exergy Analysis of Thermal, Chemical and Metallurgical Processes*, Hemisphere Publishing Corporation: New York, 1988.
3. Sankaranarayanan, K. *Thermodynamic Efficiency of the Polyolefin Industry*. MSc thesis, Delft University of Technology, Delft, 1997.
4. Ren, T.; Patel, M.; Blok, K. Olefins from conventional and heavy feedstocks: Energy use in steam cracking and alternative processes. *Energy* 2006, *31*, 425–451.
5. Industrial Technologies Program. *Exergy Analysis: A Powerful Tool for Identifying Process Inefficiencies in the U.S. Chemical Industry* (Chemical Bandwidth Study), Industrial Technologies Program: Boston, MA, 2006.
6. Kyaw, K.; Matsuda, H.; Hasatani, M.J. Applicability of carbonation/decarbonation reactions to high-temperature thermal energy storage and temperature upgrading. *Journal of Chemical Engineering of Japan* 1996, *29*(1), 119.

12 A Note on Life Cycle Analysis

In this chapter, we briefly touch on a tool called life cycle analysis (LCA) and discuss its usefulness. Motivation is given for extending this analysis to include exergy depletion, exergetic life cycle analysis (ELCA). An example is given of the so-called zero-emission ELCA, in which the analysis further includes the exergetic cost of making a product "environmentally friendly."

12.1 INTRODUCTION

In the previous chapters, thermodynamic analysis is used to improve processes. However, as pointed out in Chapter 9 (Energy Conversion), the exergy analysis did not make any distinction between the combustion of coal and natural gas and, as a result, could not make any statements regarding toxicity or environmental impact of exploration, production, and use of the two fuels. A technique that can do this is LCA. What exactly is life cycle analysis? In ISO 14040 [1], life cycle analysis[1] (or life cycle assessment) is defined as "the compilation and evaluation of the inputs, outputs and potential environmental impacts of a product throughout its life cycle."

Increased environmental awareness has led to the emergence of the concept to conduct a detailed examination of the life cycle of a product or a process. We note that this cradle-to-grave analysis is not truly a cycle, but this is the terminology which is in use. In the late 1960s and early 1970s, global modeling studies and energy audits were conducted to evaluate the costs of resources and the environmental implications of mankind's behavior. The LCAs were a natural extension of this and are useful in assessing the environmental impact of various products. They are also slated to be used in so-called eco-labels that allow the consumer to discern between various products. LCA is a potentially powerful tool that can be used in many instances. It can, for example, be used to (1) analyze products and processes and improve them, (2) assist in formulating (environmental) legislation or steer public policy [2], and so forth. In using the tool, some care must be taken. Like most tools, it must be used correctly. A tendency to use LCAs to "prove" the superiority of one product over another has cast a shadow of disrepute over the concept in some areas [3].

Thus, LCA is a tool to evaluate and analyze the environmental burden of a product throughout all stages of its life. Thermodynamic analysis, on the other hand, is not restricted to a product, but is equally well applied in the analysis of processes. A product and a process are separate entities, but they are related since the purpose of a process is to manufacture a product. Because of this, thermodynamic analysis can easily be used for a product (e.g., lost work per unit weight product, as in the polyethylene (PE) case study, Chapter 11).

DOI: 10.1201/9781003304388-15

The environmental burden covers all impacts on the environment and includes extraction of raw materials, emission of hazardous and toxic materials, land use, and disposal. In certain cases, the analysis only takes into account the burden up to the "gate" of the producing facility and, in other cases, the analysis takes into account the actual disposal of the product. In the former case, the analysis is termed a "cradle-to-gate" analysis, while in the latter case it is referred to as a "cradle-to-grave" analysis.

Thus, LCAs are useful in (1) quantifying environmental impact and comparing various process routes for the same product, (2) comparing improvement options for a given product, (3) designing new products, and (4) choosing between comparable products. In Section 12.2, we outline the general methodology of LCA, and in Section 12.3, we discuss the need to extend LCA to include exergy in the so-called exergetic life cycle analysis. Section 12.4 briefly discusses zero-emission ELCA and gives an example of how this further extension of LCA can account for environmental "friendliness." In Section 12.5, we make some concluding remarks.

12.2 LIFE CYCLE ANALYSIS METHODOLOGY

Broadly speaking, a life cycle analysis consists of the following steps:

1. Goal and scope definition
2. Inventory analysis
3. Impact assessment
4. Interpretation and action

12.2.1 GOAL AND SCOPE

An LCA typically starts by defining the goal and scope of the study. The goal and scope definitions are important since they define the level of detail in the study. The exact question to be answered using the LCA method is formulated (e.g., for product X, two process pathways exist. Which pathway has the least environmental burden?). Too often, this step is given too little time, and the LCA is launched and completed. However, without a proper goal definition, it is impossible to say whether the study was successful [4]. From [5] we can obtain the following definitions.

The goal definition element of an LCA identifies the purpose for the study and its intended application(s). This step will present reasons why the study is being conducted and how the results will be used. Scoping defines the boundaries, assumptions, and limitations of a particular LCA. It defines what activities and impacts are included or excluded and why. . . . Scoping should be attempted before any LCA is conducted to ensure:

- The breadth and depth of analysis are compatible with and sufficient to address the goal of the LCA.
- All boundaries, methodologies, data categories, and assumptions are clearly stated, comprehensible, and visible.

The goal-and-scope definition process is an integral part of any LCA study. At the outset of an LCA, before any data are collected, key decisions must be made regarding the scope and boundaries of the system being studied. These decisions are mainly determined by the goal, i.e., the defined reasons for conducting the study, its intended applications, and the target audience.

The scope definition is similar to the definition of the control volume in the thermodynamic analysis or the battery limits in process design, and for the LCA in terms of space and time (e.g., we follow the use of product X in the process from the raw materials to the time it is disposed by the consumer. Throughout the lifetime of the product, we analyze the environmental burden). The reasons for the study are also clearly defined (e.g., is the study necessary to make a decision about a process?), as well as an answer must be given as to who is performing the study and for whom. Consider the following hypothetical example:

The goal of this LCA study is to examine the environmental burden of using PE film (made with the high-pressure tubular process) for food packaging. The results will be used to improve the environmental performance of the production process by changing, where possible, process parameters. This LCA will not compare the PE film made by the high-pressure tubular process with that made with the high-pressure autoclave process.

The LCA will be performed by an in-house team of engineers and has been commissioned by Food Packaging Inc., which uses the PE film. An overseer from Food Packaging Inc. has been assigned to review the study.

The LCA is performed for Food Packaging Inc., which is based in New Jersey. The data for the study will come from the plant, which is based in New Jersey. The total size of the study is 10 person months, and the bulk of the time will be spent on gathering the data necessary for the study and will not include the ultimate disposal of the food packaging film (cradle-to-gate).

It is clear from this hypothetical example, what is going to be studied, by whom, and who commissioned it.

It is important to note that the LCA is a tool and cannot provide an all-encompassing assessment. One of the reasons is that industrial processes are interconnected globally, so that complete consideration of all these interdependencies is practically impossible. Also, the results of an LCA are approximations and simplifications of cumulative burdens to the environment and of resources used. Therefore, the LCA process does not directly measure actual environmental impact, predict effects, or represent causal linkages with specific effects. As a result, to meet the needs of the study users, it may be necessary to supplement the LCA with other tools or methods to provide a basis for decision making. These tools include risk assessment, site-specific environmental assessment, etc. As a part of the *scoping* process, it is useful to identify where and how these other tools will be used to augment the findings of the LCA [5].

A noteworthy point is that in certain cases, it is convenient to speak in terms of function. For example, the main function of the plastic film is to keep the food fresh for a specific duration of time. In cases where a comparative study is commissioned, and alternatives are required, the function definition is important to clearly state what the product is meant to do, so alternatives can be sought.

12.2.2 INVENTORY ANALYSIS

Probably the most important step in the LCA is the inventory analysis, which carefully documents the (raw) materials necessary for the product, the emissions, required energy input and environmental burden at disposal by either recycle, disposal, etc. Here it is very useful to clearly mark the system boundaries. ISO 14040 [1] defines inventory analysis as follows:

> Inventory analysis involves data collection and calculation procedures to quantify relevant inputs and outputs of a product system. These inputs and outputs may include the use of resources and releases to air, water, and land associated with the system. . . . These data also constitute the input to the life-cycle impact assessment.

Let us take the PE mentioned in the goal as an example. PE is produced from ethylene, as we saw in Chapter 11. The ethylene, on the other hand, comes from a refinery or chemical plant, and the ultimate source of ethylene is, therefore, crude oil or another fossil fuel. Electrical energy is required for the process. There are emissions at the various stages. The final PE is then used to package food. The control volume does not take into account the disposal of the plastic film, since the goal stated that this was a cradle-to-gate analysis.

The key concept of inventory analysis is that all in- and outflows of matter are carefully documented. This inventory is therefore cumulative (see Figure 12.1). For example, an inventory of the PE production can be constructed from plant data (note that compounds and figures are hypothetical in Table 12.1).

For ethylene, similar data could be constructed from either an average of ethylene producers in this geography or on-site naphtha crackers. In certain cases, databases exist for certain commonly used compounds. In other cases, data are unavailable and estimates have to be made. In case of plants that produce more than one product, the total emissions are allocated proportionally to their sales.

12.2.3 IMPACT ASSESSMENT

Once a complete inventory has been generated for the product, the impact can be assessed. According to ISO 14040, various classes of impact exist. These classes can either coincide with known groups such as acidification, global warming, and resource depletion, or be specifically defined to be in line with the goal [6]. Various characterization methods exist for these classes. For example, human toxicity potential (HTP) is a possible measure for human toxicity, whereas acidification potential is a useful measure for acidification. These measures usually allow the transformation of the results into equivalent amounts of a certain compound and allow for the transformation of the total inventory in an environmental profile. Once this environmental profile is complete, the results are normalized with respect to reference information for a certain community at a certain time. Weighting, which is an optional step in the

Emissions

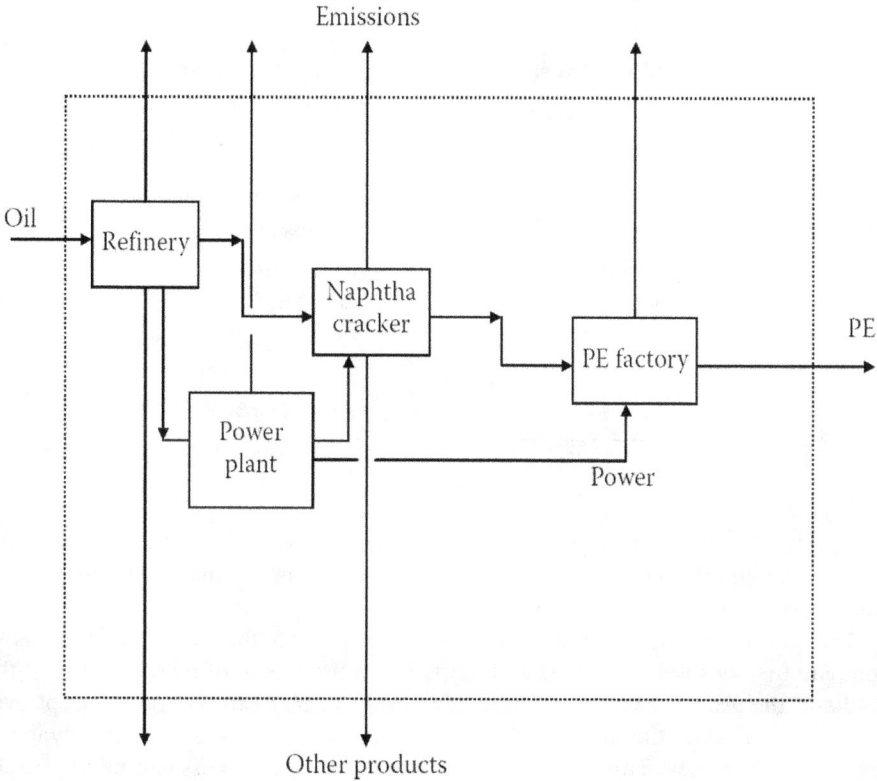

FIGURE 12.1   Example of cumulative inventory.

impact assessment, can be used to highlight certain categories by assigning weight factors.

12.2.4   INTERPRETATION AND ACTION

This step is difficult since the results have to be interpreted, and the interpretation will be subjective. For example, what is more important—the environmental burden of acidification or global warming? To answer these questions, the goal and scope definition must have been given proper care.

12.3   LIFE CYCLE ANALYSIS AND EXERGY

According to Ayres [7], exergy is "that part of the energy embodied in a material, whether it be a fuel, food or other substance, that is available to perform work," to be contained as in a compound (food) or to be dissipated as in a process (refrigeration). As such, it is also a measure for the quality of the energy under discussion. Here, engineering science offers a wonderful common measure of resource quantity and quality that cannot be supplied by either mass or energy. In this way, economists are

TABLE 12.1
Example of Inventory (Hypothetical) of PE Plant.

Polyethylene Production

Source of data	PE plant
Date	January 1, 2002
Geography	North America
Economic input	Ethylene
Economic output	PE
Emissions	
CO_2	0.5 kg/kg PE
Ethylene	12 mg/kg PE

offered "a resource accounting framework covering stocks and flows of, not only, fuels, agricultural and forest products, and other industrial materials, but also of wastes and losses."

The essence of life cycle analysis, states Ayres in another publication [8], is to compare the consolidated inputs and outputs of a process or of a product from "the cradle to the grave." Ayres sees three advantages of using the exergy concept over energy and mass in the analysis. First, it allows the estimation of thermodynamic (exergetic) efficiency. This will tell us to what extent the process can be improved. This improvement will have an immediate impact on the amount of resources employed and waste and other emissions generated. With an efficiency of 10%, the scope for improvement may be much larger than for an efficiency of, let us say, 60%. We refer to Chapter 14 for further details.

The second advantage of "exergy-enhanced" LCA is the fact that it facilitates comparing very different items such as chemical products, utilities such as electricity and heat, and waste: comparing "apples" with "oranges." Monetary units are much less satisfactory because obviously they will be a function of time and of other, often political, factors ("the oil crisis") that in the long term are not significant. Although exergy is not a panacea, it is difficult to find another common measure that in quality and performance comes even close to it. It may look inaccessible as a concept, but it is inconceivable that an economist or politician would have more difficulties with the concept of available work than with concepts such as inflation and deflation. And how reassuring to know that the available work concept is the coproduct of two fundamental laws of science, the first and second law of thermodynamics.

The third advantage, according to Ayres, is the possibility to accomplish year-to-year environmental auditing for large firms, industries, or even nations. The current approach is highly unsatisfactory, with a built-in incompetence to compare flows of different nature: "apples" and "oranges." Exergy takes away this important imperfection.

12.4 ZERO-EMISSION ELCA

One year before Ayres' publications [7,8], Cornelissen [9] completed his PhD dissertation in which he had combined life cycle analysis with exergy analysis. He called this extension of LCA exergetic life cycle analysis. He explained that ELCA should be part of every LCA because the loss via dissipation of exergy is one of the most important parameters to properly assess a process and measure the depletion of natural resources. Cornelissen even went one step further and extended ELCA to what he called zero-emission ELCA. In this extension of ELCA, the exergy required for the abatement of emissions, that is, the removal and reuse of environmentally friendly storage of emissions, is accounted for. Cornelissen illustrated his ideas with examples of the optimization of equipment, heat exchangers, and of laundry machines, to compare various production processes. We will show his results for the comparison of two consumer products: the porcelain mug and the disposable plastic cup.

12.5 EXAMPLE OF A SIMPLE ANALYSIS

Cornelissen [9] decided to compare the LCA of a disposable polystyrene cup with that of a porcelain mug, both often used for coffee in the Western world. The cup is often thrown away after using it once; the mug is usually washed with a detergent. In his analysis, he included nine environmental effects. He assumed that the mug is used about 3,000 times which corresponds to its average lifetime; he assumed that the cup is thrown away after using it once, which is common to do, and so he compared 3,000 plastic cups with one porcelain mug. The nine environmental problems included effects such as the greenhouse effect, depletion of the ozone layer, heavy metals, winter and summer smog, etc. He made a full analysis of both the cup and the mug "from the cradle to the grave." He looked at the raw materials input for both life "cycles," the emissions to air, water, and soil. He concluded that the total environmental impact, according to Eco Indicator 95, of the plastic cup is significantly higher than that of the porcelain mug, nearly three times.

Next, Cornelissen extended the LCA study to include the effect of depletion of natural resources making use of ELCA, the *exergetic* life cycle assessment. In this analysis a full mass and energy balance was made, that is, a first law analysis. Exergy values for all mass and energy streams were included in accordance with Tables 6.1, 6.2, 6.3, 7.1, 7.2, 7.3, and 7.4 in Chapters 6 and 7. This analysis clearly showed where work was available in inputs and outputs and where it was lost. He could show that the cup scored less favorable than the mug in terms of depletion of natural resources (817 MJ vs. 442 MJ).

Finally, Cornelissen performed a *zero-emission* ELCA. In such an analysis the depletion of natural resources is extended to include the real work it takes to reduce the environmental effects to acceptable levels, for example, by 80%–90%. It appeared that in terms of depletion of resources, both life cycles showed an increase of about 20%: 817 → 992 MJ for the plastic cup, 442 → 528 MJ for the porcelain mug. The plastic cup weighs about 4g. Assuming for simplicity that the cup's chemical unit is CH_2 with a molecular weight of 14, then the work available (exergy) in 3,000 plastic cups is $3{,}000 \times 4/14 \times 700 \approx 600$ MJ, assuming that the exergy of CH_2 is about 700 kJ/mol. So the production of 3,000 plastic cups in such a way that it is harmless to the environment is $(600 + 992) = 1{,}592$ MJ. This number is called the *cumulative exergy consumption* (see Chapter 7). Less than 40% (600 MJ) of this consumption is required for the product itself and is conserved in the product, more than 60% (992 MJ) is required for producing it including some 10% (992 – 817) for doing this in an environmentally friendly way, and is lost.

12.6 CONCLUDING REMARKS

In this chapter, we gave a brief outline of LCA, the extension ELCA and zero-emission ELCA. It is beyond the scope of this book to give detailed instructions on how to perform every step, but we hope that the mechanics of the procedure are clear. For more details regarding LCA, we refer to the publications of the Society of Environmental Toxicology and Chemistry and the International Standards Organization. It will be useful to analyze processes using a form of ELCA, since both methods have the potential to quantify the environmental burden. In this case, it would be preferable to use a cumulative exergy analysis to obtain estimates of the exergy required to manufacture a certain product (and of course, also the lost work), and use the life cycle analysis methodology to obtain estimates for greenhouse emissions, for instance. The example given illustrates the usefulness of this combination.

NOTE

1 Recently, terms such as life cycle inventory (LCI), cradle-to-grave-analysis, eco-balancing, and material flow analysis have come into use.

REFERENCES

1. Environmental Management—Life-Cycle Assessment—Principles and Framework. *ISO 14040*, International Standards Organization: Geneva, 1998.
2. Consoli, F.J.; Davis, G.A.; Fava, J.A.; Warren, J.L. *Public Policy Applications of Life-Cycle Assessment*, Allen, D.T. (ed.), Society of Environmental Toxicology and Chemistry (SETAC): Pensacola, FL, 1997.
3. *Life Cycle Analysis and Assessment*. http:/www.gdrc.org/uem/waste/life-cycle.html
4. Weitz, K.; Sharma, A.; Vigon, B.; Price, E.; Norris, G.; Eagan, P.; Owens, W.; Veroutis, A. *Streamlined Life-Cycle Assessment: A Final Report from the SETAC North America, Streamlined LCA Workgroup*, Todd, J.A.; Curran, M.A. (eds.), Society of Environmental Toxicology and Chemistry (SETAC): Pensacola, FL, 1999.
5. Fava, J.; Denison, R.; Jones, B.; Curran, M.A.; Vigon, B.; Selke, S.; Barnum, J. (eds.). *A Technical Framework for Life-Cycle Assessment*, Society of Environmental Toxicology and Chemistry (SETAC): Pensacola, FL, 1991.
6. Barnthouse, L.; Fava, J.; Humphreys, K.; Hunt, R.; Laibson, L.; Noesen, S.; Norris, G. et al. (eds.). *Evolution and Development of the Conceptual Framework and Methodology of Life-Cycle Impact Assessment an Addendum to Life-Cycle Impact Assessment: The State-of-the-Art*, 2nd edn., Society of Environmental Toxicology and Chemistry (SETAC): Pensacola, FL, 1997.
7. Ayres, R.U. Ecothermodynamics, economics and the second law. *Ecological Economics* 1998, *26*, 189–209.
8. Ayres, R.U., Ayres, L.W.; Martinás, K. Energy, waste accounting and life cycle analysis. *Energy* 1998, *23*, 355–363.
9. Cornelissen, R.L. *Thermodynamics and Sustainable Development*. PhD thesis, Twente University, Enschede, 1997.

Part IV

Sustainability

This world provides enough for everybody's need but not enough for everybody's greed.

—**Mahatma Gandhi**

As Chapters 9 through 12 dealt with process efficiency, Chapters 13 through 18 are all closely related to the concept of sustainability. This concept has been defined in Chapter 13 for our society as a whole, and is discussed in Chapter 14 for the chemical industry. Chapter 15 focuses on the need for CO_2 sequestration, that is, its capture, storage, and/or reuse. Chapter 16 deals with the sense and nonsense of green chemistry and fuels. Chapter 17 focuses on solar energy as a renewable, sustainable energy resource and its conversion into other energy carriers. Finally, Chapter 18 discusses many aspects of what may well be the fuel of the future: hydrogen.

DOI: 10.1201/9781003304388-16

13 Sustainable Development

So far, all our contributions and attention have been focused on the thermodynamic efficiency of processes. Theory provided us with thermodynamic tools with which we could establish the difference between the real and the minimum amounts of work required and identify this difference as lost work. We could also show the way to keep these losses to a minimum, given the various constraints to the process: thermodynamic optimization. But operating processes efficiently is not enough; an emerging requirement is that our technology also be sustainable. Most people have an intuitive notion of what sustainability is, but our examples will show that there is a need for substantiation and, if possible, for quantification of this concept. Biology shows us that nature most probably provides us with what is, in our opinion, an elementary and perfect example of sustainability. Similarly, a simple economic analysis points to the essentials of what a sustainable economy is and what this implies for operating our industry. One conclusion seems to be that technology needs to learn some essential lessons from nature. Industry, mainly based on nonrenewable resources, should be transformed into one based on renewable resources. In more general terms, a society, driven by material energy sources with, as a consequence, material emissions, should be transformed into one driven by immaterial sources of energy such as radiation, wind, geothermal energy, and so on, resulting in nonmaterial emissions.

As shown in Chapter 9, solar energy seems to be abundant compared to the energy our society requires and could potentially be a candidate for energy supply. But a closer analysis shows that even here there might be, for the moment, significant limitations.

13.1 SUSTAINABLE DEVELOPMENT

After one of our studies on the efficiency of industrial processes had been completed, the client contacted us again some time later and asked us for a definition of sustainability. More specifically, his industry was probably not sustainable but if it was not, was it close or remote? The client suspected that this was a question with a high thermodynamic content and had therefore contacted us again: "Your discipline should be able to tell us more about it." Now, some years later, our conclusion is that his intuition was partly right: Thermodynamics is of great help in structuring the concept of sustainability and in giving it a sound basis for quantification. But strictly speaking, it is not essential for understanding and defining it, as we will illustrate in the following.

According to the *China Daily* of November 3, 2001, the term "sustainable development" was catapulted into the consciousness of world leaders at the United Nations

DOI: 10.1201/9781003304388-17

FIGURE 13.1   Dr Gro Harlem Brundlandt.

Conference on Environment and Development in Rio de Janeiro in 1992. At its core was an understanding that development, namely, poverty eradication, is integrally tied to keeping the world's natural resources and ecosystems free from pollution and degradation. Agenda 21, called by some the blueprint for sustainable development, is a collection of commitments by governments on climate change, biological diversity, and forestry principles. In 2002, South Africa hosted the first world summit on sustainable development to assess and evaluate progress made in the 10 years since the Rio Declaration.

Of course, the term "sustainable development" was first launched with success in a United Nations report called "Our Common Future" [1], better known as "The Brundtland Report" after the chairperson of the commission that produced it (see Figure 13.1). In this report, sustainable development is defined as a social development required to satisfy the needs of present generations without putting at risk the needs of future generations.

Trilemma: Three Major Problems Threatening World Survival [2] points out that amid the explosive rate at which the world population is increasing, from 5 billion people in 1995 to an estimated 9–10 billion people in 2040 [3], mankind is faced with a triad of serious problems: an unprecedented economic growth; consumption of energy and resources; and all this in trying for conservation of the environment. In the authors' words, the world is facing a formidable trilemma that can only be obviated by what they call a sustainable development. Such a development can be accomplished by a serious multidisciplinary effort in which science, technology, sociology, and economy seem as yet to be the most prominent disciplines.[1]

13.1.1 THREE VIEWS

Harmsen [4] points out that there are several views with which one can look at sustainable development. One view is the *theocratic* view, prominent in Judeo-Christian religion in which the Creator stands central and needs to be honored and served by man. Nature is trusted to man, who should take care of it, while technology should help him and simplify his task. Man assumes this responsibility, much as a steward of property and capital does. This is the same role as the one many responsible company directors play: The interest of the company should be served and is central in one's performance and behavior toward the company. This view is to some extent expressed in a moving way by the text on the statue of the Japanese-American environmentalist Yasui Minoru in Denver's Sakura Center: "We are all put on this earth to leave it a better place for our having been there."

In the *ecological* view nature is central. By nature is meant all living systems, the climate, and natural resources. Human actions and activity should preserve nature's integrity, stability, and beauty. Technology should be embedded in nature. Exchange of materials should take place in closed cycles, whose conditions are so severe that it seems that only materials of plant origin can fulfill it. This view can be recognized in Yoda's nearly desperate words when he states in *Trilemma* that "Nature has lost its ability to cleanse itself."

In the *anthropological* view, man is central, which is best illustrated with the popularized definition of sustainable development from the earlier cited Brundtland Report [1]: "fulfilling the needs of the present generations without sacrificing the needs of future generations." A very "down-to-earth" definition is given by Okkerse and van Bekkum [3] in their chapter "Towards a Plant-Based Economy": "Feed double the number of people, provide them with energy and materials, let them live according to the requirements of a developed society and do not pollute the earth nor change the climate."

13.1.2 SOME OTHER VIEWS

It is interesting to read what "captains of industry," such as the leaders of prominent multinational companies, have to say about sustainable development. Sir Marc Moody-Stuart, former chairman of the Committee of Managing Directors of the Royal Dutch/Shell Group of Companies, stated in 1999 in Ref. [5] what sustainable development meant for his company: "balancing our own legitimate commercial interests with the wider need to protect and enhance the environment and contribute to social progress and stability."

Tabaksblatt, as former chairman of Unilever, emphasized in a speech [6] that within his company, "No business or policy decision can be taken before it has been tested on its impact on sustainable development."

We conclude this brief review with what has been perhaps the most rigorous and strict definition of sustainable development so far. It has been given by Dr. Angela Merkel, a physicist from Leipzig University, and present Chancellor of Germany (2009), while she was German Minister of the Environment, Nature Conservation and Nuclear Safety. Discussing "The role of science in sustainable development," she states in *Science* [7], "Sustainable development seeks to reconcile environmental protection and development; it means nothing more than using resources no faster than they can regenerate themselves, and releasing pollutants to no greater extent than natural resources can assimilate them." This definition is, in our opinion, of dramatic dimension and consequence while also nonrealistic, because of the impossibility to effectively adopt it in the short term. When prestigious scientists, as the American economist Ayres [8], cofounder of Industrial Ecology, and Japanese economist Takamitsu [9], appeared to use the terms *metabolic society* and *industrial metabolism*, it seemed to us wise to turn to some prominent books on biochemistry [10] and bioenergetics [11].

13.2   NATURE AS AN EXAMPLE OF SUSTAINABILITY

Industry and our industrial society lean heavily on material resources, raw materials, and energy or fuels. Most of these resources have a limited abundance; we consume them and do so without regeneration. From this supply basis, we produce electricity, mechanical energy, heat, chemicals, and other specified materials, sometimes of great purity or complexity. *Nature*, in particular or more specifically living or animate systems, does the same but in a remarkably different way that appears to send us a clue for what we should understand by sustainability. A closer analysis shows that man and animals do not seem to be essential for the maintenance of sustained life, but plants and microorganisms are. Man and animals are ultimately dependent on plants. Plants and microorganisms are essential for life and cannot live without each other. They live on air, water, minerals, and sunlight,[2] and based on these they make complex compounds necessary to sustain life. Energy can be converted into chemical, mechanical, and electrical energy and work for transportation and concentration of matter and for transporting signals through the nerve system. At the synthesis side, photosynthesis is the start of building a complex material system. At the respiration side, oxygen breaks down complex molecules to transfer chemical energy to all other forms of energy to sustain life. Ultimately, all originally absorbed solar energy will be transformed into heat that is radiated into the universe. Figure 13.2 clearly illustrates what we have just described. It pictures in a primitive and limited simplification how matter is cycled in a closed system of the biosphere and is neither absorbed nor emitted. The biosphere acts as an open system, however, for energy, taking in solar energy and emitting this energy in the degraded form of heat into the universe. Solar energy absorbed is, in contrast to fossil fuel, an immaterial source of energy, and after this has permeated through the biosphere, it is emitted as an immaterial form of energy: heat. Meanwhile the biosphere extracts this solar energy partly from its useful part for conversion into many forms of energy. The essence is that matter is recycled in this "cycle of life," without net material intake or output, partly stripping incident solar energy from its available work. Only a very small fraction (<1%) of the incident solar energy is utilized in this way. The major part of the incident energy is directly emitted as heat. Ultimately, the minor part will also be emitted as heat after it has been degraded via the living systems. So there is a strictly unidirectional flow of energy from the sun to the earth and into the universe. Only a very small part of this flow will be delayed via a detour through living systems.

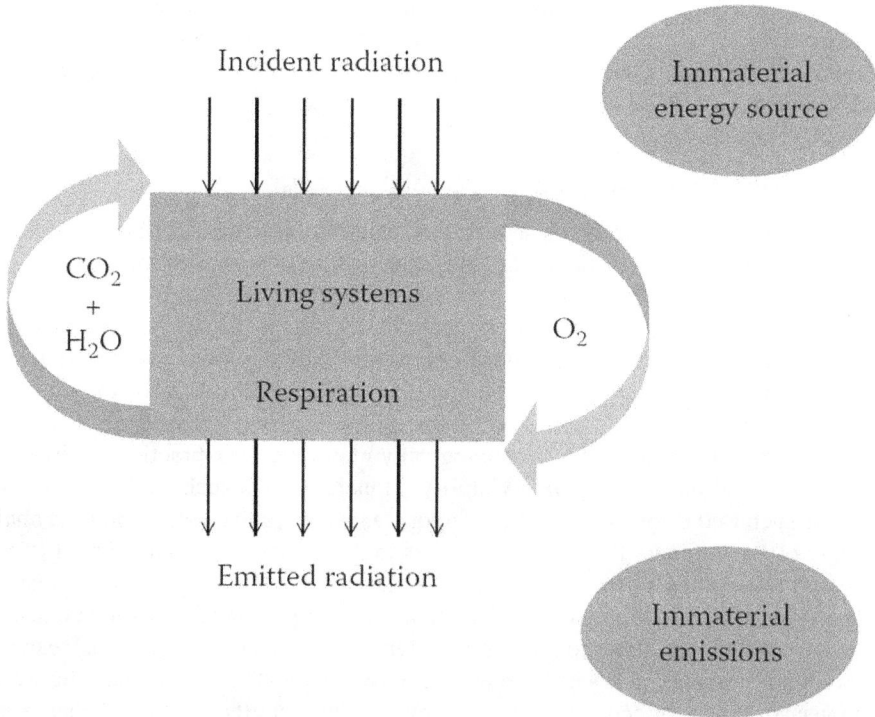

FIGURE 13.2 Cycle of life.

As for the microorganisms, they appear to be vital in the degradation process. Nature does not seem to know waste, as some microorganism will always take care of it and lead it back to CO_2 and water or other building stones for the more complex molecules of life. Microorganisms are by far the most abundant form of life; in fact, nearly all life is made up of them. On one hand, they are able to fixate CO_2 and form oxygen in the process of synthesis; on the other hand, they are elementary in the degradation of living matter to the elementary building blocks.

The general conclusion of this analysis seems to be that the *biological* definition of sustainability is that use should be made of an immaterial permanent source of energy, like solar energy or the heat of the inner earth, and that matter should move in closed cycles.

13.3 A SUSTAINABLE ECONOMIC SYSTEM

Economics is the science of economy, economy being "the management of a household or state, its income and expenditures," and more particularly, "the *careful* management of its wealth and resources" [13]. Ecology is concerned with the relations between living organisms and their environment. An important aspect of these relations is the exchange and conversion of energy and matter. And so, it is not too surprising that many consider thermodynamics, being the ultimate science of the transformation of energy and matter and always dealing with the exchange between systems *and* their environment, as essential for inclusion into any economic and ecological analysis.

13.3.1 Thermodynamics, Economics, and Ecology

In appearance, thermodynamics seems to be nothing more or less than a nice collection of abstract mathematical relations between the properties of matter valid for the various states in which this matter may prevail. It becomes more substantial when thermodynamics is applied, as in process technology. The extent to which one form of energy (e.g., heat) can be converted into another (e.g., work) or to which one form of matter (e.g., methane) can be converted into another form of matter (e.g., methanol or hydrogen) is traditionally governed by thermodynamics. But even if such conversions appear to be "technologically" feasible, their practical realization may still depend on the *economic* viability. Monetary units such as the dollar and concepts such as the cost of production factors (e.g., labor and capital) enter the analysis and often dominate the outcome. Interestingly enough, labor and capital relate more and more often to the environment: What is the cost to repair the burden of technology on the environment? More sophisticated, the following question is raised: What part of the cost relates to the cost of natural capital such as forest, the ocean, or the soil and of natural services such as waste assimilation? So the relation between and interwovenness of economy and ecology are clear, but the relation of these disciplines with thermodynamics is less obvious. However, as we will see below, when analyzing economic systems and their interaction with the environment, it becomes clear that economic systems, whatever their details, appear to be driven in the same direction as the industrial or agricultural processes that constitute them, that is, the direction prescribed by the second law, namely, the direction of entropy production, the formation of waste, or the dissipation of useful work. In the subsequent sections, we endeavor to illustrate how close the relationship often is among thermodynamics, economics, and ecology.

This "ecothermodynamic approach," leads to the same definition of sustainability as the biological definition given above and stemming from the cycle of life—it comes from *economics*. In the view of classical economics there is an "equilibrium" among production, consumption, and capital (Figure 13.3). Necessarily this equilibrium has a dynamic nature, much as the equilibrium in thermodynamics where on a macroscopic scale, there may be "rest" but where on a microscopic or rather molecular scale, there is "unrest." An economic system, however, much like thermodynamic systems, interacts with its environment. The first and second laws govern the evolution of a system in its environment toward equilibrium and can account for the nonequilibrium dynamic state, as in living systems, and its steady state. An economic system, much like a thermodynamic system, also has an environment, and, as has become clear in recent times, this environment is of global dimensions. The economic system is driven by a finite amount of resources. Similarly, the economic system has its by-products—waste—and the environment has no infinite absorptive capacity for this waste. So supply of resources and absorption of waste cannot be

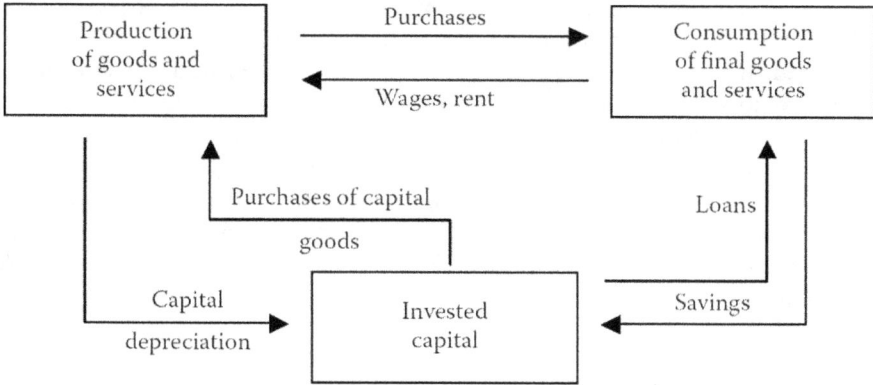

FIGURE 13.3   Classical economics.

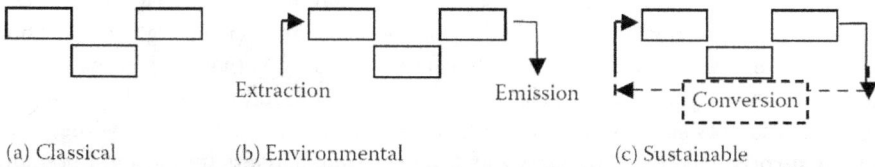

FIGURE 13.4   Schematic description of economic systems, following Figure 13.3: (a) Classical, (b) environmental, and (c) sustainable economic systems.

taken for granted, and the classical picture of an economic system, in Figure 13.3 (or Figure 13.4a), has to be extended as in Figure 13.4b. This picture is characteristic for so-called environmental economics or, even better, *ecological economics*.

Ecological economists thus recognize features of the economic system that are most common in thermodynamics where the main laws express the prevailing inter-action between the system under consideration and its environment. The extent to which matter and energy are exchanged dominates this representation. An economic system, therefore, much resembles a dynamic nonequilibrium system as discussed in the last section, much like living systems. Food sustains the living system; material resources and energy sustain the economic system. It is for this reason that some economists [14,15] state that the production factors, labor and capital, are outdated and archaic concepts and that the real production factors are matter and energy, or even better exergy and information, serving man, machine, and nature.

Figure 13.4b emphasizes the finite nature and strong irreversibility of an eco-nomic system. The stock of energy and resources will eventually run out and so will the absorptive capacity of the environment for waste. An obvious extension of Figure 13.4b, therefore, is the one represented by Figure 13.4c. Just like in nature, waste has to be recycled. In nature, there is no real waste. Every form of waste is a resource for a living system. This living system is very small and called a *microbe*. Microbes make sure that all matter recycles in nature. Man needs to assume this humble but valuable and important role of microbes in the economic system and make sure that the material cycles get closed. Therefore, energy (or rather work) is required. But obviously this work should not be supplied from a nonrenewable source, like fossil fuels, but rather from a renewable source like the sun. Figure 13.4c therefore seems to be characteristic for a sustainable economic system and agrees remarkably with the definition of sustainability from biological systems: A nonequilibrium dynamic system is sustained through the cycling of matter in the system and its environment, driven by the supply of an immaterial, renewable energy source. For the definition of what should be understood by a (non-)renewable source, we refer to Lems et al. [16] and Section 13.6 in this chapter.

13.3.2 ECONOMICS AND ECOLOGY

The authors of this book are all engineers and do not claim much more than the layman's understanding of both economics and ecology. In addition, they rely on some distinct authors and their publications [15,17–20] and, not in the least, on their own common sense, which has developed in many years of engineering practice, both in industry as well as in academia. Their understanding is that economics is about the production of goods and services (Figure 13.3), production realized by so-called production factors such as capital and labor, and about monetary flows while, much to their surprise, the flows of energy and matter are much less important. Economic activities are captured in more or less precise models. Conventional economists have a strong belief in price and market mechanisms. All economic activities are expressed in monetary flows. Economic performance and growth are measured, in monetary terms, at current prices in the so-called gross national or domestic product (GNP or GDP), which acts as a macroeconomic indicator and is considered to be a universal yardstick. Conventional economists seem to have great optimism about resource supply and absorptive capacity of the environment and an infinite faith in human ingenuity to solve problems whenever they occur and whatever their nature, for example, environmental problems or questions about the so-called carrying capacity of this planet.

As mentioned earlier, environmental or ecological economists disagree largely with this picture and attitude and insist on some major adjustments of prevailing economic ideas. For one thing, ecological economists consider economic activity constrained by the regenerative capacity of the ecosphere, namely, that part of the planet containing living organisms. The economy is contained by the ecosphere [20] and as such is dependent on the services supplied by this ecosphere (Figure 13.4b). However, market prices mask the role and contribution of these services and the natural capital that supplies these services. Services rendered by natural capital are not accounted for and thus, in our monetary world, not valued. The price of a depleting resource, like oil, may go down to less than $10, as happened a few years ago, which does not account for oil's longer-term scarcity, but at most for its current temporary abundance, or go up to $150 as in 2008 from fear of the rising needs of countries like China. Equally, market and price mechanisms do not reflect ecofunctional scarcity, rather short-term demand. In the sharp words of Georgescu-Roegen [21]:

The savage deforestation which at one time menaced all the woods in the world was the result of the fact that the prices were *right*. And it was not brought to a halt by the price mechanisms, but only by the introduction of some quantitative restrictive rules.

And if the cheetah is now an endangered species, it is because the price of a cheetah pelt is just right for some people to hunt one animal after the other.

(p. 17)

The case of sustainable development or, as some prefer, the development toward sustainability is illustrative for the general observation that there is a conflict between short-term and long-term values. To quote the famous sociobiologist Wilson [17],

To select values for the near future of one's area of responsibility is relatively easy; to select values for the distant future of the whole planet is relatively easy too—in theory at least. To combine the two visions to create a universal environmental ethic is very difficult. *But combine them, we must.*

(pp. 65–75)

Wilson claims that environmentalism is much more than a special-interest hobby. Its essence has been defined by science. The earth is not in physical equilibrium; its non-equilibrium condition needs to be sustained with the help of an extraneous energy source that is driving numerous cycles of matter and energy, emulating each other in complexity. The fragility of these self-organized cycles is for many of greater concern than the finite capacity of the environment to supply resources or absorb waste [15]. A major understanding of this biosphere is required for many purposes. One such purpose is the proper analysis of economic activities. Apart from establishing the ultimate carrying capacity of this planet (10 or 16 billion people?), it is necessary to include the cost of ecosystem services in any economic analysis. After all, such an analysis is aimed not only at helping us to sustain our own prosperity in the near future but also at helping our offspring to achieve prosperity in the distant future, one of the main goals defined in the Brundtland report [1].

One cannot help but reflect man's position in nature. Up to a certain period in the past, man fitted nicely in nature's cycles and did not disturb its course. Then man stepped out of the prevailing cycles and, while entering the Industrial Revolution, may have started to disturb this course. Is it naive to conclude for the moment that the only way to repair this situation is to become part again of nature's cycles?

Another important objection to conventional economic wisdom is monetary appraisal as in project appraisal or evaluation. There is a prejudice to the future by discounting; the present value of future benefits may be exceeded by the immediate short-term benefits [20]. Application of what some have called a revolution in investment analysis (i.e. investment under uncertainty [22]) may alter this, for example, by accounting for possible increased scarcity of ecological services [19]. Discounting makes nature appear less valuable, the farther we look into the future, although future generations may need at least the same amount of ecological goods and services we need today, whatever the discount rate may be. Should we not account for this whenever natural capital (a piece of land) is transformed into human-made capital (a shopping center) [20]?

So expressing economic activities in terms of monetary flows has a number of serious disadvantages, the main ones being that goods and services from natural capital are ignored and that it is, for the moment, impossible to express these contributions in terms of money. But let us first look at the components of these resources and estimate their financial size.

13.3.3 NATURE'S CAPITAL AND SERVICES

In 1997 Costanza et al. [19] published an article that came as a revelation. Their starting point was the concern that ecosystem services are not fully "captured" in commercial markets or well compared with manufactured capital and economic services and thus given little weight in policy decisions. They studied 16 marine and terrestrial biotic communities such as oceans, deserts, and forests and tried to estimate the current economic value of 17 renewable ecosystem services. Services comprise goods, such as food, and services, such as waste assimilation. These services, which consist of flows of materials, energy, and information, as contained in natural capital stocks, and albeit not visible in GNP or balance sheets, are combined with manmade goods and services to produce human welfare. Major categories of ecosystem services considered were food production, nutrient recycling, waste treatment, and recreation, among others. Estimating the global gross national product to be around $18 trillion (= 18×10^{12}), the authors estimate the minimum current economic value of these ecosystem services to be in the range of $16–$54 trillion per year. The authors conclude that if one were to try to replace these services—supposing this is possible, which of course it is not—the global GNP would nearly triple without any increase in welfare. If ecosystem services were paid for, the global price system would be very different, in particular for products using these services. Another conclusion is that although economic welfare has significantly increased in terms of conventional GNP, correcting for ecosystem services shows that not much improvement has taken place since 1970. Finally, the authors point out that the impact that such estimates will make, if they are included in project appraisals, is considerable. It would take away the earlier objection against discounting procedures in economic analysis, as these are biased against the future. In this new analysis, natural capital and ecosystem services are likely to become more scarce in the future and thus more valuable rather than less valuable.

13.3.4 ADJUSTMENT OF THE GROSS NATIONAL PRODUCT

There have been some prudent other efforts to make a start with correcting the GNP for factors such as the deterioration of the quality of life and the natural environment. In the Green GNP, deductions have been introduced for goods and services required for environmental protection [2]. Such corrections can easily lead to a 10% reduction in the GNP or a 50% reduction in economic growth. In Japan, the Economic Planning Agency (EPA) formulated a new economic indicator called National Net Welfare (NNW). The EPA's estimates were that the NNW is 12% less than Japan's GNP [2].

Nevertheless, the Japanese question how to quantify environmental damage and how to translate this into monetary value. It is obvious that a global approach is required; a country taking measures on its own may be easily "out-competed" by countries that do not take measures at all. Other aspects are that pollution (e.g., in the form of waste) can be "exported" and productive land "imported" (e.g., citrus fruit imports), which does not add to the solidarity that the world's citizens must eventually develop. We would like to conclude this section with a reference to a book excerpt that has appeared in the February 2002 issue of *Scientific American*. The book, *The Future of Life*, is written by Wilson, whom we cited at the end of the last section [17].

13.3.5 Intermezzo: Thermodynamics and Economics—A Daring Comparison and Analogy

In Chapter 5, we discussed the maximum power that can be obtained from a heat engine driven by heat from a source at a rate of $\dot{Q}_{in}$ J/s, rejecting heat to the environment at rate $\dot{Q}_{out}$ and producing work at a rate of $\dot{W}_{out}$. A simplified scheme is given after De Vos [23] in Figure 13.5 and shows that the engine is assumed to run *endoreversibly*, by which we mean that all irreversibilities are assumed to be concentrated in the interaction between the engine and its environment (i.e., when heat enters or leaves the engine). Under these assumptions, a value is found for the maximum power $\dot{W}_{out}^{max}$:

$$\dot{W}_{out}^{max} = \dot{Q}_{in}\left(1 - \sqrt{\frac{T_0}{T_1}}\right) \tag{13.1}$$

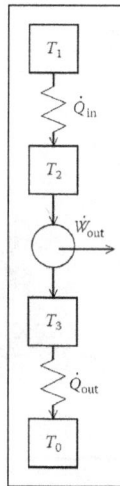

FIGURE 13.5 The endoreversible heat engine; irreversibilities are assumed to take place only in the engine's exchange of heat with the environment. (Taken from De Vos, A., *Energy Convers. Manage.*, 36, 1, 1995.)

instead of the Carnot value

$$\dot{W}_{out}^{max} = \dot{Q}_{in}\left(1-\frac{T_0}{T_1}\right) \tag{13.2}$$

Due to the irreversible heat exchange between the four heat reservoirs (for $T_0 = 300$ K and $T_1 = 600$ K), the real maximum power is found to be close to a value of 0.3 rather than the ideal value of 0.5. The maximum power is found at an optimal flow rate $\dot{Q}_{in}^{opt}$ corresponding to an optimal set of temperatures T_2 and T_3, satisfying the Carnot relation

$$\dot{W}_{out}^{max} = \dot{Q}_{in}^{opt}\left(1-\frac{T_3}{T_2}\right) \tag{13.3}$$

Figure 13.6 schematically gives $\dot{W}_{out}$ as a function of $\dot{Q}_{in}$ and at the same time, the corresponding thermodynamic efficiency $\eta = \left(\dot{W}_{out}/\dot{Q}_{in}\right)$ has its highest value, the Carnot value, for an infinitely slow operation of the engine at a zero heat input

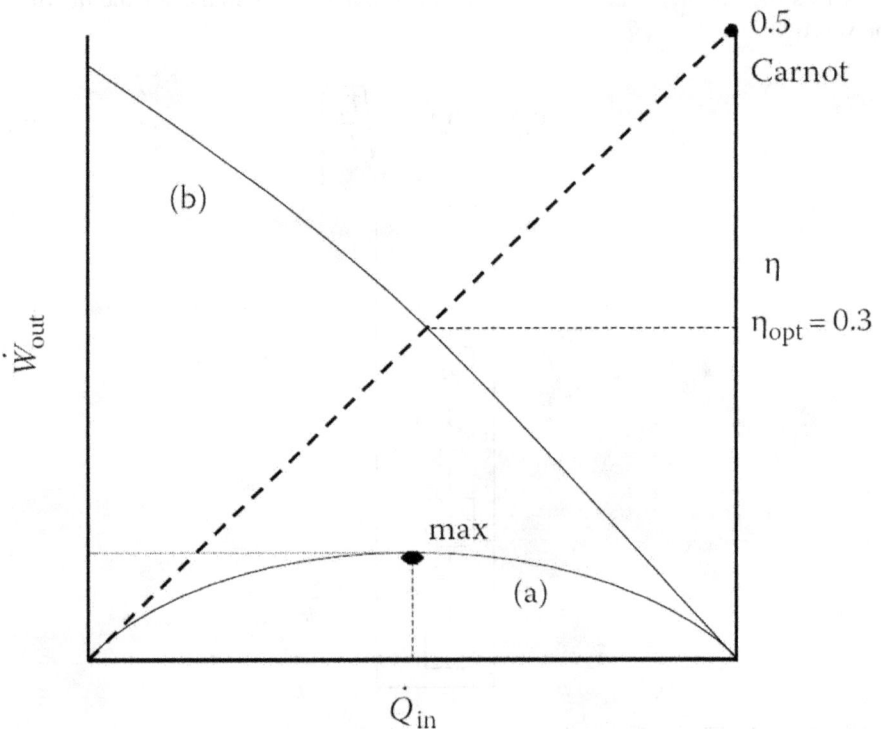

FIGURE 13.6 Power output $\dot{W}_{out}$ (curve a) and thermodynamic efficiency η (curve b) as a function of heat input rate $\dot{Q}_{in}$ for a ratio of $T_0/T_1 = 0.5$.

rate $\dot{Q}$ but also at zero power output. Note that then $T_2 \rightarrow T_1$ and $T_3 \rightarrow T_0$. The thermodynamic efficiency η is zero when $\dot{W}_{out}$ is zero, but now at the maximum possible heat input rate that the engine can absorb, namely, when $T_2 = T_3$. Somewhere between these extremes the power output is at a maximum value for which the heat input rate and the efficiency are at optimal values.

In short, this simplified thermodynamic analysis for the operation of a power station shows that, to a certain approximation, there is an optimal rate of heat input for which the power output is maximum. The thermal efficiency in this mode of operation is lower than the Carnot efficiency.

Next we use a similar approach to show further, after De Vos [24], that this *thermodynamic* optimum is not necessarily the *economic* optimum. After all, the higher the delivered power $\dot{W}_{out}$, the faster the investments for building the power plant will be returned, but the higher the efficiency η, the better the expenses for processing the primary fuel will be recovered. So, clearly, the economic optimum for the operation of the plant is in between the point for maximum power and the point for maximum efficiency, or

$$0 < \dot{Q}_{in,econ}^{opt} < \dot{Q}_{in,therm}^{opt} \tag{13.4}$$

De Vos assumes that the rate of running costs of the plant, $\dot{C}$, is made up of two parts. One part is the capital cost, which is assumed to be proportional to the investment and therefore proportional to the size of the plant. The other part consists of the fuel cost and is therefore proportional to the heat flow rate $\dot{Q}_{in}$. It is then assumed that $\dot{Q}_{in}^{max}$ is an appropriate measure for the size of the plant and thus

$$\dot{C} = a\dot{Q}_{in}^{max} + b\dot{Q}_{in} \tag{13.5}$$

a and b are in economic units (ecu) per Joule, ecu/J. The quantity to be maximized is now

$$p = \frac{\dot{W}_{out}}{\dot{C}} \tag{13.6}$$

in which p is the profit in terms of Joules produced per unit of money spent. De Vos then derives a relation between η^{opt} and the fraction of total costs $\dot{C}$ incurred on fuel, the so-called fractional fuel cost f defined as

$$f = \frac{b\dot{Q}_{in}}{a\dot{Q}_{in}^{max} + b\dot{Q}_{in}} \tag{13.7}$$

Table 13.1 gives the fractional fuel cost f for various energy sources in Belgium, De Vos's home country. They are, of course, a function of the technology applied.

The result is given in Figure 13.7 for an assumed ratio of $T_0/T_1 = 0.5$, that is, an ordinary power station. For a renewable fuel f is close to 0 and the optimal efficiency

TABLE 13.1

Fractional Fuel Cost f for Various Energy Sources.

Fuel	f
Renewable	0
Uranium	0.25
Coal	0.35
Gas	0.5

Sources: De Vos, A., *Energy Convers. Manage.*, 36, 1, 1995; De Vos, A., *Energy Convers. Manage.*, 38, 311, 1997.

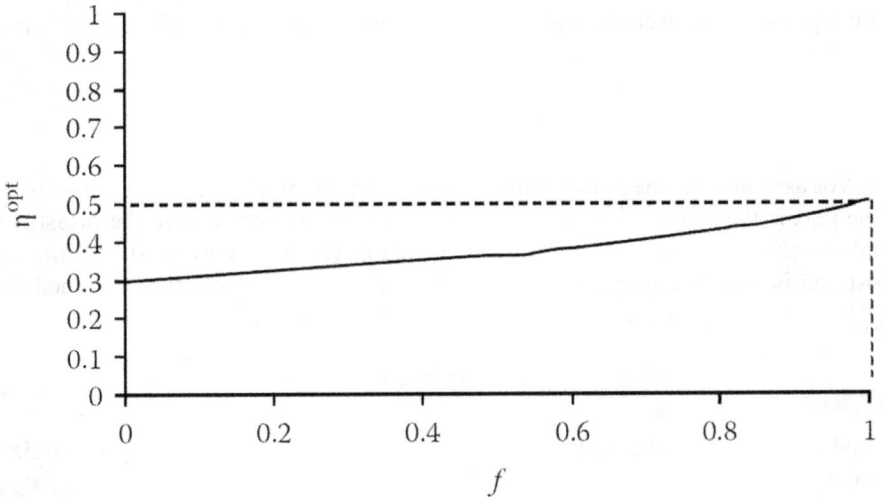

FIGURE 13.7   Optimal efficiency as a function of fractional fuel cost f for $T_0/T_1 = 0.5$.

is 0.3 as opposed to the Carnot value of 0.5. For a costly nonrenewable fuel such as natural gas with $f = 0.5$ (i.e., 50% of all costs are spent on fuel), the optimal efficiency is 0.35, so around 15% better, although possible environmental costs related to the emission of waste have been ignored (although we should not exclude environmental costs for renewable fuel beforehand). Figure 13.8 depicts how the situation improves when T_1, the temperature of the heat source, increases. Nevertheless, the trend that nonrenewable fuels are more favorable appears to persist, however, under the same restriction as just mentioned. By the way, this optimum, which we call the *economic* optimum, is also known as the *thermoeconomic* optimum and the analysis with which it was obtained is known as *thermoeconomic* analysis.

Finally, De Vos makes the daring step from a heat engine that produces work (as in the thermodynamic analysis) or money (as in the thermoeconomic analysis) to an

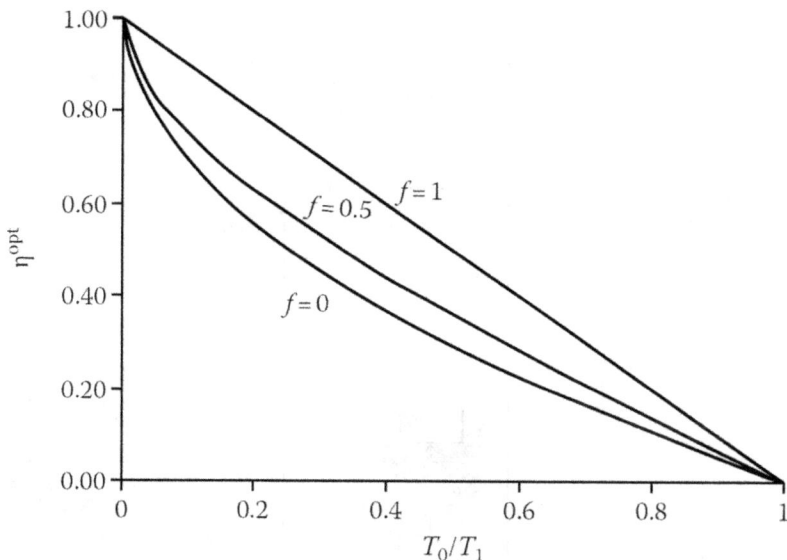

FIGURE 13.8 Optimal efficiency as a function of the temperature ratio T_0/T_1.

economic engine that produces tax revenues [24]. Figure 13.9 pictures the essentials of this engine. V_1 and V_0 are the high- and low-value reservoirs, respectively. Goods flow upward at rates $\dot{N}$, from V_0 to V_1, whereas money flows downward at rates $\dot{N}V$ from customer to salesperson. V_2 and V_3 are values in exchange or prices paid by the consumer and received by the supplier, respectively. De Vos introduces two axioms equivalent to the two laws of thermodynamics: conservation of matter and conservation of money.

A tax rate η is introduced according to

$$\eta = \frac{\dot{W}_{out}}{\dot{N}_2 V_3} \tag{13.8}$$

which, applying the axioms, appears to be

$$\eta = \frac{V_2 - V_3}{V_3} \tag{13.9}$$

Under reversible conditions, η equals the Carnot value

$$\eta_C = \frac{V_1 - V_0}{V_0} \tag{13.10}$$

Reversible conditions imply that economic activity takes place at an infinitely slow rate; thus, $\dot{N} = 0$ and $\dot{W}_{out} = 0$, but η has the maximum value of η_C. Applying the

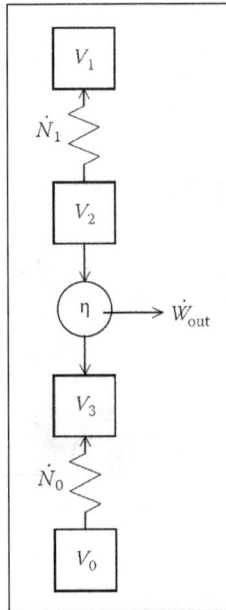

FIGURE 13.9   The money engine. V stands for value, $\dot{N}$ for flow rates of goods, η for tax rate, and $\dot{W}_{out}$ for the rate of tax revenue. Arrows pointing up represent flow rates of goods, and arrows pointing down represent flow of money. (From De Vos, A., *Energy Convers. Manage.*, 38, 311, 1997.)

same techniques as earlier for the heat engine, an optimal rate of goods $\dot{N}^{opt}$ can be determined for which the rate of tax revenues is at its maximum value. In conclusion, De Vos has shown that economics and thermodynamics, under certain assumptions, show a remarkable analogy in which the indirect tax revenue rate of the economic engine plays the same role as the power produced by a heat engine!

13.4   TOWARD A SOLAR-FUELED SOCIETY: A THERMODYNAMIC PERSPECTIVE

If we adopt the model for sustainability as we have developed in the previous two sections, then one obvious necessity is to transform a mainly fossil-fueled society into a society fueled by a more sustainable fuel such as solar energy. Solar energy is abundant and "renewable" on a time scale of millions of years. Its availability, estimated at 2.8×10^6 ExaJoules/annum (EJ/a = 10^{18} J/a) exceeds by far the needs of our society, 800 EJ/a even in 2040 (Table 13.2). In recent publications of Shell International [25], the world's energy needs are estimated to be a factor 2.5–3 times larger in the year 2050 than in the year 2000. Not all of the 2.8×10^6 EJ/a is accessible; estimates are that "only" 300–3,000 EJ/a or 0.01%–0.1% can be captured.

Fossil fuel, like oil, coal, and gas, which takes care of more than 90% of industry's energy needs, is not renewable on this time scale, at least if we adhere to the

TABLE 13.2
Situation Analysis 1995 and Prognosis for 2040.

Critical Global Data	1995	2040
Population	5×10^9	9×10^9
Energy consumption	350 EJ	800 EJ
Energy consumption/cap	2200 W	3000 W
Agricultural land	3.4×10^9 ha	2.8×10^9 ha
Organic materials	300×10^6 ha	1000×10^6 ha

Sources: Okkerse, C. and van Bekkum, H., *Starch 96—The Book, Carbohydrate Research Foundation,* Zestec, The Hague, the Netherlands, 1997; Shell Energy Scenarios 2050, 2008.

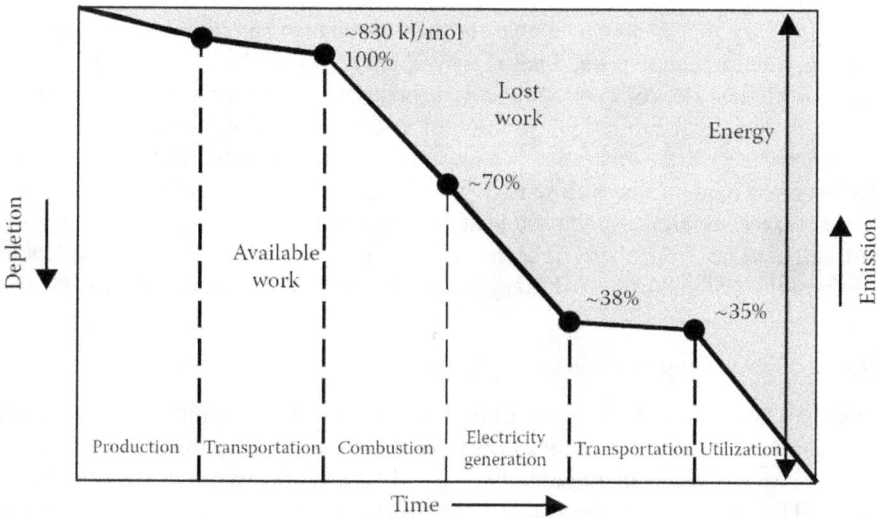

FIGURE 13.10   The fate of work from natural gas.

definition that a fuel is not renewable when its consumption rate exceeds its production rate. With fossil fuel so prominent in our industrial society, let us make a thermodynamic analysis of a fossil-fueled power station.

13.4.1   THERMODYNAMIC ANALYSIS OF A POWER STATION

Our choice of fuel is natural gas, for reasons of simplicity taken as pure methane. Natural gas has been called by some the fuel of the twenty-first century because of its abundance, relative purity, and low CO_2 production per unit of electricity. Figure 13.10 depicts the fate of the work available in natural gas, its exergy, from source to sink (i.e., all the way from the gas reservoir to the environment into which its emissions of heat and combustion gases are released). The production of natural gas requires exergy and so does its transportation to the power station. The exergy of methane is about 831 kJ/mol, so strictly speaking less work is available for the power station due to the incurred exergy cost of production and transportation. Our reference should be natural gas as it is present and at rest in our environment. Nevertheless, for the discussion to follow and for the observations to be made, our starting point will be methane, at P_0 and T_0, at the power station. On our way to electricity, the first loss incurred is that of the spontaneous combustion of methane. From earlier chapters we know that close to 30% of methane's exergy is lost in this process, due to the reaction velocity and affinity, in other words, the strictly irreversible nature of this spontaneous process (see Chapter 4). Then the power station transforms the remaining 70% of the original exergy into electricity, and we assume for the *thermodynamic*, exergetic, efficiency of this conversion process 50%, which compares to a *thermal* efficiency of 55% for some of the best power stations in the world. Next, we consider the loss for transportation of the produced electricity to the consumer and estimate this loss at about 6%. So we arrive at a number close to one-third of the original exergy that is now available for the customer to run his or her refrigerator, washing machine, and so on. That is why Yoda [2] in his book *Trilemma* writes a chapter with the title: "66% of our energy is being wasted." It is indeed shocking that due to our technical limitations we are not able yet to explore the exergy available for more than 35%. At this point, we must draw attention to the fact that instead of 35% we often read numbers close to 50%. This is a consequence of the fact that such sources take as reference point the heat available in the combustion gases and start their calculations from there, ignoring the exergy loss incurred in the combustion step from the original chemical exergy of methane to the products of its combustion.

13.4.2   SOME OBSERVATIONS

Let us have another look at Figure 13.10. It allows us to make a number of interesting observations. First of all, it is an illustration of the first law. By introducing 1 mol of methane, we introduce an amount of energy, chemical energy, to the amount of some 830 kJ. This amount is fully available for work, but in the process that follows we are not able to recover all this energy for work. Instead, this *available work* is partly converted into work dissipated, *lost work*, apparent as heat transferred to the environment at T_0 and P_0. However, the sum of available work and lost work remains constant, the law of conservation of energy. This is what Baehr [26] formulated as, "The sum of Anergy and Exergy is constant." Second, the curve depicting the decrease in available work during the process is an expression of the second law, showing that the quality of energy (see Chapter 6) is decreasing over the process in the direction of the process. Baehr's definition of the second law was, therefore, "Exergy can always be converted into Anergy. Anergy can never be converted into Exergy."

A third observation is that the use of a fossil fuel contributes to the *depletion* of our resources, as their supply is finite. Because this energy source is a material source of energy, embodied in compounds mainly made up of carbon and hydrogen, we are bound to material *emissions*, of which CO_2 is the most prominent product. After our excursion in Section 13.2 to biology, it strikes us that the material flows of emission and extraction do not participate in closed cycles, at least not to the extent that regeneration can keep up with consumption. Last but not least, the observation can be made that, for the *purpose* of the power station, which is to produce electrical energy, strictly speaking, no material source of energy is required. Our conventional way to produce electricity is to use a material fuel, but, as a simple photovoltaic cell illustrates, radiation, as an immaterial source of energy, can produce electricity as well, and this without material emissions.

13.4.3 FROM FOSSIL TO SOLAR

Figure 13.11 tries to express that the work originally *available* in fossil fuel is used to "turn the wheel" for most of our global energy needs and is *lost* in time, while the mass in which this work was embodied is conserved and emitted. In the process, the mass and energy are conserved, but the quality of this energy diminishes, expressed in the fainting "color" of the energy flow.

In contrast, Figure 13.12 expresses that for the same global exergy dissipation, the sun as the energy source can be exploited without the need for material emissions. Of course, this assumes that the solar exergy reaching our earth exceeds that of the earth's exergy needs. Although this appears to be the case [3], some serious

FIGURE 13.11   The fate of fossil fuels.

FIGURE 13.12   A solar-fueled globe.

reservations have been made [20] by those who have looked into man's so-called ecological footprint. We return to this subject later in this chapter.

It is interesting to realize that, eventually, all exergy in the sun's radiation to the earth is dissipated. Heat released to the environment, without exergy, will be radiated into the environment, being a source of exergy for any body lower in temperature. In this way, the earth's steady-state temperature is maintained [23]. If the original solar exergy were partly tapped for exploiting it for our industrial activities, the same amount of entropy would eventually be produced, but for this part with some delay in time.

13.5　ECOLOGICAL RESTRICTIONS

The previous sections have pointed to the important role that solar energy will play in a sustainable society. Whether we talk of a living system, an industrial society, or an economic system, it seems that ultimately solar energy should drive it: "The sun is our nuclear reactor, remote and safe" [3]. In that context, it makes much sense to pay attention to an inspiring and consciousness-raising monograph with the title *Our Ecological Footprint* [20,28].

13.5.1　ECOLOGICAL FOOTPRINT

In 1996, Dr. Wackernagel and his PhD supervisor Professor Rees published a unique monograph entitled *Our Ecological Footprint* [20]. This publication has made a dramatic impact on our thinking with regard to the intelligent management of our earth in terms of resources and emissions.

Let us first turn to the definition that the authors have given to the Ecological Footprint [19, p. 9]:

> A resource accounting and management tool that measures how much land and water area a human population requires to produce the resources it consumes and to absorb the wastes it generates, taking into account prevailing technology.

Figure 13.13 shows a cartoon description of the ecological footprint. In this definition, you will recognize much of the definitions of sustainability that we gave earlier in this chapter, in particular, the one given by Angela Merkel, Germany's Chancellor. On the Web site www.footprintnetwork.org/ that we just mentioned, you can see that the organization "Global Footprint Network" [19] reports the ecological footprint that man makes on a global and on a national scale. In Table 13.3 we notice that in 2003, man's global ecological footprint was over 20% larger than what the planet can regenerate, 2.2 ha versus 1.8 ha. It takes more than one year and two months for the earth to regenerate what each global citizen consumes in a single year. The excess is generated by consuming the planet's natural resources (capital). From this table, it is also clear what should be understood by sustainability: "Sustainability is when Man's global demand on Nature's resources, his footprint, is in balance with Nature's capacity to meet that demand (biocapacity)."

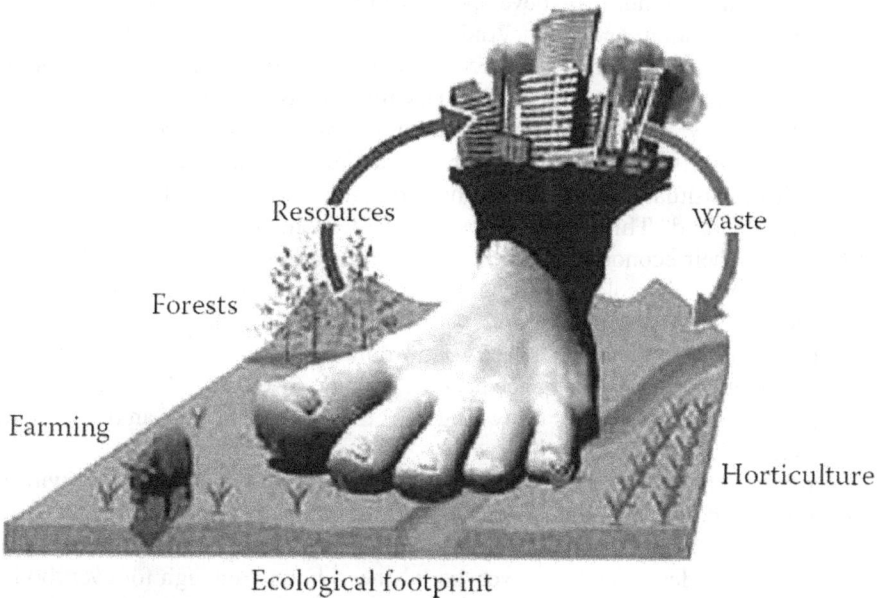

Ecological footprint

FIGURE 13.13 Cartoon of ecological footprint.

TABLE 13.3
Ecological Footprint (ha).

Population (×10⁶)	Footprint	Biocapacity	Surplus
Australia (19)	6.6	12.4	+5.9
Canada (31.5)	7.6	14.5	+6.9
China (1293)	1.6	0.8	−0.8
India (1033)	0.8	0.4	−0.4
Japan (127)	4.4	0.7	−3.7
Russia (145)	4.4	6.9	+2.5
Europe (454)	4.8	2.2	−2.6
The United States (288)	9.6	4.7	−4.9
World (6148)	**2.2**	**1.8**	**−0.4**

Table 13.3 also shows that although a large country as the United States has a large biocapacity (4.7 ha/capita), its footprint (9.6 ha/capita) is nearly double that capacity and instead of a surplus, this nation has a deficit. Although we all know that Americans consume energy excessively, few of us will have expected that this large country has a deficit in terms of its footprint. A country like Australia has also an enormous footprint but can afford it, having an impressive biocapacity. The world as a whole shows a deficit and there is a lot to be worried about, because countries like

China and India have not fully developed yet in terms of living standard, transportation, accommodation, and so on. And turning to China in this respect: The *ratio* of China's ecological footprint (resource demand) and biocapacity (resource supply) has changed from 0.8 in 1961 to 2.0 in 2003. China needs twice its productive area to meet sustainability. Its deficit is compensated by import of resources or liquidating ecological capital. In 40 years time, China's situation has completely reversed.

How will this situation be in 2050 when so many Chinese people have improved their living standard? This is not only a concern for China but for all countries that use China for their economic output: "China, the world's factory." Most developed countries import their ecological footprint from outside and therefore have a national ecological deficit. A developed country as the Netherlands has a footprint that is nearly five times its national biocapacity. How remarkable it is that our economic and ecological interests and performance become so interwoven on a global scale. There seems to be no national solutions any longer, as these problems take an international or even a global dimension.

From this analysis, it is clear that our world does not live in harmony with its environment and our "appetite" for material things appears to be too big. The great Indian statesman Mahatma Gandhi (1869–1948) once made the following statement: "This world provides enough for everybody's need but not enough for everybody's greed." Sixty years before nobody talked about sustainability, but then Gandhi made this visionary statement. How impressive. World Watch Institute, an independent research organization, based in Washington, DC (http://www.worldwatch.org), claims to work for an environmentally sustainable and socially just society. In its "State of the World 2004" report, World Watch Institute states that "12% of the world's population is responsible for more than 60% of the world's consumption" and concludes that

> The world is consuming goods and services at a rate that is not sustainable, threatening the well-being of the planet and its people. Consumer appetite is undermining Nature's capital and services and makes it harder for the world's poor to meet their basic needs. Nearly half of the world's population has an income of less than 2$/day.

Therefore, Sher Khan [27] observes, "The greatest challenge in the modern world is for people to give up the materialism that surrounds them." This may appear easier than it seems because World Watch Institute, in another report with the title "Richer, fatter and not much happier," reports that a survey in 65 countries has shown that life-satisfaction (happiness) grows with income up to a certain level, and then flattens off. A high quality of life is not so much obtained by "wealth," the accumulation of material goods, but by the experience of "well-being" in terms of freedom, health, security, information, and so on.

A last word on the "footprint." Canadians and Australians by far exceed the footprint that the average global citizen is allowed to make, but their countries can afford it. The department head of one of China's most famous universities once remarked, however, that he taught his students to be global citizens first, then Chinese citizens. The world's greatest problems have global dimensions, and it seems important that for solving such problems we need to be global citizens first.

By the way, everyone who is interested in his or her personal footprint should visit the "Global Footprint Network" Web site and do the Footprint Quiz.[3]

Of course, a concept with such dramatic implications for the state of the world has been the subject of great scrutiny, and an illustration is the discussion that has been published in the *Journal of Ecological Economics* [28]. Experts have expressed both serious criticism and support on the assumptions made and the analysis applied. But we believe that the concept remains one of great illustrative power and cannot be ignored in any discussion on sustainable development. In early 2002, the famous sociobiologist Wilson [17,29] discusses this concept in his book *The Future of Life* [29], which has been reviewed in the *Scientific American* [17]. He mentions an ecological footprint of 1 ha/cap in developing nations, but 9.6 ha/cap in the United States, while the average footprint worldwide is 2.1 ha/cap. He concludes that with existing technology, we would require four more planet earths to reach present U.S. levels of consumption. The least one can say is that the ecological footprint can contribute to creating wide public awareness and act like a warning light. Wackernagel and Rees [20] conclude by stating, "What we lack is intellectual and emotional acceptance of the fact that humanity is materially dependent on nature and that nature's productive capacity is limited." Therefore, dematerialization of economic goods and services must proceed faster than economic growth. They have estimated that a drastic (4–10-fold) reduction in material and energy intensity per unit of economic output is required for global sustainability.

In the preceding studies [16,19,27] it has become apparent that energy consumption is an important part of the ecological footprint. For example, for Canada nearly 60% of the footprint stems from the need for energy. Therefore, the authors [20] also studied the productivity of various energy sources in a sustainable economy. From these studies it appeared that solar energy as in photothermal and photovoltaic conversions or as in wind energy is close to 10 times more efficient than fossil fuel. This source of energy would therefore fit well within scenarios for sustainable development but also calls for new technology.

13.5.2 WASTE

The laws of mass and energy conservation and the second law combined give a sharp insight into the industrial process, its resources, waste, and losses (Figure 13.14). We wish to illustrate this in a simplified example [30]. Suppose we want to produce metal X from its oxide XO, which is readily found in nature as a resource. The most elementary reaction to handle this transformation is

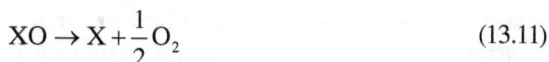

$$XO \rightarrow X + \frac{1}{2}O_2 \tag{13.11}$$

In a handbook such as Szargut et al. [31], these values are tabulated and the exergy values of XO and X are found, for example, to be 50 and 440 kJ/mol, respectively. So this reaction requires an amount of exergy of at least some 390 kJ/mol product. Usually, this is supplied by a nonrenewable fuel such as coal, oil, or gas (the exergy of oxygen is small and has been neglected). From literature we learn that a suitable

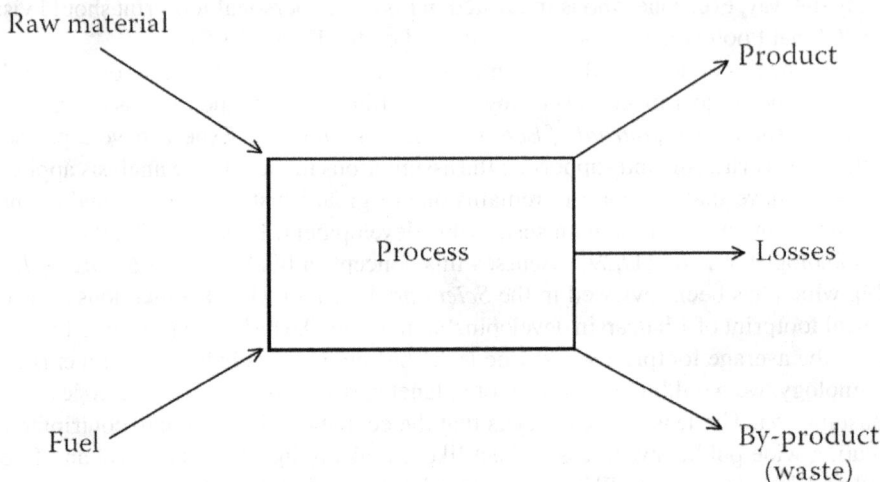

FIGURE 13.14 Characteristics of a process.

reactant, at the same time a supplier of exergy, is carbon from coal (with an exergy value of 410 kJ/mol) and that the exergy value of CO_2 is 20 kJ/mol

$$XO + \frac{1}{2}C \rightarrow X + \frac{1}{2}CO_2 \tag{13.12}$$

This equation expresses the law of the conservation of mass, overall and for each chemical element separately. The combustion reaction as written in Equation 13.12 is still short of $(440 + \frac{1}{2} \times 20 - 50 - \frac{1}{2} \times 410) = 195$ kJ. So theoretically, the work available in another ½ mol of C is required, corrected for the work available in another ½ mol of CO_2. Theoretically, the amount of waste is now 1 mol of CO_2 per mole of product. But suppose that the thermodynamic efficiency of the overall scheme is 50%. In that case, 1 mol more of C will be required, resulting in total of 2 mol of CO_2, that is, a fourfold of the CO_2 emission of the reaction in Equation 13.12.

We observe that we have drawn from nonrenewable resources and emitted waste at four times the rate corresponding to that for the reaction. Of course, this is all a consequence of choosing a material, nonrenewable resource, coal, as the reactant and supplier of required chemical work: The use of fossil fuel and the inefficiency of processes are responsible for most of the waste in industry. Figures 13.15 and 13.16 give an impression of the size of waste [30], where we should note that in the chemical industry, things quickly get worse if we move, for example, from base chemicals to fine chemicals to pharmaceuticals.

Relief may be offered if we move from nonrenewable fuels to a renewable exergy source such as sunlight. But this would require completely new technology at acceptable efficiencies. One might envisage, for example, the photolytic dissociation of water to produce hydrogen, hydrogen taking the role of carbon in reducing

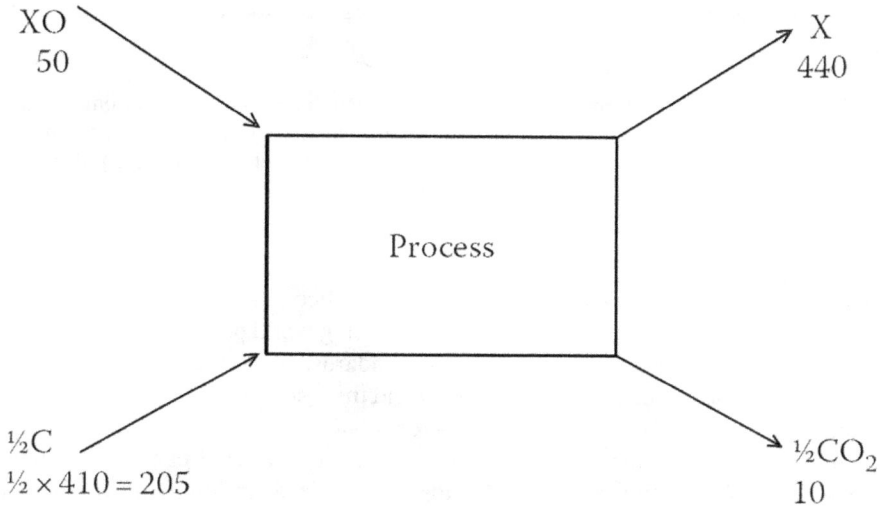

FIGURE 13.15 Thermodynamic analysis of a process.

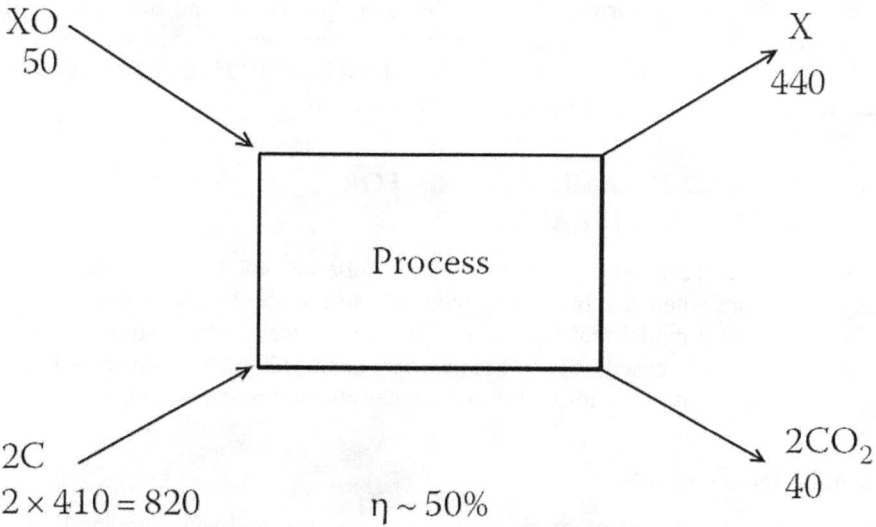

FIGURE 13.16 Waste generation with efficiency of approximately 50%.

XO toward the pure metal X. But the efficiency of photolysis so far is a few percent [32], so considerable amounts of exergy will be required to manufacture all equipment required for operation. Most likely, this exergy has to be "advanced" by fossil resources, which have tempted some to point out that solar energy applications and, perhaps, a more general other renewable energy will be based mainly on fossil fuel consumption (see, e.g., [33]), which tempted Georgescu-Roegen to state the following in 1979:

> The simple fact that no enterprise is at present manufacturing collectors only with the aid of the energy produced by collectors is sufficient proof that the corresponding technology is not viable. Consequently, at present, all direct uses of solar radiation represent parasites of the fossil fuel technology.

> (p. 22)

This example of waste analysis illustrates why some economists are concerned about the prospect of economic growth. With the increasing world population and an anticipated worldwide increase in average living standards, economic growth is a necessity. But given the inefficiency of technology and the resultant drain on resources and production of waste, these may exceed the economic output by more than a factor. But the finite size of our planet, and as a result the restricted capacity to supply resources and absorb emissions, may hamper the required economic growth. And, in full support of what we wrote in Section 13.2, Ayres states that "there is every chance that the supply of environmental services will dwindle as the demand, generated by population and economic growth, grows exponentially."

Things get even worse if one looks at concerted efforts to establish the thermodynamic efficiency of various countries like Sweden, Japan, and Italy [34]. If the inefficiencies in the use of consumptive exergy toward final services are included, such studies confirm earlier estimates by Ayres and Kneese [35] that the overall efficiencies of advanced economies are only a few percent.

13.6 THERMODYNAMIC CRITERIA FOR SUSTAINABILITY ANALYSIS

Sustainability is a critical issue, and the word is often bandied about. However, how can we measure whether a process is genuinely sustainable? This section relates to the evolution of a model that can be used to help decide on the sustainability of a given practice and is based on two of our earlier papers [16,36]. Key advances include the quantification of renewability beyond a renewable/not renewable basis.

13.6.1 INTRODUCTION

Industry is under increasing pressure from governments and environmental groups to improve the sustainability of its processes. However, how this higher level of sustainability should be achieved is not really clear, and even the definition of sustainability is often only qualitative, as we have seen earlier. There is a need for a tangible *quantitative* description that allows the sustainability of technological processes to be systematically evaluated, compared, and improved as Lems [16] argued.

In an earlier work by De Wulf et al. [36], we showed that different aspects of process sustainability can be quantified by using thermodynamic principles. Indeed, the thermodynamic concept of exergy is used as the basis for the construction of sustainability parameters, which conveniently express particular aspects of process sustainability on a scale of zero to one. Elements of this work were used to analyze the sustainability of several industrial processes, and new insights have led to some meaningful improvements.

All real processes must consume exergy to proceed, and this means that all our technological activities are ultimately limited by our ability to supply exergy to our processes. The only exergy that can be considered as truly sustainable is the exergy supplied by solar radiation, because this solar exergy will be available on a very large time scale and its immaterial nature allows processes using it to operate within closed material cycles on earth.

However, our potential to use solar exergy is limited. Although the total amount of exergy reaching the earth as solar radiation is enormous, this exergy is dispersed over a very large area and its effective harvesting is limited to relatively few sites. These sites are often also the areas needed for agriculture, living space, and industrial activity and too intensive harvesting of solar exergy can be disruptive to the natural environment. For instance large-scale introduction of solar panels can alter the natural heat absorption of the surface and it thereby has the potential to alter local climates.

It is clear that ultimately our only limitation to sustainable production is obtaining the exergy to run the production processes and to drive closed material cycles. In view of this, exergy can be considered the ultimate scarce resource in our technological processes, and exergy flows to or from these processes are therefore vital elements of a method aiming to quantify process sustainability. Our first attempts to use exergy flows in the construction of sustainability parameters were published by De Wulf et al. [36]. This original quantification method is the basis for our newly developed method, which expresses the sustainability of technological processes in a set of three independent sustainability parameters. These three parameters deal with the resource utilization, the energy conversion, and the environmental compatibility of the process, respectively. The second and third parameters are revised elements of the work of De Wulf et al. [36], while the definition of the first parameter is entirely different. The parameters are discussed separately in the following sections.

13.6.2 SUSTAINABLE RESOURCE UTILIZATION PARAMETER α

One of the major factors undermining the sustainability of a production process is the depletion of the resources it uses. A quantification of process sustainability should therefore include a parameter that deals with the sustainability of resource utilization, and the construction of such a parameter begins with defining a quantitative measure for the depletion of an individual resource. One way of doing this is to classify each resource as either renewable or nonrenewable, as was done by De Wulf et al. [36]. The distinction made between renewable and nonrenewable resources is that renewable resources are created at least as fast as they are consumed (e.g., solar energy), while nonrenewable resources are consumed faster than they are created (e.g., crude oil).

However, there are two major problems with this renewability concept. First, the idea of a resource being either renewable or nonrenewable seems somewhat artificial.

Even the sun is depleting a finite amount of nuclear fuel, and it is therefore dubious to regard solar energy and derived energy sources such as wind- and hydro-energy as completely renewable. Also, so-called nonrenewable resources such as fossil fuels or mineral deposits are being formed naturally to some extent, and therefore they are not entirely nonrenewable. Obviously, the renewability of resources is more gradual than suggested in the black-and-white representation of renewable versus nonrenewable, and a parameter expressing the sustainability of resource utilization should account for these more subtle differences in resource renewability.

A second objection that can be made against the renewability concept is that renewability is only a part of sustainable resource utilization, since it does not involve the natural reserves of resources. Although the concept of renewability rightfully views the consumption rate of a resource in relation to its production or regeneration rate, this gap between consumption and regeneration rate should in turn be viewed in relation to the size of the natural reserves of that resource. In fact, a temporary discrepancy between the consumption and the regeneration rate of a resource does not necessarily threaten the sustainability of a process when the resource is plentiful. Of course, not having certain material cycles closed can have a profound effect on the natural environment (e.g., via harmful emissions), but this is beyond the scope of this particular sustainability parameter, which deals only with resource depletion.

Hence, instead of considering the renewable versus the nonrenewable resources in a process, it is preferred to quantitatively combine the consumption rate ($\varphi_{m,consumption}$), the regeneration rate ($\varphi_{m,production}$), and the extent of natural reserves ($M_{reserves}$) of a resource, as they are known at this time, in a resource depletion time (τ) defined as

$$\tau = \frac{M_{reserves}}{\phi_{m,consumption} - \phi_{m,production}} \tag{13.13}$$

Besides taking into account any gap between the consumption and regeneration rate of a resource, the depletion time relates this gap to the extent of the natural reserves. The depletion time is then a measure for the rate at which the currently known reserves of a resource are being depleted. For example, a depletion time of 100 years means that currently 1% of the known reserves is being depleted yearly; likewise, a depletion time of 1,000 years means that yearly 0.1% is being depleted. It should be stressed that the depletion time as defined in Equation 13.13 does not attempt to predict resource depletion in the future; it merely indicates how fast a known supply of a resource is *currently* being depleted.

This definition means that the depletion time of a resource is time dependent. The consumption rate of a resource may increase over time as a result of increased industrial production, or it may decrease if alternative resources are increasingly being used. Likewise, the regeneration rate may increase when more resources are recycled, and reserves may shrink after prolonged utilization or may expand when new natural deposits are found. In addition, more accurate data may become available, for example, on the extent of currently known reserves or on the natural formation rates of certain deposits. In any case, the depletion time τ is variable, and it reflects the rate of resource depletion only in the present situation.

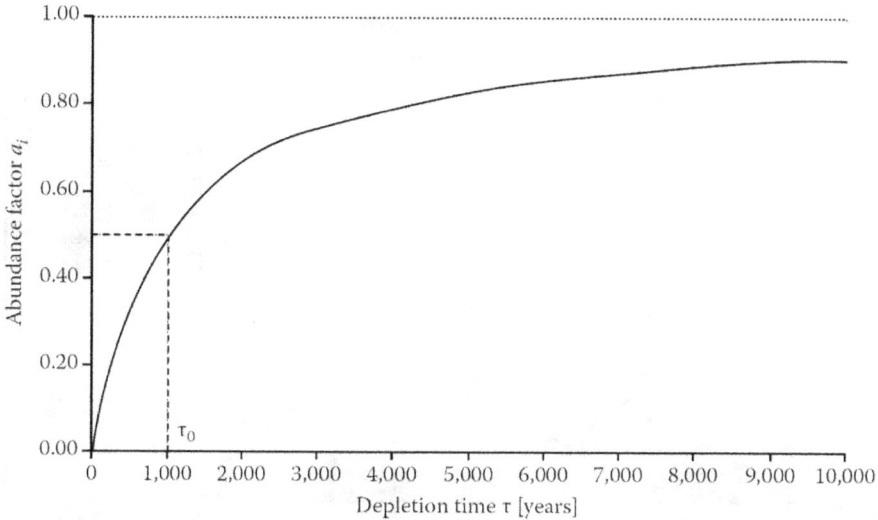

FIGURE 13.17 Graphical representation of Equation 13.14. Abundance factor as a function of depletion time.

The depletion time of a resource obtained with Equation 13.13 cannot be used directly in the construction of the sustainability parameter and, therefore, it is first converted to a factor that can be viewed as an expression of resource abundance on a scale of zero to one (Figure 13.17). For a resource i with a depletion time τ_i, the abundance factor a_i is defined as

$$a_i \equiv \left(\frac{\tau_i}{\tau_i + \tau_0} \right) \qquad (13.14)$$

The reference time τ_0 in Equation 13.14 represents the resource depletion time at which the abundance factor a_i is at the value of exactly one half (i.e., 50%). This reference time τ_0 must be given an appropriate value at which the factors a_i adequately reflect the differences in the abundance of real resources. For example, a reference time τ_0 of 1,000 years will give fossil fuels abundance factors roughly ranging from 0.1 to 0.5, while sunlight, probably available for billions of years, will have an abundance factor approaching unity.

Important to note is the asymptotic behavior of Equation 13.14, which is clearly seen in Figure 13.17. This nonlinearity causes the abundance factor a_i to be more sensitive at small depletion times and less sensitive at large depletion times, which is in accordance with our intuitive judgment. When comparing two different natural resources, the difference between 100 and 1,000 years of depletion time is much more important than the difference between 1 billion and 10 billion years. The latter time scales are both so large that it is almost equally unlikely that the depletion of either one of these resources will cause a process using them to become unsustainable. Hence, the nonlinear behavior of Equation 13.14 brings some common sense to the quantification method.

Renewability indicator α

Focus on exergy inputs:

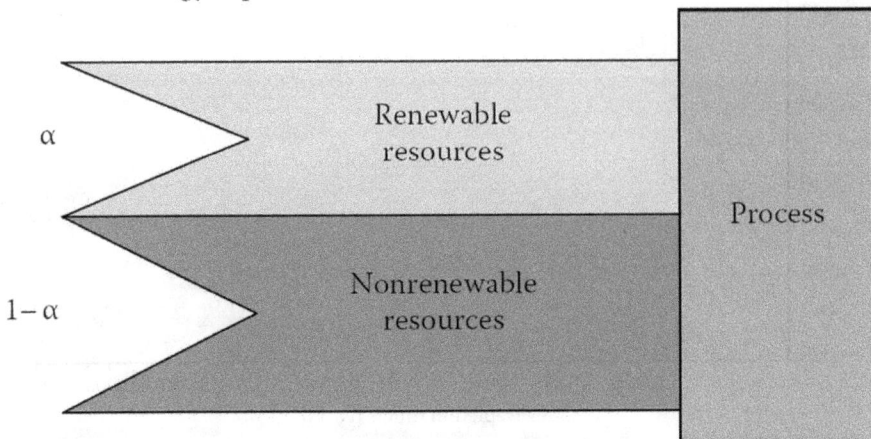

FIGURE 13.18 The sustainability parameter α is based on the average abundance factor a^{average}, which considers all the abundance factors a_i and exergy flows Ex_i of the individual resources used in the process.

With the abundance factors as quantitative measures for the abundance of individual resources, the focus can now be on expressing the sustainability of multiple-resource utilization in a particular technological process. The first step is to determine how available are, on average, all the resources used in the process. This average abundance (a^{average}) is obtained by averaging the abundance factors (a_i) of the individual resources while using their exergy flows ($\text{Ex}_{\text{in},i}$) to the process as averaging weights (Figure 13.18):

$$a^{\text{average}} = \frac{\sum_i a_i \cdot \text{Ex}_{\text{in},i}}{\sum_i \text{Ex}_{\text{in},i}} \tag{13.15}$$

Equation 13.15 uses the exergy flows of the resources because these values express the minimum amount of work required to produce these resources from compounds that are thermodynamically in equilibrium with the natural environment. As, ultimately, only exergy is scarce, the exergy flows of resources are better indicators of relative importance than their mass or volume flows.

Although very useful, the average abundance of resources (a^{average}) alone is not capable of adequately expressing the sustainability of resource utilization. As a result of the averaging procedure, small abundance factors, which come from resources that are rapidly being depleted, can be (partly) compensated by large abundance factors, which come from resources that are depleted relatively slowly. In this way, the weakest link in the chain, namely, the resource being depleted at the highest rate, does not act as a barrier to the sustainability of the resource utilization in a process.

This opposes common sense because a process already becomes unsustainable when only one of the resources it requires is depleted.

This problem can be solved by looking not only at the *average* resource abundance, but also at the *minimum* resource abundance. Minimum resource abundance (a^{min}) can be defined simply as the smallest of the resource abundance factors a_i relevant in the process:

$$a^{min} = \underset{i}{min}[a_i] \qquad (13.16)$$

With a^{min}, we have identified the weakest link in the chain of resources used in the process, and this can be used to limit the effect of the average abundance factor ($a^{average}$). Hence, the parameter for sustainable resource utilization (α) can then be defined as the product of the average ($a^{average}$) and the minimum (a^{min}) abundance of resources:

$$\alpha = a^{average} \cdot a^{min} \qquad (13.16b)$$

In this way, the sustainability parameter α includes the depletion rates of all resources used in the process (via $a^{average}$), while, at the same time, it is directly limited by the resource with the highest depletion rate.

The proposed method for quantitatively describing the sustainability of resource utilization in a process has several advantages. First, it considers the *degree* of resource renewability, which allows even subtle differences in the depletion of different resources to be accounted for. Second, it also includes the (natural) reserves of resources, making the method yet more refined. Finally, the concept of depletion times and their translation to abundance factors allow the resource sustainability parameter α of a process to be limited by its most rapidly depleting resource, which is quite realistic.

EXAMPLE 13.1

Consider the following hypothetical process. Suppose a process can be driven by three different sources of exergy: oil, coal, and solar energy. Using their current consumption rate, regeneration rate, and extent of natural deposits, the depletion times of oil and coal are calculated (Equation 13.13) to be 150 and 1,000 years, respectively. For solar energy, the depletion time equals the lifetime of the sun, which is approximately 5 billion years (Table 13.4).

Taking the reference depletion time τ_0 at 1,000 years, the depletion times of oil and coal yield abundance factors of 0.13 and 0.50, respectively; the very large depletion time of the nuclear fuel in the sun leads to an abundance factor for solar energy that approaches unity (note that the abundance factor is less sensitive at high depletion times and that this is convenient because higher depletion times can usually be determined less accurately).

With Equations 13.14 through 13.16, and with the abundance factors for oil, coal, and solar energy as determined earlier, it is possible to determine the parameter

TABLE 13.4

Depletion Times and Abundance Factors of Oil, Coal, and Solar Energy.

Exergy Source	Depletion Time [years]	Abundance Factor [–]
Oil	150	0.13
Coal	1,000	0.50
Solar energy	5 billion	–1

TABLE 13.5

Relevant Data on Sustainability of Resource Utilization in Processes Using Oil, Coal, and Solar Energy.

Process	Oil [%]	Coal [%]	Solar [%]	$a^{average}$	a^{min}	α
1	100	0	0	0.13	0.13	0.02
2	0	100	0	0.50	0.50	0.25
3	0	0	100	–1	–1	–1
4	50	50	0	0.32	0.13	0.04
5	20	50	30	0.58	0.13	0.07
6	10	0	90	0.91	0.13	0.12
7	0	10	90	0.95	0.50	0.48

for sustainable resource utilization for a process that uses solely one or more of these three resources. Table 13.5 lists the relevant data for several such processes, each extracting different percentages of the total required exergy from oil, coal, and solar energy.

First of all, Table 13.5 lists three processes that each extract their exergy entirely from one of the three resources (processes 1, 2, and 3). In these situations, there is only one relevant abundance factor, and its value then solely determines the values of the average and minimum abundance factor. Also, the sustainability parameter is then simply this value squared. The results show that the process using only oil (process 1) is the least sustainable in its utilization of resources, the process using only coal (process 2) is more sustainable, and the process extracting all exergy from solar energy (process 3) is practically entirely sustainable in terms of resource utilization.

Table 13.5 also lists three processes (processes 4, 5, and 6) that extract a decreasing percentage of the total required exergy from the least durable resource oil, and an increasing percentage from the more durable resources coal and solar energy. Correspondingly, the average abundance factor increases, also causing the value of the sustainability parameter to rise. However, the minimum abundance factor remains the same since all three processes use oil to some extent, which has a relatively small depletion time of only 150 years. The values of the parameter rightfully indicate that using more durable resources helps to increase the sustainability parameter, but that its effect is limited, because the weakest link, namely, the least durable resource, is not completely replaced. In fact, process 7 in Table 13.5 illustrates that when the use of oil in process 6 is replaced by coal, the sustainability parameter indicates a much more sustainable use of resources.

The effect of the minimum abundance factor $a^{\min}$ on the sustainability parameter α may seem counterintuitive. However, it must be realized that the sustainability parameter α only expresses that part of process sustainability that involves the availability of the resources used. The fact that process 6 in Table 13.5 may be more sustainable than, for example, process 2 in terms of environmental impact is not relevant to this particular aspect of process sustainability. Purely in terms of resource utilization, process 2 is more sustainable than process 6, because process 2 does not depend on a rapidly depleting resource like oil.

13.6.3   NOTES ON DETERMINING DEPLETION TIMES AND ABUNDANCE FACTORS

A special situation in the determination of depletion times and abundance factors occurs when a resource used in a process is actually a half-product; for example, iron metal is not a naturally occurring resource but is produced from iron ore in a blast furnace. The abundance factor of such a half-product must then be derived from the smallest depletion time among the resources used during its production. Since the abundance of, for example, iron metal is limited by the abundance of heavy oil fractions, which are used to convert coal into coke, the abundance factor of iron metal can be based on the depletion time of oil.

Another special situation arises when one resource has its origin in two different processes. For instance, when a process uses the electricity provided by an energy company, it is possible that this electricity is generated partly by burning coal and partly by burning natural gas. In this case, the process should be considered to use two different types of electricity: electricity from coal and electricity from natural gas. Both types of electricity then have their own derived depletion time and, based on how much they contribute to the total amount of electricity supplied, their own exergy flow.

Finally, regarding the concept of depletion times, as it is used in the construction of the sustainability parameter, it is also suitable to include the sustainability of production of facilities needed to harvest resources. For instance, solar energy may in principle have a depletion time of billions of years, but the production of photovoltaic cells usually requires some very scarce elements for which the depletion times are much smaller. The sustainability of using solar energy harvested with photovoltaic cells should then be based on these latter depletion times and not on the depletion time of solar energy itself.

13.6.4   EXERGY EFFICIENCY η

The efficiency parameter focuses on the conversion of energy in the process itself (Figure 13.19). Since exergy rather than any other resource is the ultimate limiting factor to production activities, a process is most sustainable when it uses the exergy of its ingoing resources most efficiently. For this reason, the exergy efficiency of the process is important enough to be considered as a separate sustainability parameter:

Efficiency indicator η

Focus on exergy conversion:

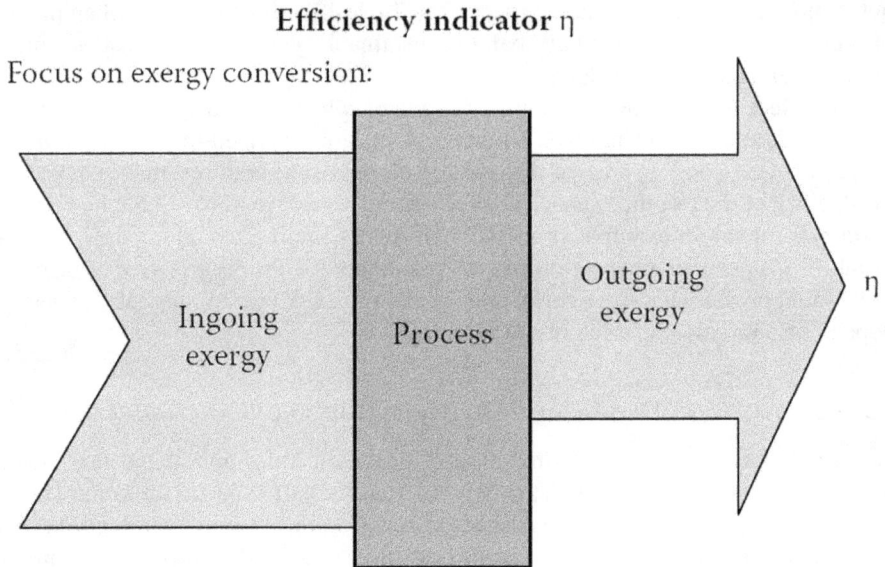

FIGURE 13.19 The sustainability parameter η is the efficiency with which the exergy of resources is transferred to the products of the process.

$$\eta = \frac{\sum Ex_{out,useful}}{\sum Ex_{in,process}} \qquad (13.17)$$

Equation 13.17 explicitly mentions the *useful* exergy flows coming out of the process because exergy can be lost in two different ways. First, exergy is lost in any real process as a result of irreversibility in the process itself, and such losses are called *internal* exergy losses. Second, exergy can be lost via waste streams that are not yet at equilibrium with the natural environment. Examples of such *external* exergy losses are the release of hot flue gases or high-pressure gas to the atmosphere. Both the internal and the external exergy losses are in principle inefficiencies, and the exergy used efficiently in the process is therefore only the exergy of products and the exergy of waste products, provided they are made useful in other processes.

13.6.5 THE ENVIRONMENTAL COMPATIBILITY ξ

A third main aspect of process sustainability is not damaging the natural environment, and the environmental parameter ξ as defined by De Wulf et al. [36] elegantly quantifies this part of process sustainability. Its basic idea is that negative effects of a process on the environment must be abated before they can do damage to the natural environment. The extra exergy required to achieve this abatement then reflects the environmental incompatibility of a process to the natural environment (see Figure 13.20).

Environmental indicator ξ

Focus on exergy input for abatement:

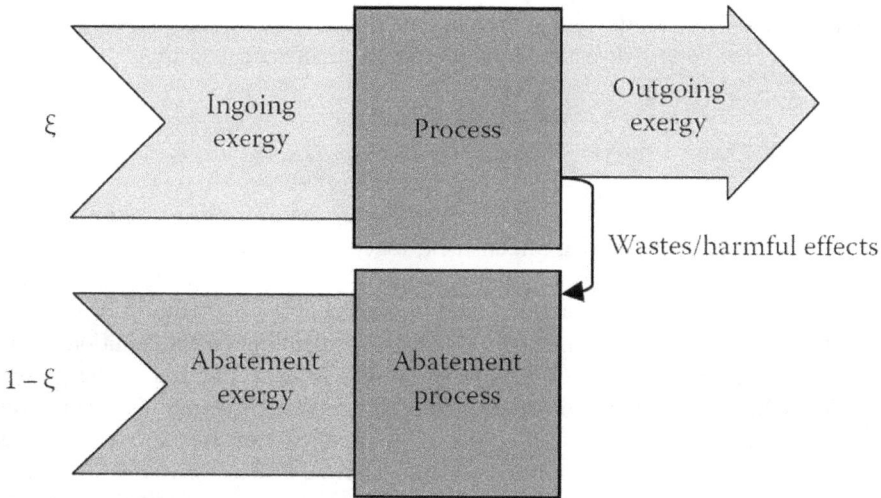

FIGURE 13.20 The sustainability parameter ξ expresses the exergy required to abate all the effects of the process that are harmful to the natural environment.

It should be noted that abating negative effects is a very broad concept. Abatement not only applies to emissions, as considered by De Wulf et al. [36], but can also apply to all kinds of negative effects, including thermal pollution or even extraction of resources from the environment. Examples of harmful effects of resource extractions are erosion of soil after cutting down forests or subsidence of ground after extraction of natural gas or oil from underground deposits. Abatement is required for all effects considered harmful to the environment that occur either during the process or during the eventual destruction of the products after their use.

Based on the exergy required for abatement, the sustainability parameter expressing the ecological compatibility of a process is defined as given in Equation 13.18:

$$\xi = \frac{Ex_{in,process}^{total}}{Ex_{in,process}^{total} + Ex_{in,abatement}^{total}} \tag{13.18}$$

The parameter ξ relates the exergy required to run the process to the exergy required to run the process in an environmentally sound way, which includes the extra exergy required for abating the harmful effects on the environment. A large exergy demand for abatement leads to a small value for the environmental parameter ξ, indicating that the process has a small compatibility with the natural environment. Only if a process requires no exergy for abatement, the environmental parameter ξ = 1 and the process is considered completely compatible with the natural environment. Now, for a process to be economically viable, it would seem that

$$\text{Ex}_{out} \bullet P_{Ex,out} \geq \text{Ex}_{in} \bullet P_{Ex,in} + K \tag{13.19}$$

where

> P denotes the economic value of the in- and outgoing exergy streams
> K denotes the capital investment, appropriately amortized over time

EXAMPLE 13.2

The main characteristics of a biomass conversion process are illustrated in Figure 13.21. The biomass feedstock, together with added water, is pumped and heated to a temperature and pressure not too different from water's critical point. The conversion process taking place at these conditions results in a transport fuel-type product, that still needs upgrading by the addition of certain chemical compounds. Even then, only a modest fraction of the product is of high enough grade to be considered suitable as a fuel. The remaining fraction's grade still requires considerable chemical treatment. At this stage, an exergy analysis was performed and this showed a thermodynamic efficiency η of nearly 70% (Figure 13.22). Figure 13.23 shows a more realistic picture of the process. This picture takes into account that for environmental compatibility, wastewater needs to be treated, requiring work most likely from a fossil resource. At the same time, the utilities also require work that, equally, is likely to be from fossil origin. Figure 13.24 shows that as a result, the fraction of renewable input exergy is $\alpha = 0.6$ and the environmental compatibility factor ξ has been brought from 0.8 up to the desirable level of 1.0; the thermodynamic efficiency has dropped, however, to $\eta = 47\%$. Figure 13.25 shows that α has been made 1.0 by replacing all fossil input by biomass together with, out of necessity, low grade product, but Figure 13.26 shows the high price for this: a drop in the thermodynamic efficiency from 47% to 20%.

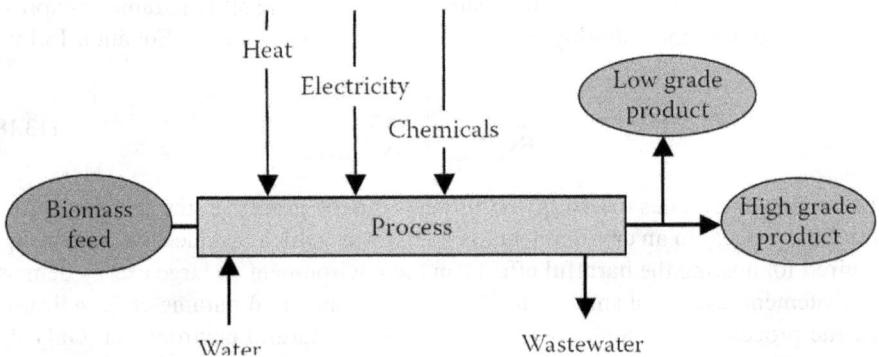

FIGURE 13.21   Main characteristics of biomass conversion process.

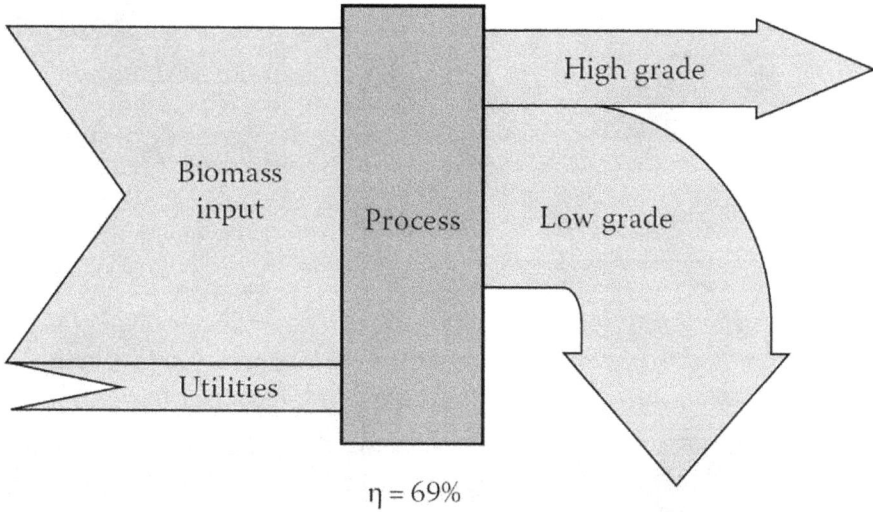

FIGURE 13.22 Schematic Grassmann diagram of the process shown in Figure 13.21.

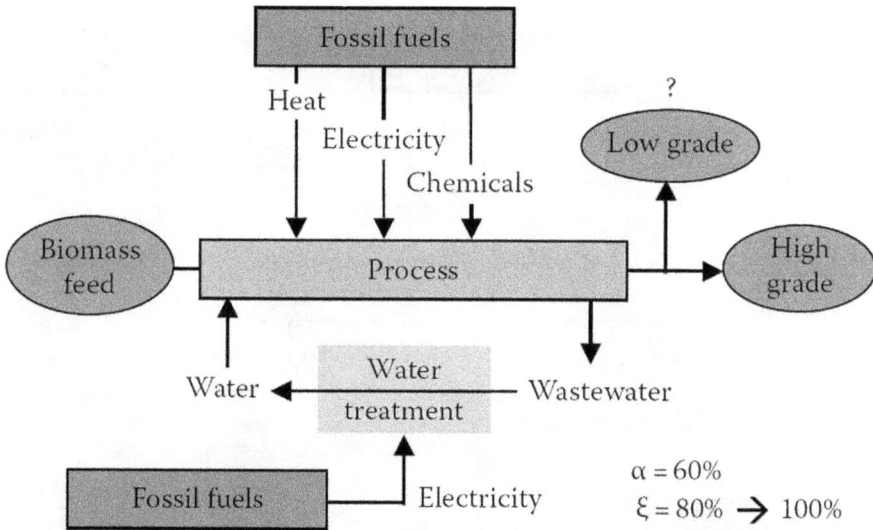

FIGURE 13.23 More realistic version of process shown in Figure 13.21 with mitigating steps to ensure that $\xi = 80\% \rightarrow 100\%$.

Equations 13.17 and 13.19 have to decide whether this value of η is enough for the economic feasibility of the process. Although the final result is disappointing, we should realize that this process or an alternative process offers a wide scope for improvement. Research will undoubtedly develop many new leads to improve the final product quality, to increase its high grade fraction and so on.

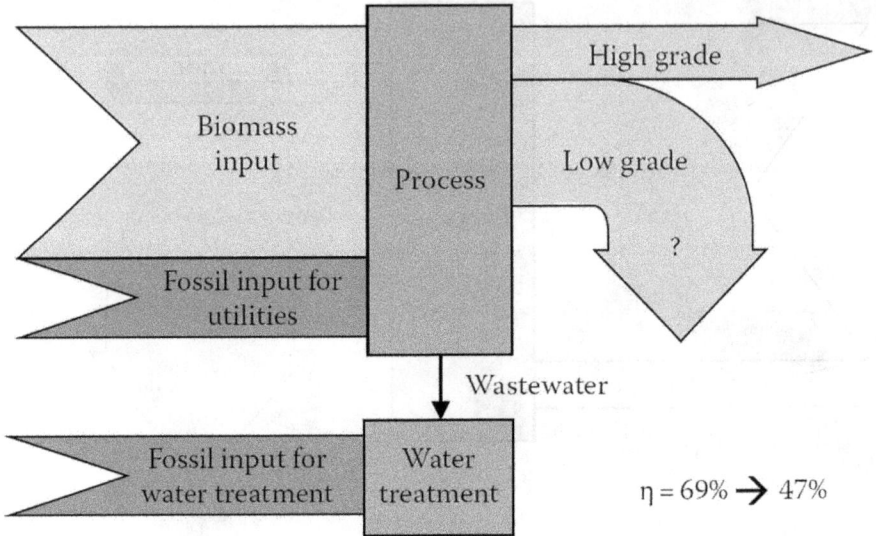

FIGURE 13.24 Schematic Grassmann diagram of process shown in Figure 13.23.

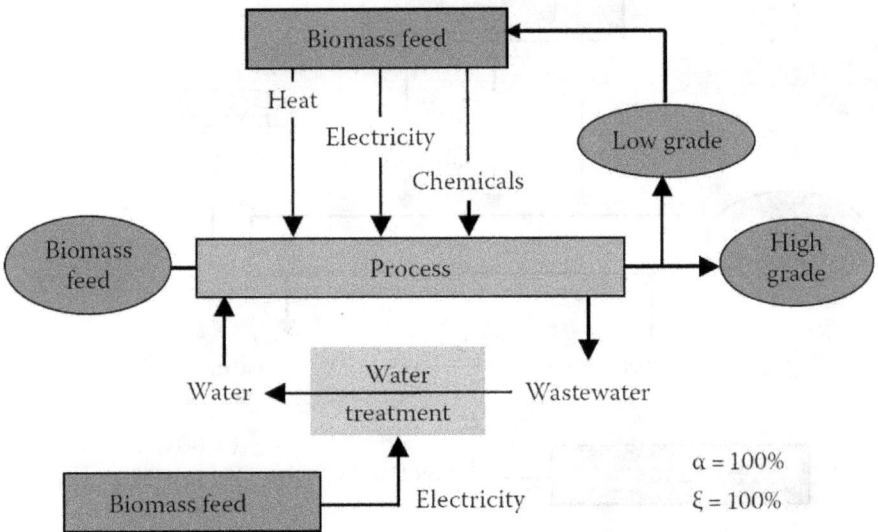

FIGURE 13.25 Modified realistic version of process shown in Figure 13.23 with mitigating steps to ensure that $\alpha = 60\% \rightarrow 100\%$.

13.6.6 DETERMINING OVERALL SUSTAINABILITY

De Wulf et al. [36] combine individual sustainability parameters to form one overall sustainability coefficient S. However, although a single expression for the sustainability of a technological process may seem appealing, it has some serious disadvantages.

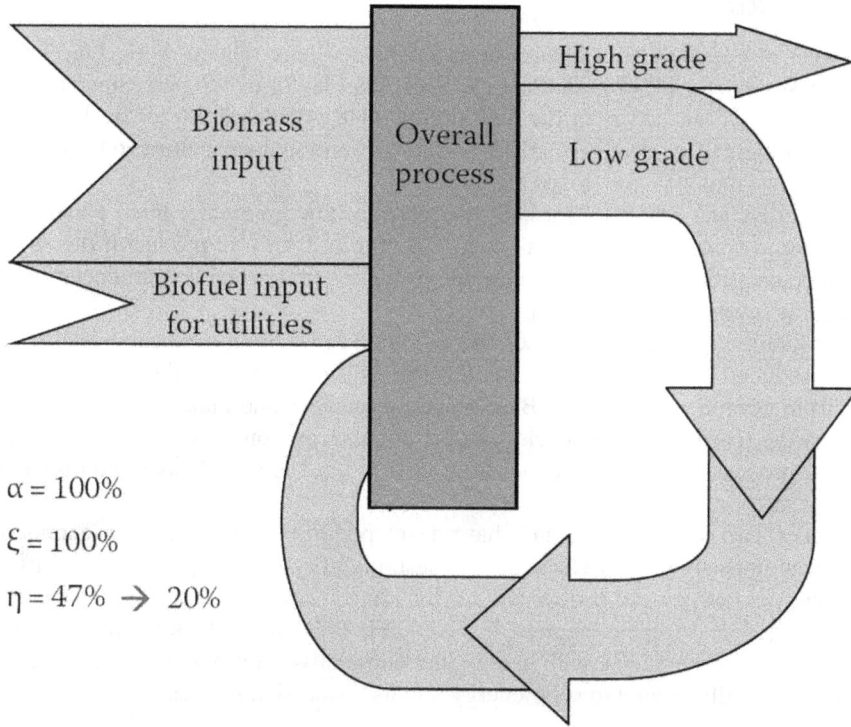

FIGURE 13.26 Schematic Grassman diagram of process shown in Figure 13.25.

First, one sustainability coefficient falsely suggests that sustainability is one-dimensional, and this undermines the idea that there are different aspects to a complex concept such as sustainability. Second, a lot of valuable information is lost when merging the individual parameters into one overall coefficient; the combination basically means that concrete and meaningful expressions are merged into an overall expression without a tangible meaning. The overall coefficient merely expresses something vague as the overall sustainability of a process. Third, the method used to combine the three individual parameters is highly subjective. Many methods of combining parameters may be applied, such as several different averaging techniques or a least-squared method, but every method unavoidably involves a value judgment on the relative importance of the different aspects of sustainability.

Hence, to keep the quantitative expression of process sustainability as meaningful as possible, separate sustainability parameters should not be combined into one overall sustainability coefficient. Expressing the sustainability of a process quantitatively is always a trade-off between completeness and surveyability; one aims to lose as little valuable information as possible, but at the same time the expression of process sustainability must be concise and well organized. It is felt that the three sustainability parameters as defined in this chapter do exactly that and that the sustainability of a process is therefore best evaluated by independently considering these quantifications of three fundamentally different aspects of process sustainability.

13.6.7 RELATED WORK

We wish to conclude this section by referring to some related work. Our former student Wassenaar [37] has defined the fossil load factor as the percentage of fossil exergy input of the exergy of the final product. He calculated 9% for fresh potatoes from "ecological" agriculture and 13% from conventional agriculture, 64% for fresh French fries, and 80% for frozen French fries.

Gerngross and Slater [38] asked the question how green are green plastics and show that in some cases less fossil energy is required per unit product if this is produced conventionally from nonrenewable resources than when it is produced from a renewable resource, such as corn.

Berthiaume et al. [39] introduce the so-called renewability indicator relating the work produced from solar energy to the work required to restore the degraded products from nonrenewable origin. Based on their analysis and making use of concepts such as the thermodynamic cycle, exergy, and exergy consumption, they conclude that the process to produce ethanol from corn is not sustainable as it requires more work of restoration than is produced.

Finally, Brown and Ulgiati [40] have developed indicators to monitor economies and technology on their performance in sustainability. They make extensive use of the concepts *emergy* and *transformity*, which have been introduced by two pioneers in ecological engineering, namely, E. G. and H. T. Odum [41]. Both concepts show a remarkable resemblance with exergy and cumulative exergy consumption but are more specifically related to solar energy and its absorption in ecosystems.

13.7 CONCLUSION

As the growth of the world population and of the average living standards are certainties, so are economic growth, the growing need for resources, and the increasing burdening of the environment. The environment is neither an infinite supplier of resources, nor an infinite absorber of waste. Prominent people and reports therefore call for a global sustainable development where society develops in harmony with the earth, with nature. But it seems that however good the intentions of the advocates of such a development are, they are not always aware of the implications and consequences thereof. This becomes clearer if we analyze nature itself in its cyclic behavior.

Nature seems to provide us with a true example of sustainability: to make use of renewable resources, solar energy or heat from the inner earth, closing material cycles by making waste a feedstock for another actor in the cycle and doing all this with a remarkable efficiency. The "cycle of life" is a thermodynamic cycle producing mechanical, chemical, and electrical energy, driven by an immaterial source of exergy while making use of a "working fluid," which changes in composition depending on its position in the cycle. The cycle produces solely immaterial emissions, that is, radiating waste heat into the universe. This process is strictly irreversible, dissipative, and unidirectional, transforming what is useful—work—into what is useless—heat of the environment—and delayed by the creation, maintenance, activity, and degradation of living systems.

Economic and industrial systems share many features with living systems but differ markedly from them in the sense that useful work is extracted from nonrenewable, material resources and useless heat is emitted together with material waste products.

As the earth is a closed system for matter, this shall ultimately lead to the exhaustion of resources and the accumulation of waste products. These systems should therefore rely increasingly on renewable resources while simultaneously closing material cycles. Even then, the constraint of the limited availability of ecologically productive areas may exist, as the originators of *The Ecological Footprint* have propagated.

A last conclusion is that thermodynamics as the ultimate accountant of the conversion and storage of energy and matter can provide the fundamental tools to assess to what extent an industry and even an economy is sustainable.

NOTES

1 Suppose a scientist, say a physical chemist, discovers a new principle by which solar energy can be converted into electricity. The engineer is required to make this principle work in practice. The economist will be needed to assess the economic feasibility, which should include ecological aspects. But even then the sociologist may be needed to explain why people don't buy it. If the innovation serves a global interest, a multidisciplinary effort with participants also skilled in communicating the essence of their discipline to each other will enhance the successful introduction and acceptance of the innovation.
2 In a wonderful article [12] the late Stephen Jay Gould discusses that life based on solar energy and photosynthesis may be the exception rather than the rule. Equally, he disputes that the bulk of life's biomass should reside in the wood of trees in our forests and argues that the mass of subterranean living material as that of bacteria is comparable to that at the surface and is possibly in excess of it.
3 In a course on this subject that one of the authors gave to a group of Dutch Energy Business Managers, one of the participants conducted the Quiz and was shocked by her result. Her Footprint was 6.55 ha, compared to the Dutch average of 4.7 ha and compared to the 1.8 ha that she is allowed as a global citizen. She decided to use her car less, and use the bicycle instead, to eat more regional vegetables and fruits, less frozen food, and less meat and readymade meals.

REFERENCES

1. Brundtland, G.H. *Our Common Future, The World Commission on Environmental Development*, Oxford University Press: Oxford, 1987.
2. Yoda, S. (ed.). *Trilemma: Three Major Problems Threatening World Survival*, Central Research Institute of Electric Power Industry: Tokyo, 1995.
3. Okkerse, C.; van Bekkum, H. Towards a plant based-economy? In *Starch 96—The Book, Carbohydrate Research Foundation*, Zestec: The Hague, 1997.
4. Harmsen, G.J. *Inaugural Address*. Delft University of Technology, Delft, 1998.
5. *Shell Report*. http://sustainabilityreport.shell.com/2008/servicepages/downloads/files/shell_report_1999.pdf, 1999.
6. Tabaksblatt, M. Sustainability as a challenge. In *Speech at the Occasion of the Unilever Research Awards*, 1999.
7. Merkel, A. The role of science in sustainable development. *Science* 1998, *281*, 336–337.
8. Ayres, R.U.; Simonis, U. *Industrial Metabolism, Restructuring for Sustainable Development*, United Nations University Press: Tokyo, 1994.
9. Takamitsu, S. (ed.). *In Harmony with the Earth*, Central Research Institute of Electric Power Industry: Tokyo, 1996.
10. Nelson, D.L.; Cox, M.M. *Lehninger Principles of Biochemistry*, 3rd edn., Worth: New York, 2000.
11. Lehninger, A.L. *Bio-Energetics*, 2nd edn., W.I. Benjamin: Menlo Park, CA, 1973.
12. Gould, S.J. This view of life: Microcosmos. *Natural History* 1996, *105*(3), 20–23.

13. *Webster's New Twentieth Century Dictionary*, 2nd edn., Simon & Schuster: New York, 1983.
14. Ayres, R.U. Eco-thermodynamics: Economics and the second law. *Ecological Economics* 1998, *26*, 189–209.
15. Baumgartner, S. *Ambivalent Joint Production and the Natural Environment*, Physica Verlag: New York, 2000.
16. Lems, S.; van der Kooi, H.J.; de Swaan Arons, J. The sustainability of resource utilization. *Green Chemistry* 2002, *4*, 308–313.
17. Wilson, E.O. The bottleneck, feature article. *Scientific American*, February 2002, 82–91.
18. Brown, L.R. *Eco-Economy. Building an Economy for the Earth*, Earthscan: London, 2001.
19. Costanza, R. et al. The value of the world's ecosystem services and natural capital. *Nature* 1997, *387*, 253–260.
20. Wackernagel, M.; Rees, W. *Our Ecological Footprint*, New Society Publishers: Vancouver, 1996 (See also the forum on this subject in *Ecol. Econ.* 2000, *32*, 341–394).
21. Georgescu-Roegen, N. Myths about energy and matter. *Growth Change* 1979, *10*(1), 16–23.
22. Dixit, A.K.; Pindyck, R.S. *Investment under Uncertainty*, Princeton University Press: Princeton, NJ, 1994 (See also from the same authors, The options approach to investment. *Harvard Bus. Rev.* May/June 1995, 105–115).
23. De Vos, A. Endoreversible thermoeconomics. *Energy Conversion and Management* 1995, *36*, 1–5.
24. De Vos, A. Endoreversible thermoeconomics. *Energy Conversion and Management* 1997, *38*, 311–317.
25. Energy Needs, Choices and Possibilities. *Scenarios to 2050*, Global Business Environment, Shell International: London, 2001.
26. Baehr, H.D. *Thermodynamik*, 6th edn., Springer-Verlag: New York, 1988.
27. Brian, H. Man and the future environment. *The European Review* 2004, *12*(3), 273–292.
28. Forum: The ecological footprint. *Ecological Economics* 2000, *32*, 341–394.
29. Wilson, E.O. *The Future of Life*, Alfred A. Knopf: New York, 2002.
30. Baumgärtner, S.; de Swaan Arons, J. Necessity and inefficiency in waste generation. A thermodynamic analysis. *The Journal of Industrial Ecology* 2003, *7*, 113–123.
31. Szargut, J.; Morris, D.R.; Steward, F.R. *Exergy Analysis of Thermal, Chemical and Metallurgical Processes*, Hemisphere Publishing Corporation: New York, 1988.
32. Heyduk, A. F.; Nocera, D.G. Hydrogen produced from hydrohalic acid solutions using a two-electron mixed-valence photocatalyst. *Science* 2001, *293*, 1639.
33. Georgescu-Roegen, N. Myths about energy and matter. *Growth Change* 1979, *10*, 16–23.
34. Wall, G. Exergy conversions in Japanese society. *Energy* 1990, *15*, 435–444.
35. Ayres, R.U.; Kneese, A.V. Externalities: Economics and thermodynamics. In *Economics and Ecology: Towards Sustainable Development*, Archibugi, F.; Nijkamp, P. (eds.), Kluwer Academic Publishers: Dordrecht, 1989.
36. De Wulf, J.; van Langenhove, H.; Mulder, J.; van den Berg, M.M.; van der Kooi, H.J.; de Swaan Arons, J. Illustrations towards quantifying the sustainability of technology. *Green Chemistry* 2002, *2*, 108–114.
37. Wassenaar, J.A. *Sustainability of the Potato Production Chain. A thermodynamical Approach*. Master's thesis, Delft University of Technology: Delft, 2000, p. 38.
38. Gerngross, T.U.; Slater, S.C. How green are green plastics. *Scientific American*, August 2000, 37–41.
39. Berthiaume, R.; Bouchard, C.; Rosen, M.A. Exergetic evaluation of the renewability of a biofuel. *Exergy, An International Journal* 2001, *1*(4), 256–268.
40. Brown, M.T.; Ulgiati, S. Emergy based indices and ratios to evaluate sustainability. *Ecological Engineering* 1997, *9*, 51–69.
41. Odum, H.T. Self-organization, transformity and information. *Science* 1988, *242*, 1132–1139.
42. *Shell Energy Scenarios 2050*, 2008. http://www.shell.com/home/content/aboutshell/our_strategy/shell_global_scenarios/dir_global_scenarios_07112006. html

14 Efficiency and Sustainability in the Chemical Process Industry

14.1 INTRODUCTION

The process industry is a large consumer of raw materials, which are utilized both as feedstock for its numerous products and as an energy source to drive its various processes. In the scope of sustainable development, consensus on the limited availability of our natural resources, and on the need for closed cycles in our ecosphere, has grown. Hence, the current approach of our process industry is in question, in terms of both efficiency and sustainability. First, we need to address the subject of efficiency. There is a need for quantitative figures on the efficiency with which natural resources are consumed. Such quantitative figures can be provided from a thermodynamic analysis of the process indicating the discrepancy between the ideal thermodynamic process and the state of the art in current process technology. Next, we will address the question of sustainability. Is it possible to indicate in quantitative terms to which extent a process or an industry is sustainable? This chapter has been based largely on one of our earlier publications [1].

14.2 LOST WORK IN THE PROCESS INDUSTRY

In Part I, we have defined the concept of *lost work* and identified it with the *entropy production* in the process. With the help of irreversible thermodynamics, we have found its origin in the finite *driving forces* of the process. We have also introduced in Chapter 6 the concept of *minimum work* and shown how this concept leads to a convenient thermodynamic tool called *exergy*, or *available work*. It was also shown that *exergy analysis*, also known as *thermodynamic analysis*, can give an insight into the amount and origin of lost work and thereby into the thermodynamic efficiency of a process. Simple illustrations were given in Chapters 6 and 8, while real case studies were dealt with in Chapters 9 through 11.

Exergy analysis has so far drawn most attention in the energy-system area, where heat is converted into power or electricity. It has penetrated less into the chemical process industry, perhaps because of its greater complexity. In real material conversion processes, primary materials, natural resources, are converted into consumer materials

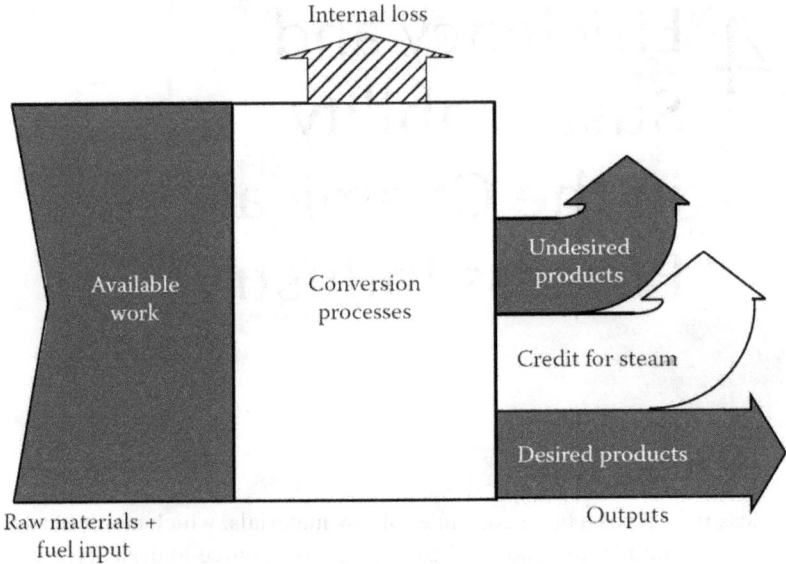

FIGURE 14.1   Grassmann diagram of a typical conversion process.

and heat. These processes do not proceed ideally, so part of the work available in the primary materials will be lost. To obtain a "feeling" for lost work in the process industry, the production processes of several "large-quantity" products have been analyzed. This chapter only focuses on input and output streams of the conversion process; production of the natural resources, transportation, and storage are excluded from the system boundary, because most often the largest part of lost work is incurred in the conversion step. Elaborate lost work analyses are given, for example, by Hinderink et al. [2] and Wall [3]. Figure 14.1 illustrates the general result of a lost work analysis of a material conversion process in a Grassmann diagram. A lost work analysis reveals internal losses due to process imperfections, whereas an energy analysis deals with losses in physical streams such as those of waste material and heat leaving the process. It should be noted that the U.S. Department of Energy has commissioned a study to quantify process efficiencies in the U.S. chemical industry [4]. It is not clear to the authors, however, how the study was exactly performed, since for certain cases, the ratio of lost work to exergy input exceeds 100%. It may well be that the chemical exergy component was not taken into account, as this seems to not be mentioned in the report.

The input side of such a process is represented by natural primary resources. No distinction has been made between resources used as feedstock and resources applied as fuel; they are all quantified by their exergy value, the universal measure. The output side is represented by the exergy of the desired product(s) and of recovered useful heat, in the form of a steam credit. By comparing the total amount of exergy entering and leaving the process, the loss in available work is revealed, which is due to either process inefficiencies or material/heat release to the environment, so internal and external losses are lumped. All data have been taken from published literature (Table 14.1).

TABLE 14.1

Results of Global Lost Work Analyses of Several Important Production Processes.

Final Product	Molecular Weight	Raw Materials	Data Taken From	Technology Level	Available Work (kJ/mol Final Product)					Thermodynamic Efficiency[a] (%)
					Raw Materials	Final Product	Steam Credit	Lost Work		
Hydrogen	2	Natural gas/air	[5]	1990	409	236	28	145		58
Hydrogen	2	Natural gas/air	[6]	2007	317	236	0	81		74
Ammonia	17	Natural gas/air	[7]	1980	763	338	85	340		44
Aluminum	27	Bauxite	[8,9]	1990	4703	888	n.a.	3815		19
Methanol	32	Natural gas/air	[10]	1985	1136	717	80	339		63
Oxygen	32	Air	[11]	1980	64	4	n.a.	60		6
Urea	60	Natural gas/air via ammonia	[7,12]	1980	1590	686	150	754		43
Nitric acid	63	Natural gas/air via ammonia	[7,13]	1975	995	43	151	801		4
Copper	63.5	Copper ore	[9,14]	1980	1537	130	n.a.	1407		9
Methane	16	—	—	—	830	—	—	—		—

[a] Excluding steam credit.

14.3 THE PROCESSES

Table 14.1 gives a "thermodynamic blueprint" of some large-scale production processes. The numerical values presented refer to the technology level of the 1970s or 1980s and are based on primary, natural, resources only, such as natural gas and air. Because only primary resources are allowed to enter the processes, several subprocesses can be present inside the system, for instance for the generation of intermediate products, or for the generation of steam or electricity. Lost work involved with the latter type of subprocesses is handled by using commonly applied thermodynamic efficiencies (e.g., 50% for power production *via* cogeneration).

An example of a process with an intermediate product is the urea process. The second step, starting from ammonia, is over 90% efficient, whereas the total process, having ammonia just as an intermediate product, shows an efficiency of only half of this value. For the nitric acid process, Figure 14.2, the second step is the least efficient as indicated by the simplified Grassmann diagram in Figure 14.3 for this process. However, an important message that Figure 14.3 conveys is that large losses occur in the conversion processes, namely, the transformation of CH_4 into NH_3 and NH_3 into nitric acid. Nitric acid has a low exergy content, and it is partly responsible for the large loss incurred in the conversion step. Losses, therefore, can also be a useful metric in the evaluation of processes and can be caused by process inefficiencies or a final product that is simply low in exergy.

Results of lost work analyses strongly depend on the system boundary considered and the credit that is given to coproducts and by-products. Therefore, the analysis results can vary per author.

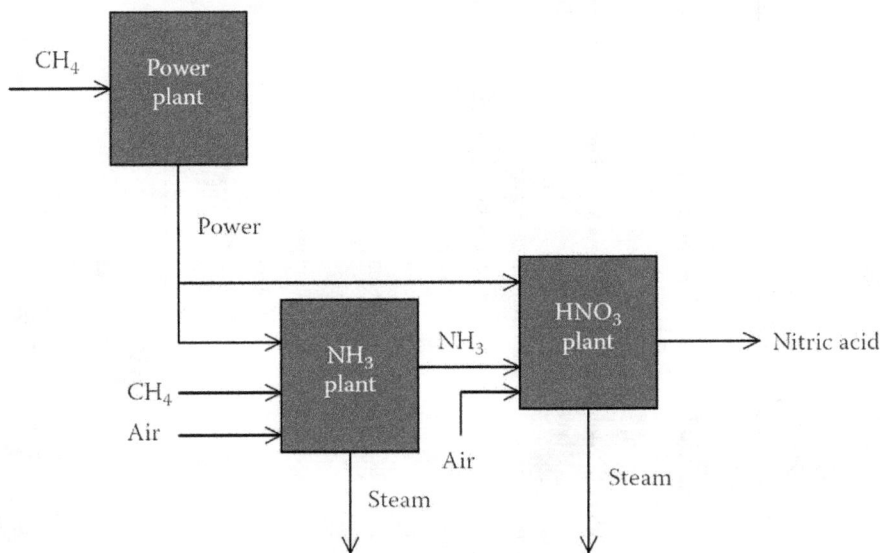

FIGURE 14.2 Flowchart of a nitric acid process.

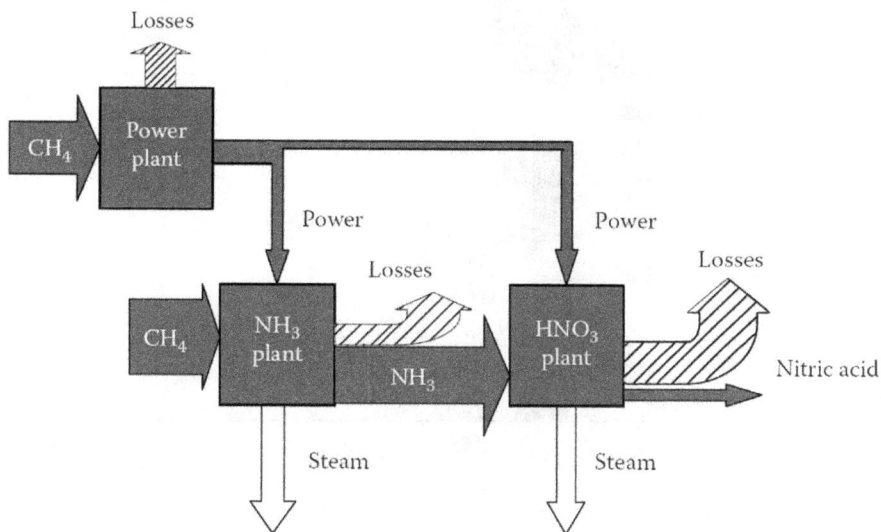

FIGURE 14.3 Simplified Grassmann diagram of the nitric process shown in Figure 14.2.

14.4 THERMODYNAMIC EFFICIENCY

Although efficiency values can be misleading because they can be defined in numerous ways [15,16], they are easy to handle. The definition of thermodynamic efficiency applied here is the ratio between the available work or exergy of the desired products, excluding useful heat, and the primary resources applied. Usually, thermodynamic efficiencies based on lost work analysis do not exceed 70% when starting from primary resources. At the higher side of the efficiency range, the production of organic products can be found, while inorganic and metallurgical processes are at the lower side of the efficiency range. A combination of low efficiency and high input of exergy indicates, in general, the need for process improvement. It is important to distinguish between efficiency and lost work. Lost work refers to the process, and is related to the driving forces, whereas efficiency has to include the exergy of the desired product. In the case of nitric acid, as we saw earlier, it is low. Lost work is an absolute metric, as opposed to efficiency, which is a relative metric. An overview of absolute lost work figures for the process industry, such as shown in Figure 14.4, is more distinct and can be of use to determine which products and/or which processes need to be reconsidered in view of sustainable development.

14.5 EFFICIENT USE OF HIGH-QUALITY RESOURCES

Figure 14.5 shows lost work figures for the utilization of natural gas for various purposes. From this picture, it can be concluded that it is best to use natural gas for those chemical processes in which its exergy eventually ends up in the desired products. The basic rule behind this conclusion is that the degradation of exergy has to be

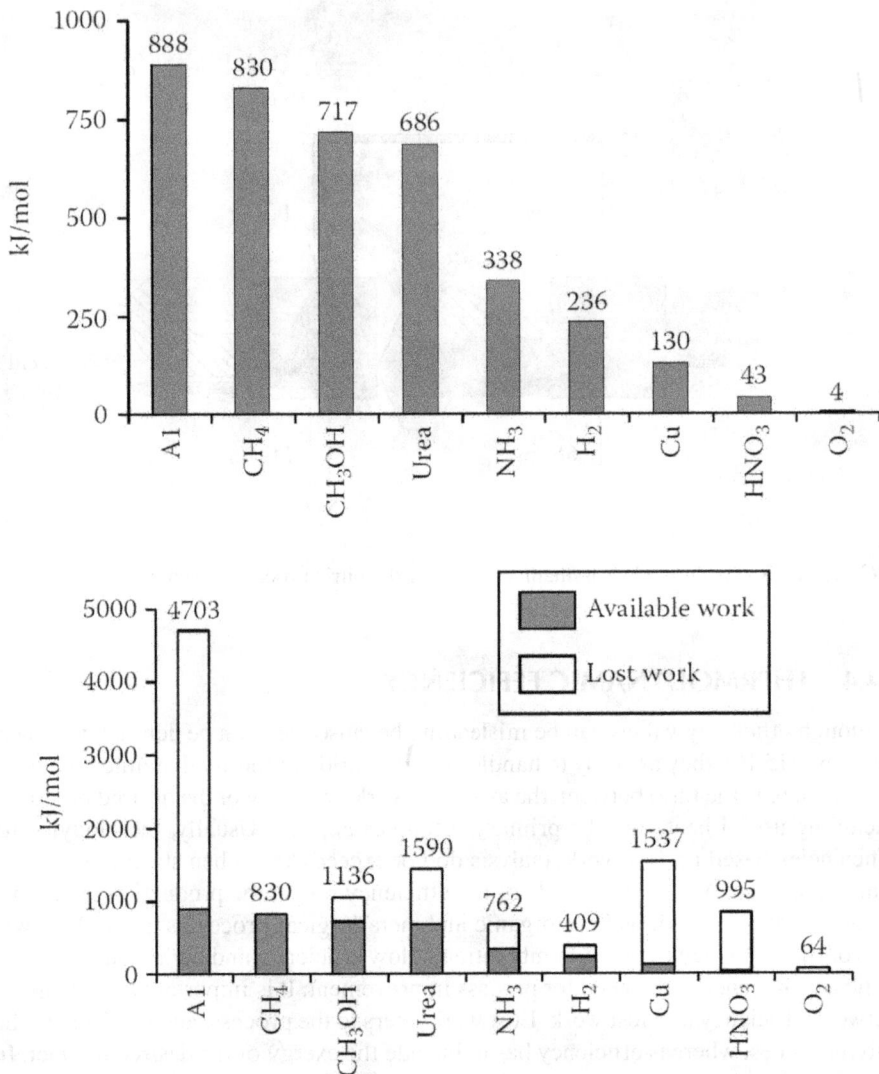

FIGURE 14.4 Available and lost work for typical chemical processes.

delayed as long as possible. This rule facilitates the choice of chemical routes and raw materials. For the nitric acid process, this rule is broken; the very high-quality raw material natural gas contributes negligibly to the exergy content of the low-quality final product nitric acid. This suggests that an incredible drop in exergy takes place per atom N (NH$_3$ → HNO$_3$) and perhaps an alternative chemical route should be sought, or the drop in exergy should be utilized. Also, the direct use of high-quality natural gas for low-quality heating purposes does not coincide with the idea of sustainability.

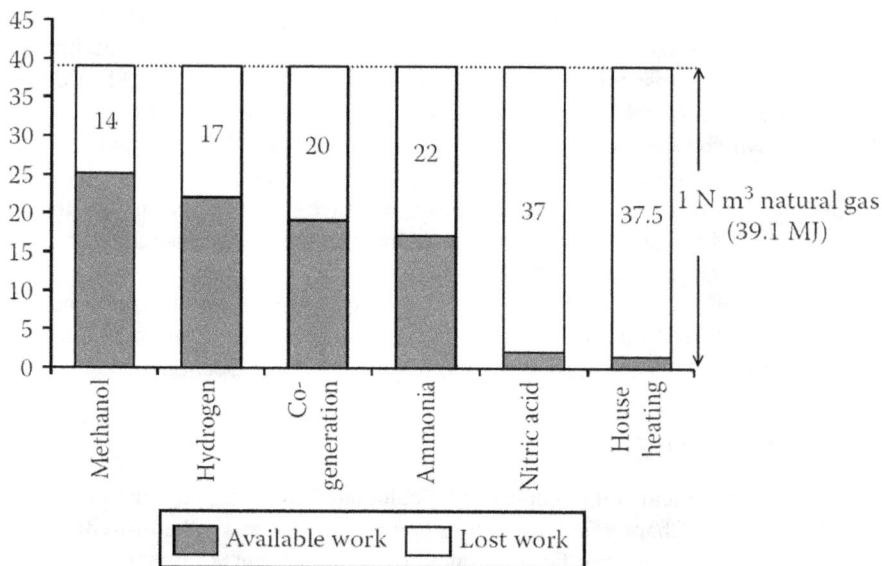

FIGURE 14.5   Lost work figures for the utilization of natural gas.

14.6   TOWARD SUSTAINABILITY

In the discussion about how we should set up our chemical process industry in the near future, the sustainability issue is of prime importance. Sustainability in the ecological sense means that we do not place an intolerable load on the ecosphere and that we maintain the natural basis for life. The complexity of the chemical industry with its numerous products has made us lose sight of the associated ecological impact of these products' life cycles: When you produce something, you also produce long-term effects. In economics, this concept is known as *joint production* [17].

The improvement of thermodynamic efficiency is frequently but erroneously considered as the only contribution to sustainability. An increase of thermodynamic efficiency, however, has to do with a lowering of the rate with which our nonrenewable natural resources are consumed, whereas sustainability implies more than that, as we have discussed in Section 13.6.

Although the figures presented in this chapter are quite indicative, they do not reflect the degree of sustainability of the process but rather its efficiency. In other words, the extent to which the process uses renewable materials is not given explicitly. Actually, a chemical conversion process can be 100% efficient when all the originally available work ends up in the desired product(s). In real processes, exergy is lost. Hence, more exergy enters the process than leaves it. This excess of exergy entering the process to make it proceed has to originate from renewable sources, such as solar exergy, in order to contribute to sustainability.

Such a "balanced" process industry can still lead to exhaustion of our natural resources if the exergy content of the nonrenewable end products is not utilized at the end of the product life. If we go one step further, our starting material should be renewable, for example, CO_2 and water. Real sustainable systems/chains need to be circular with respect to matter; outputs should become inputs. The driving force for such a sustainable system must stem from solar energy, as we discussed in Chapter 13, which is essentially available in a large quantity. Wall [15] and Gong and Wall [16] give elaborate discussions on exergy and sustainability.

Looking into this matter more carefully, we have come to the insight that the extent to which a process contributes to sustainability can be characterized by three parameters. A first parameter is the thermodynamic efficiency of the process. A second parameter needs to reflect the extent to which use has been made of renewable resources. Finally, a third parameter is needed to indicate the extent to which cycles have been closed. This has been discussed at great length in Section 13.6.

14.7 CHEMICAL ROUTES

The exergy of chemical components can be calculated from thermodynamics, as has been illustrated in Chapter 7. This exergy can be considered as the minimum work needed to synthesize the specific component from constituents of its surroundings. It has been shown that in practice the production of desired chemicals requires far more work than indicated by the exergy of this desired product. In other words, the exergy entering and leaving such processes does not balance, so work is lost. The challenge of our process industry is to limit the losses, while still being able to let our processes run with sufficient speed.

In past decades, increasing energy efficiency was accomplished mainly by complex heat integration within existing chemical processes, requiring considerable investments. The thermodynamic or exergy analyses described by Hinderink et al. [2], however, show that the chemical reaction step largely determines the overall thermodynamic efficiency. Chemical reactions have been found to be a notorious source of lost work. If chemical processes are developed from scratch by state-of-the-art methods—namely, by structured process synthesis procedures—attention can be paid to the core of the process, that is, the chemical reactions or the chemical reactor. Then, a significant improvement in energy efficiency and process economics can be achieved simultaneously [17].

Losses resulting from chemical reactions can be viewed in the same way as losses resulting from heat exchange as we discussed in Chapter 4. The driving force for heat transfer is the temperature gradient, which determines not only the rate of transfer, but also the degree of devaluation of exergy. Chemical reactions also proceed along a gradient from high to low chemical affinity. On flowing along this gradient, heat is released and exergy is lost for a spontaneous chemical reaction. The relation between the Gibbs energy of reaction and lost work is linear. This relationship has been established by Denbigh [18] and is discussed by Hinderink et al. [19]. This insight is of prime importance for the development of future chemical routes [20].

For more sustainable chemical routes, chemical gradients should be reduced, or should be counterbalanced by chemical reactions proceeding against their gradient, a principle of which biological systems make extensive use [21]. In this view, there is an analogy between heat pinch and reaction pinch. Examples of reactions proceeding with a large gradient are the production of nitric acid by the partial oxidation of ammonia and the conversion of H_2S into elemental sulfur and steam. The development of new chemical processes should thus focus on approaching exergy-neutral reactions. In addition, one should also look for sophisticated utilization of the Gibbs energy of reaction [22], for example, by using it directly to drive a separation, such as reactive distillation. Finally, although Figure 14.4 is fairly indicative, the graph lacks information on which part of the primary exergy is renewable. If, for example, the energy needed for the metallurgical processes comes from hydro-energy, our conclusions with respect to the improvement potential of these processes will have to be adapted.

14.8 CONCLUDING REMARKS

Some general observations that can be made from lost work analyses are

- The amount of lost work differs from product to product and from process to process and depends largely on how skilled we are in process design.
- Processes showing a large steam credit, albeit useful, should be distrusted, because this implicitly means that more primary exergy is applied than is actually needed.
- All basic efforts to reduce lost work have to come from postponing as long as possible the devaluation of exergy; preserve the quality of energy and matter.

With regard to process sustainability, the following observations can be made:

- The best source for exergy is a renewable resource.
- A more complete thermodynamic analysis of processes deals not only with the efficiency but also with the extent to which renewable resources have been used and to which extent material cycles have been closed.
- A nonbasic approach toward efficiency and sustainability is in our view unacceptable given the importance and urgency of these issues. Therefore, it is advisable to benefit from the quantitative power of the extended thermodynamic analysis proposed.

REFERENCES

1. Hinderink, P.; van der Kooi, H.J.; de Swaan Arons, J. On the efficiency and sustainability of the process industry. *Green Chemistry* 1999, G176–G180.
2. Hinderink, A.P.; Kerkhof, F.P.J.M.; Lie, A.B.K.; de Swaan Arons, J.; van der Kooi, H.J. Exergy analysis with a flowsheeting simulator. Part 2: Application: Synthesis gas production from natural gas. *Chemical Engineering Science* 1996, *51*(20), 4701–4715.
3. Wall, G. Energy flows in industrial processes. *Energy* 1988, *13*(2), 197–208.
4. Ozokwelu, D.; Porcelli, J.; Akinjiola, P. *Chemical Bandwidth study, Exergy Analysis: A Powerful Tool for Identifying Process Inefficiencies in the U.S.*, Chemical Industry: Beijing, 2006.
5. Giacobbe, F.G.; Iaquaniello, G.; Loiacono, O. Increase hydrogen production. *Hydrocarbon Processing* 1992, *3*, 69–72.

6. Feng, W.; Ji, P.; Tan, T. Efficiency penalty analysis for pure H_2 production processes with CO_2 capture. *The AIChE Journal* 2007, *53*(1), 249–261.
7. Cremer, H. *Thermodynamic Balance and Analysis of Syngas and Ammonia Plant. ACS Symposium Series*, Vol. 122, ASME: Washington, DC, 1980.
8. Habersatter, K. *Ökobilanz von Packstoffen stand. Schriftereihe Umwelt*, Vol. 132, Buwal: Bern, 1991.
9. Szargut, J.; Morris, D.R.; Steward, F.R. *Exergy Analysis of Thermal, Chemical, and Metallurgical Processes*, Hemisphere Publishing Corp.: New York, 1988.
10. Supp, E. Improved methanol production and conversion technologies. *Energy Progress* 1985, *5*(3), 127–130.
11. *Ullmann's Encyklopaedie Der Technische Chemie* (Band 20), Verlag Chemie: Weinheim, 1981.
12. Pagani, G. New process gives urea with less energy. *Energy Progress* 1985, *5*(3), 127–130.
13. Lowenheim, F.A.; Moran, M.K. *Industrial Chemicals*, Wiley: New York, 1975.
14. Boustead, F.; Hancock, G.F. *Handbook of Industrial Energy Analysis*, Wiley: New York, 1979.
15. Wall, G. Exergy—A useful concept within resource accounting. *Report No.* 77–42: *Institute of Theoretical Physics*, Chalmers University of Technology, Göteborg, 1977.
16. Gong, M.; Wall, G. On exergetics, economics and optimization of technical processes to meet environmental conditions. In *Proceedings of the International Conference on Thermodynamic Analysis and Improvement of Energy Systems, TAIES'97*, Ruixian, C. (ed.), Beijing, 1997, pp. 453–560.
17. Baumgärtner, S. *Ambivalent Joint Production and the Natural Environment*, Physica Verlag: New York, 2000.
18. Denbigh, K.G. The second law efficiency of chemical processes. *Chemical Engineering Science* 1956, *9*(1), 1–9.
19. Hinderink, A.P.; Kerkhof, F.P.J.M.; Lie, A.B.K.; de Swaan Arons, J.; van der Kooi, H.J. Exergy analysis with a flowsheeting simulator. Part I: Theory; Calculating exergies of material streams. *Chemical Engineering Science* 1996, *51*(20), 4693–4700.
20. Ratkje, S.K.; de Swaan Arons, J. Denbigh revisited: Reducing lost work in chemical processes. *Chemical Engineering Science* 1995, *10*, 1551–1560.
21. Lehninger, A.L. *Bioenergetics*, 2nd edn., W.I. Benjamin: Menlo Park, CA, 1973.
22. Harmsen, G.; Hinderink, A.P. We want less: Process intensification by process synthesis methods. In *Third International Conference on Process Intensification for the Chemical Industry* (Publication No. 38), Green, A. (ed.), Antwerp: BHR Group Conference, 2004, pp. 23–28.

15 Plastics Recycling

15.1 INTRODUCTION

It was in 1862 that Alexander Parkes introduced the first man-made plastic at the London International exhibition [9]. The product was not a commercial success, though; it took another half century or so, and two world wars to see the age of plastic truly commence. Currently, the annual production of plastics is around 400 million metric tons [5] and seems to be growing faster than linear, if not exponentially (see Figure 15.1). It is expected to be close to 600 million metric tons by 2050 [6] [4]. As of the end of 2022, the global thermoset capacity is around 48 million metric tons [3].

Indicative of the breakdown of plastic types, as of 2017 [2], is the following chart (Figure 15.2):

Figure 15.2 shows that around 60% or so of all plastics are essentially hydrocarbons, and 40% have oxygen, chlorine as part of their backbone. Additives may add different hetero atoms. The importance of this will become clear later.

Plastic was indeed a material that was cheap, light, and disposable and helped society maintain a high standard of living. Fast forward another 70 years, and it is indeed some of the same features that led to plastic being ubiquitous in the world, but also creating various problems. Plastic unfortunately would find its way to the environment and oceans and accumulate, impacting humans and wildlife. This accumulation is both visible but also invisible to the naked eye. The consumers in the western sphere were aware, but perhaps content that they were doing their bit by collecting and recycling the plastic.

In the early 1990s, the economic development and rise in living standards in China led to an increased demand for plastics. The local production was unable to meet these demands, and China started importing waste plastics to recycle these, which were also available at a lower price. In the early twenty-first century, China became the second largest producer of plastics, but also the largest importer of plastics [1].

Consumers in many countries would collect the plastics, and unbeknownst to them, these would be shipped to China. Unfortunately, the quality of recyclable materials gradually declined, and large amounts of waste was mixed with food, garbage, and other pollutants, which burdened the Chinese infrastructure.

China banned the import of certain types of plastics, and growing consumer pressure and awareness of sustainability put the industry at a crossroads, and sparked efforts to recycle plastics and introduce circularity.

DOI: 10.1201/9781003304388-19

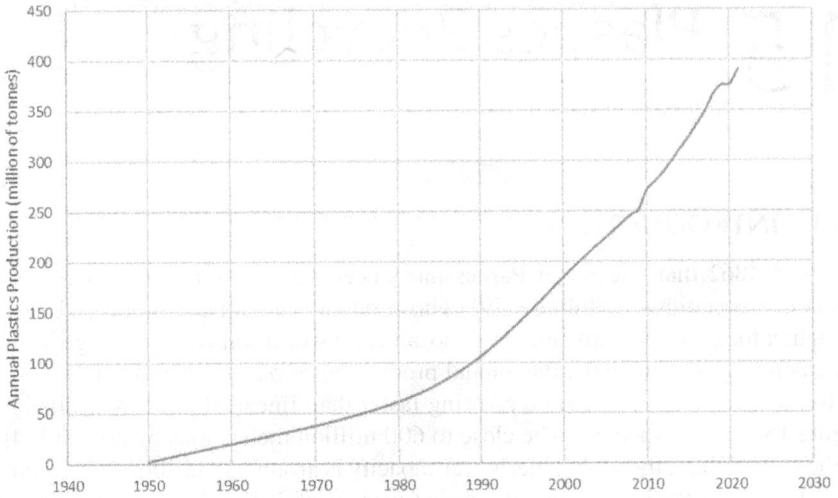

FIGURE 15.1   Annual production of plastics since 1950.

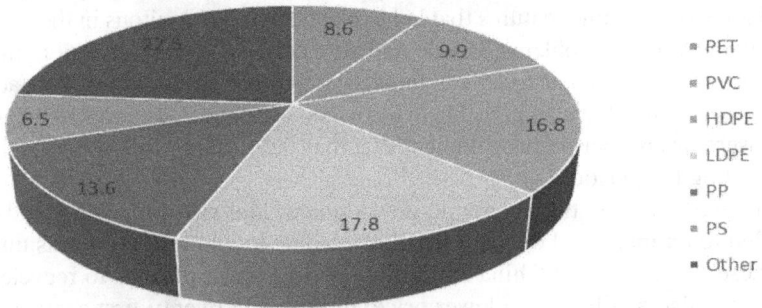

FIGURE 15.2   Worldwide production of plastics by type.

15.2 SORTING OF PLASTIC

As alluded to before, there are many different types of plastic, but generally, they are grouped as follows (Table 15.1). Note, that in this classification, only thermoplastics are captured, and thermosets, which in the vernacular are also referred to (though incorrectly) as plastics, fall outside of this classification. Their total volume as stated earlier is approximately 10–15% of worldwide plastic production.

TABLE 15.1

Classification of Plastics [7].

Type	Classification
PET. Poly ethylene terephthalate. It is used for food and drink packaging. It is recycled considerably.	1
HDPE. High Density Polyethylene. It is recycled considerably.	2
PVC. It is hardly recycled.	3
LDPE. Low Density Polyethylene. Plastic bags, plastic wrap. It is often not recycled.	4
PP. Polypropylene. It is used in a variety of applications	5
PS. Polystyrene. It is not recycled considerably.	6
Other types. This is a catch all.	7

As stated earlier, at present, the world generates around 400 million tons of plastic, of which about 10–15% is recycled, which means that roughly 360 million tons is either landfilled, incinerated, or leaks to the environment [8]. Unlike other materials which are commonly recycled such as glass and aluminum, plastics are *not* indefinitely recyclable, and generally a degradation in properties occurs, which is ameliorated by additives, mixing with virgin plastic, or choosing less demanding applications (downcycling).

At this point, it is useful to pause, and examine the differences between recycling, upcycling, and downcycling (Neutrall.us, 2023).

- **Recycling**: The process of collecting waste materials and processing them to obtain new items. Examples are the recycling of glass and aluminum which can be recycled indefinitely without losing quality or purity. Plastics, in general, do not allow for infinite recycle.
- **Upcycling:** Upcycling involves reusing and repurposing an item that would otherwise end in the trash and transforming it into something new and more valuable.
- **Downcycling:** As the word suggests, downcycling decreases the quality and value of the final products. It is generally a process consisting of breaking a product down after use into its component materials and then making a new product of lesser value.

Upcycling conserves resources and prolongs the life of materials. Downcycling diminishes the value, but also conserves resources to a certain degree.

A key challenge that we face is that in order to be successful in recycling the plastics, we need to sort and collect them, starting at the household level. This means that we need to clean the materials, the best we can, and then send them for further collection. In some countries, there are different collection outlets for the different types, while in others there is simply one, and it is mixed with paper.

15.3 WASTE FRAMEWORKS

The so-called Waste Framework Directive in the European Union (European Commission, 2023) sets out basic requirements and concepts to define a waste hierarchy. The preferred option is to prevent waste, whereas disposal to landfill or incineration is the least preferred option:

- Prevention
- Preparing for reuse
- Recycling
- Recovery
- Disposal

This framework has important considerations when designing products as the full life cycle should now be taken into account. In this chapter we will discuss the recycling aspects.

As we learned in Chapter 10, Separations, the process of separation (or sorting) requires exergy, and simply collecting all waste (= mixing) and then separating out the various plastic types is an intensive process, and will require some rethinking of how consumers play a role in this value chain.

Effective waste collection is extremely important to enable efficient recycling in the circular economy. By separating waste correctly at the point of collection, the recycling process is more efficient and will increase the quality and quantities of recycled products. Improved waste collection positively impacts the waste streams and their suitability for downstream pre-treatment, sorting, and recovery operations (Plastics Europe, 2023). In the ideal case, all the plastic waste is suitably segregated, and can be processed further.

In Figure 15.3, it can be viewed as the fate of exergy, with losses at each step. It is logical to assume that to go back, one must invest exergy or effort. Also clear from the diagram is that reducing spent plastic to e.g. fuels is not an example of downcycling but simply a delay of degradation of exergy.

In general the sorting is based on gravity, size, as well as other properties. (Figure 15.4)

In addition to the gravity sorting, smart sorting techniques, based on identification using IR detection or hyperspectral imaging, etc., are used, but it remains a tedious process.

As digital technologies are evolving, incorporation of digital watermarks is becoming more and more common, and various companies have joined the Digital Watermarks Initiative Holy Grail 2.0 (Naitove, 2020). Even with these high-tech sorting techniques, there will be a bound on the amount of mono streams suitable for mechanical recycling, whereas mixed streams will remain, and will have to be depolymerized, or pyrolyzed.

15.4 MECHANICAL RECYCLING

Mechanical recycling (Lange, 2022) of classes 1, 2, 4, 5 (PET, PE, PP) is based on the principle that once a suitably pure and clean mono-stream is obtained, the stream is

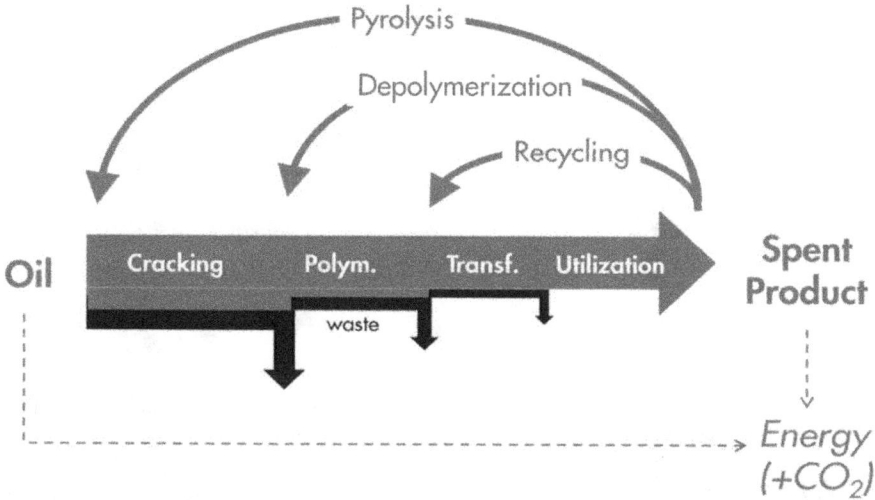

FIGURE 15.3 Options for recycling plastic waste. (With permission from Lange, Sustainable Development: Efficiency and Recycling in Chemicals Manufacturing, 2002.)

granulated, mixed with a virgin polymer of the same family, and additives are added, to ensure the quality is up to spec. It is key to understand that a melting step is present in this process, so this concept only really works for the thermoplastics, which melt upon heating (albeit with perhaps some product degradation upon repeated heating, hence the additives).

Depending on the application, and content of recycled content, the materials are recycled or downcycled, with less demanding specifications. Recycling does alter the compositions of the molecules, and hence it is not possible to recycle indefinitely in a mechanical fashion: the ultimate fate will be material which is no longer suitable for mechanical recycling.

Moreover, small amounts of additives are added *depending on the application*. Recycled material will therefore have some amounts which are "average" of the marketplace, and not a direct fit for the application—downcycling is practical.

It is also important to recognize that some applications do not allow for mechanically recycled materials, such as food packaging.

Thermoset materials, that is materials such as polyurethane, rubber tires, etc., do not melt upon heating, but will decompose or oxidize. These therefore require a different strategy such as downcycling and use as a filler. However, these approaches, of course, do not remove the need to have full molecular deconstruction.

15.5 CHEMICAL RECYCLING

Broadly speaking, chemical recycling comes in two flavors, namely depolymerization or pyrolysis (Lange, 2022). In depolymerization, the product is converted back to the original source molecules which can readily be polymerized into the new

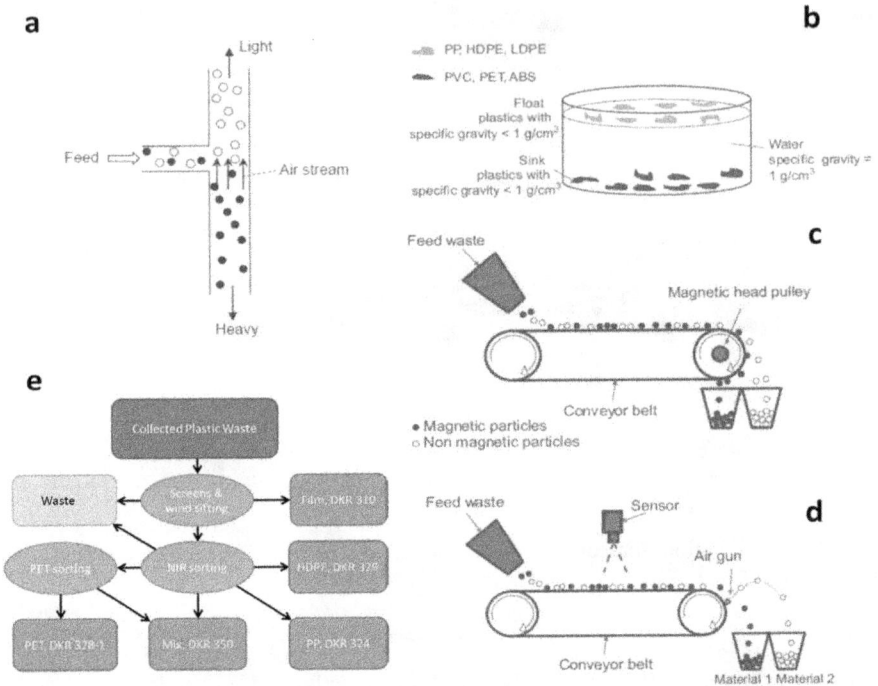

FIGURE 15.4 High level overview of sorting techniques: (a) air classifier, (b) sink-float sorting, (c) magnetic sorting, (d) sensor based sorting, (e) DKR plastic fractions. (With permission from Lange, Managing Plastic Waste—Sorting, Recycling, Disposal and Product Redesign, 2022.)

product again. In Pyrolysis, upon exposure to heat, a mixture of hydrocarbons, not unlike pyoil, is obtained, which can be suitably upgraded and added into the mix of feedstocks.

Figure 15.5 is instructive to decide how to chemically recycle certain classes of polymers. On the x axis, the heat of polymerization is shown, with a cut-off of high and low heat of depolymerization (dH = 70). The high heat of polymerization denotes the addition polymers, and the low heat of polymerization denotes the condensation polymers. The y axis denotes the cumulative amounts of resources that are consumed when producing the polymer. There are four quadrants, where the ideal is the bottom left, with low thermal energy requirements, and low waste. The bottom right lends itself best to cracking to feedstock such as pyoil, whereas the top right lends itself best to reuse and mechanical recycle. PVC is one of the polymers in the top right corner, and generally does show up in mixed plastic streams. The presence of chlorine is typically what causes headaches in the processing.

The top left quadrant lends itself well to depolymerization, though re-use and mechanical recycling are good options there as well considering the high waste generation.

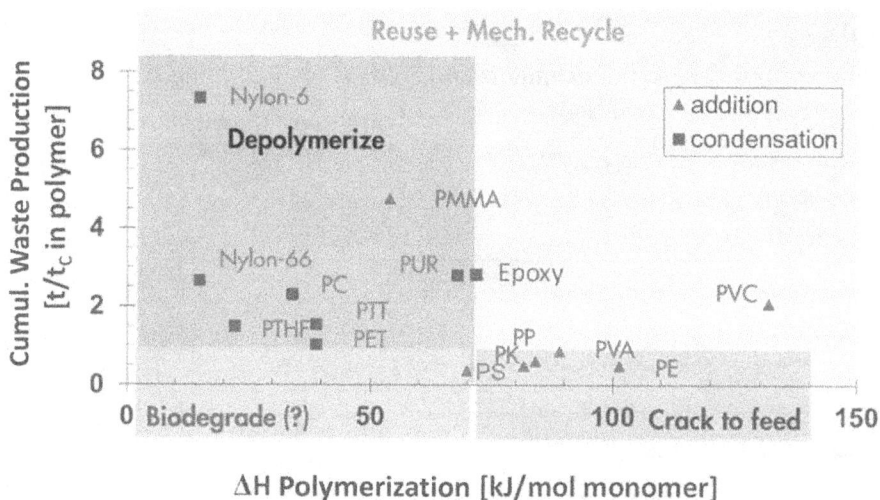

FIGURE 15.5 Options for plastic recycling (PC, p-carbonate; PTHF, p-tetrahydrofuran; PTT, p-trimethyleneterephthalate; PET, *p*-ethyleneterephthalate; PMMA, *p*-methylmetacrylate; PUR, p-urethane; PS, p-styrene; PP, *p*-propylene; PK, p-ketone; PVA, p-vinyl alcohol; PE, *p*-ethylene; PVC, p-vinyl chloride). (With permission from Lange, Managing Plastic Waste—Sorting, Recycling, Disposal and Product Redesign, 2022.)

15.6 MELT PYROLYSIS OF POLYOLEFINS[1]

Polyolefins cannot be depolymerized back to their monomeric constituents. Depolymerization requires harsh pyrolysis conditions and generally leads to a complex mixture of hydrocarbons, i.e., a general feedstock (lower right quadrant in Figure 3). The pyrolysis of polyolefins produces paraffinic/olefinic waxes under moderate temperature (450 °C) conditions, an aromatic product at more severe conditions, and olefin-rich gas and char at the highest temperature (~700 °C). Such a hydrocarbon product can be processed into a synthetic fuel, but fractions of aliphatic products produced under mild conditions can also be used as chemical feedstock and cracked into lower olefins, generally after removing the heaviest product and hydrotreating the desired distillate fraction. Steam cracking of plastic pyrolysis oil is expected to deliver olefins and aromatic base chemicals with ~65 wt % yield, with a coproduction of ~10 wt % fuel gas, ~10 wt % aromatic gasoline, and 15 wt % aromatic fuel oil.

Pyrolysis is no new technology. Oil refineries have been applying it on a large scale in various forms for decades for upgrading heavy oil fractions into gas and distillates. These technologies are then called thermal crackers, visbreakers, or cokers. Pyrolysis has also been explored for processing plastic waste by major chemical producers for some 30 years. Although technically successful, these technologies were not commercialized because they could not compete with cheap crude oil. The rise in oil price in the early 2000s encouraged start-up companies to

TABLE 15.2

Sample of Industry Announcements and Pledges for Recycling.

Company	Announced Capacity	Year
SABIC (chemical parks in Europe, 2020)	200 kTA	2025
BASF (2021)	250 kTA	2025
ExxonMobil (2021)	500 kTA	2022

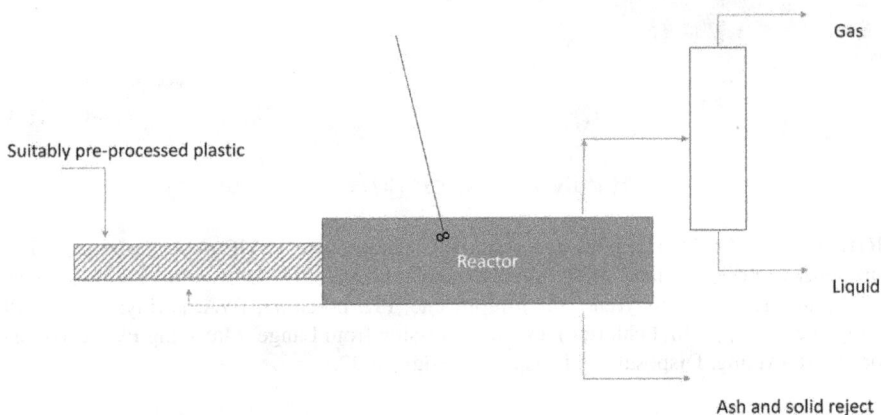

FIGURE 15.6 Typical flow diagram of melt pyrolysis.

revisit plastic pyrolysis, which led to a plethora of technology providers today. More recently, the oil and chemical majors have also joined the effort, not so much with more pyrolysis technologies but rather with plans to process the resulting pyrolysis oil in their steam cracker.

For instance, Shell announced in November 2019 that it was processing the first truck of plastic pyrolysis oil in its cracker at Norco, US, and announced its ambition to ramp up the volume of plastic recycling fed to the cracker to 1 Mt/a by 2025. Numerous chemical producers have embarked on this journey and made similar pledges as is shown in Table 15.2.

Polyolefins are generally pyrolyzed by so-called melt pyrolysis. A conceptual process scheme is provided in Figure 15.6 (top).

Accordingly, the plastic is fed and melted into an extruder, optionally with recycled liquid, then fed to a large vessel that is heated to 450–500 °C and is mechanically agitated. The cracked vapor is removed at the top of the vessel and, subsequently, condensed to liquid pyrolysis oil. The incondensable gases can be used for heating the reactor. The char is removed at the bottom of the vessel and disposed of. A key contaminant is potentially Cl containing products, or other heteroatoms.

The pyrolysis of polyolefins produces a waxy liquid product with 75–80 wt % yield and consumes about 1.5–2 GJ of energy per ton of liquid product, which represents less than half the energy available in the 10–15 wt % of gas produced during pyrolysis. Hence, the pyrolysis process can be run on its own by-product. Tight temperature control and extensive agitation seem essential to minimizing coke deposition and achieving high oil yields. Kinetic modeling and the role of heat and mass transfer in pyrolysis are discussed extensively in the literature.

As mentioned earlier, pyrolysis of plastics is similar to the pyrolysis of hydrocarbons, and is essentially heat and mass transfer limited—the process requires heat to be put into the system, to break down the polymers, and the constituent products need to find their way out. As such, some of the designs are conceptually similar to visbreakers.

These requirements seem to limit the scale of pyrolysis reactors at some 15–20 kt/a, a scale that is very small when compared to the 3 Mt/a of liquid that steam crackers are processing. The small scale is clearly harming the economic competitiveness of the process as it leads to high capital and operating costs.

The resulting pyrolysis oil and waxes can further be cracked to olefins and aromatics (so-called high value chemicals) with about 65 wt % yield. But not all steam crackers are capable of processing feedstock with such a broad range in boiling points. Some may indeed be limited to processing the naphtha fraction of the pyrolysis oil. This 1.5–2 GJ/t of energy consumed by the pyrolysis step is dwarfed by the 14–17 GJ/t of energy needed to further crack the pyrolysis to high value chemicals in a steam cracker and comparable to the energy needed to produce the standard hydrocarbon feed from crude oil. Hence, recycled olefins are not disadvantaged over virgin ones. Interestingly, over the last decade or so, the shale gas revolution led to the construction of many gas crackers which are feedstock advantaged over liquid naphtha crackers. However, only liquid crackers can effectively use the pyrolysis oil. This will undoubtedly have consequences on the corporate asset strategy of companies.

The pyrolysis technology is not fully omnivorous either. It is particularly suitable to process polyolefins, but it produces less oil and an oil that is more aromatic when the feed is contaminated with other polymers such as PS, PET, and PA. The pyrolysis of polymers other than polyolefins will be briefly discussed in a later section. Small amounts of PVC (or even the dyes and other performance enhancers) in the feed are particularly detrimental, as it liberates HCl that corrodes the equipment and makes the oil unsuitable for feeding into a steam cracker. One element of mitigation consists of heating the plastic waste in the feeding system and recovering the HCl-rich gas prior to feeding the plastic melt to the reactor. Another and complementary approach is to feed caustic elements such as $CaCO_3$ to the reactor to trap and neutralize remaining chloride. The resulting salt is then removed together with the coke.

At present, the companies investing in chemical recycling are procuring their pyoil from third parties, as they have not yet made the leap to backwards integrate into the plastic waste collection business, though this obviously provides for business opportunities by securing supply.

The observation that chemical recycling uses building blocks from refining does mesh nicely with the macro trend that fuels demand is likely to fall, thus sparing capacity which can be used as feed preparation for the chemicals portion of the complex.

Once the pyrolysis oil is obtained, it can be co-fed, albeit with hydrogenation and appropriate feed preparation, into the liquid crackers where it can follow the conventional upgrade life cycle.

At present, the industry is starting to become more circular, and we are quite far from producing the 400 million tons of capacity by means of recycling. We do have a way to go.

NOTE

1   Section 15.6 is adapted from (Lange, Managing Plastic Waste—Sorting, Recycling, Disposal and Product Redesign, 2022).

REFERENCES

1. Brooks, A.L.; Wang, S.; Jambeck, J.R. The Chinese import ban and its impact on globabl plastic waste trade. *Science Advances* 2018, *4*(6), 2375–2548.
2. Geyer, R.J. Production, use and fate of all plastics ever made—Supplementary information. *Science Advances* 2017, 25–29.
3. Global Thermoset Market Outlook to 2027. *Report ID: BQ0130*, BlueQuark Research & Consulting, Hyderabad, 2022.
4. Iea.org, 2023. www.iea.org/data-and-statistics/charts/production-of-key-thermoplastics-1980-2050
5. Stasticia. *Plastics—The Facts 2022,* PlasticsEurope (PEMRG): London, 2022, p. 16.
6. *Statistica*, 2023. www.statista.com/statistics/664906/plastics-production-volume-forecast-worldwide/
7. The 7 different types of plastic. *Plastics for Change*, January 12, 2023. www.plastics-forchange.org/blog/different-types-of-plastic
8. World Economic Forum. January 12, 2023, weforum.org. www.weforum.org/agenda/2022/06/recycling-global-statistics-facts-plastic-paper/
9. History of plastic. *www.plasticsindustry.org/*, December 8, 2022. www.plasticsindustry.org/history-plastics#:~:text=It%20was%20in%201862%20that,substitute%20for%20shellac%20for%20waterproofing

16 Project Economics, Taxes, and Subsidies for Sustainability

16.1 INTRODUCTION

We all know that investments are made to achieve a certain return, be it monetary or otherwise. Project decisions are made based on multiple factors, some of which are better known than others, but include project cost, expected return based on anticipated demand and price of product and raw material, geopolitical stability of region (will the plant operate long enough to pay back its investment), transparency of policies, etc. In this chapter, we will take a high level view of what it takes, and see how sustainability can be priced in using conventional established approaches.

16.2 WHY PROCESS AND PROJECT ECONOMICS?

Capital projects typically have a life cycle, starting with inception where things are not very well defined, all the way to start-up once the plant has been constructed. In general, spending goes up as the project progresses through a stage-gate system, where the commitment of the executing company increases, as well as the definition of the project.

Depending on the starting point of a project and the amount of work necessary, this process can take months (existing plant, near future) to years (grass roots build, novel technology). The assumptions that go in the project economic model obviously can have a huge impact on the overall projected profitability of the plant. In general, planning bases or ranges are used for feedstock, product, utilities, etc., to assess the overall economic viability, and comparisons with incumbents, as appropriate.

16.3 TECHNOLOGY STRATEGY

Projects are placed in strategic quadrants depending on market conditions, where they are deemed economically viable under certain conditions. For illustration, a hypothetical technology portfolio of a company is shown in Figure 16.1.

Under conditions of low natural gas prices, natural gas gasification processes are deemed more favorable than coal based technologies, *assuming* geopolitical availability of these resources (Figure 16.1). Similarly, when natural gas prices are high,

DOI: 10.1201/9781003304388-20

Coal Price

High	Natural Gas Gasification	Renewable Electricity and Electrification
Low	Natural Gas Gasification	Coal Gasification

                  Low   High   Natural Gas Price

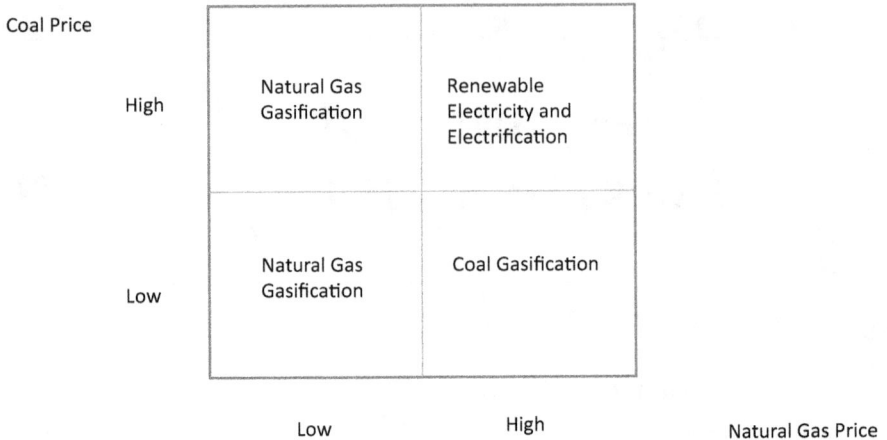

FIGURE 16.1 Technology Strategy Grid denoting economic viability of portfolio.

and the price of coal is low, coal based technologies are more economically favorable. For low natural gas prices, and high coal prices, natural gas gasification is favorable, and finally, when natural gas and coal prices are high, and then renewable electricity is available, electrification projects become favorable.

Of course, the preceding is an over simplified representation of how projects could be classified, but it is nonetheless illustrative that it is important to identify under what conditions a project will make sense.

Now, the critical reader probably would have started wondering by now, how the price of carbon, or carbon taxes and subsidies play a role. To understand this, it is important to pause and have a look at the basic economic concepts of demand and supply curves.

16.4   TAXES, CAP-AND-TRADE, AND SUBSIDIES

Demand curves are an aggregate behavior of the market in response to prices: the lower the price, the more demand will exist (Figure 16.2).

The supply curve, on the other hand, is an aggregate of the industry's ability to provide quantities (production) at certain prices. For simplicity, it is shown as a continuous curve, but in reality it is discrete: For a given price of production or minimum netback price, a certain capacity will come online (Figure 16.3).

Determining the capital cost of a plant, and cost of production, is a separate field and depending on the stage a project is in, can be done on the back of the envelope (with large error margin), or requires months of estimation with considerably smaller error margin.

In general, when considering the overall profitability of a plant, or future project, the price the market is willing to pay is important. This comes, in macroeconomic terms, from the intercept of the demand and supply curves (Figure 16.4).

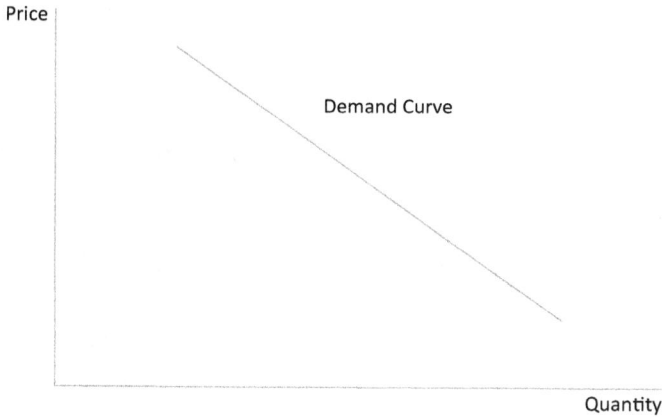

FIGURE 16.2 Sample demand curve.

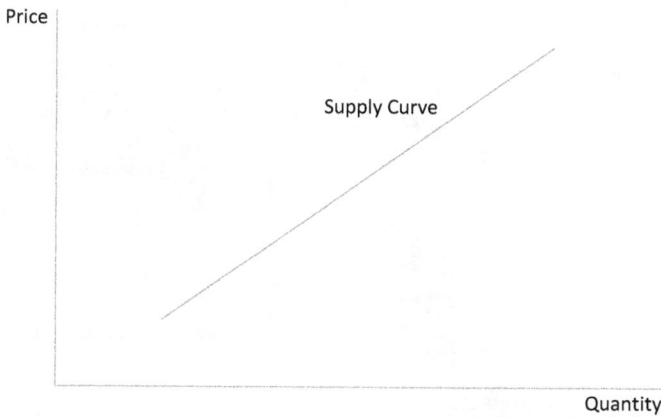

FIGURE 16.3 Sample supply curve.

Plants which have a netback price which exceeds p_0 will in general not produce goods as this would not be profitable by conventional metrics.

An introduction of a tax, which provides government revenue, will have the following effect on the price, but also, market demand (Figure 16.5):

As can be discerned from the graph, the price moves from p_0 to p_1 and the quantity moves from q_0 to q_1. It is important to recognize that the price increase is not equal to the tax increase, but that the consumer shoulders this portion of the tax increase: the consumer burden. One may wonder where the remainder of the tax went. The answer is that the producer takes this burden, as Figure 16.6 illustrates.

This simple analysis shows how taxes can change both the quantities as well as the pricing, and this is how CO_2 or other emissions taxes can guide transition.

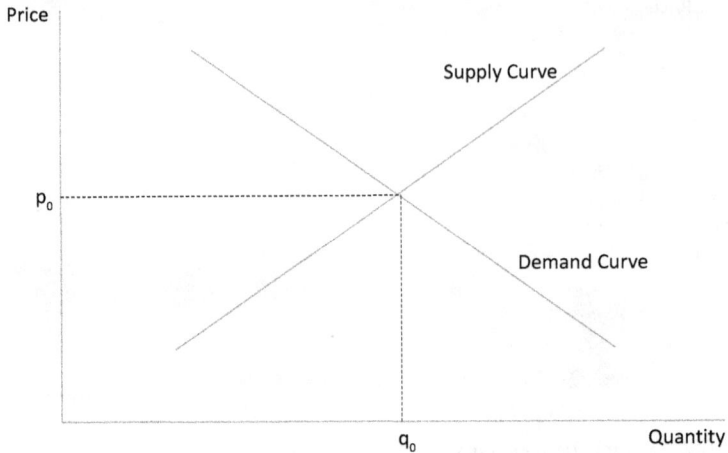

FIGURE 16.4　Demand and supply curve intercepts to determine market price and quantity.

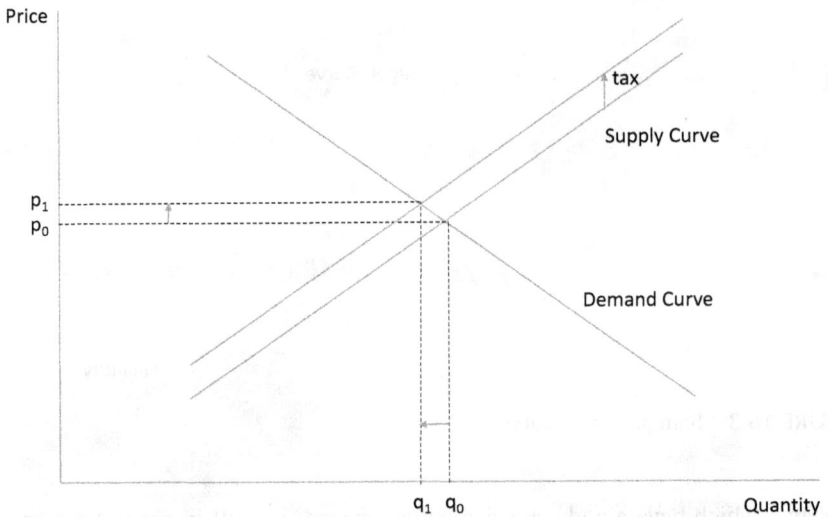

FIGURE 16.5　Effect of tax on price and quantity.

In the case of carbon tax, the money clearly is a source of revenue for the government. From a consumer's perspective, the price is p_1.

Another mechanism that is frequently discussed is cap-and-trade. Figure 16.7 shows the effect and prices and quantity.

Cap-and-trade implies that the supply is capped, or limited, regardless of the price, beyond a target quantity, which is based on some emissions metric. Hence the kink in the curve. The effect on the consumer is once again the same, the price goes from p_0 to p_1, and there is an effective tax, and consumer and producer burden, as before.

Price

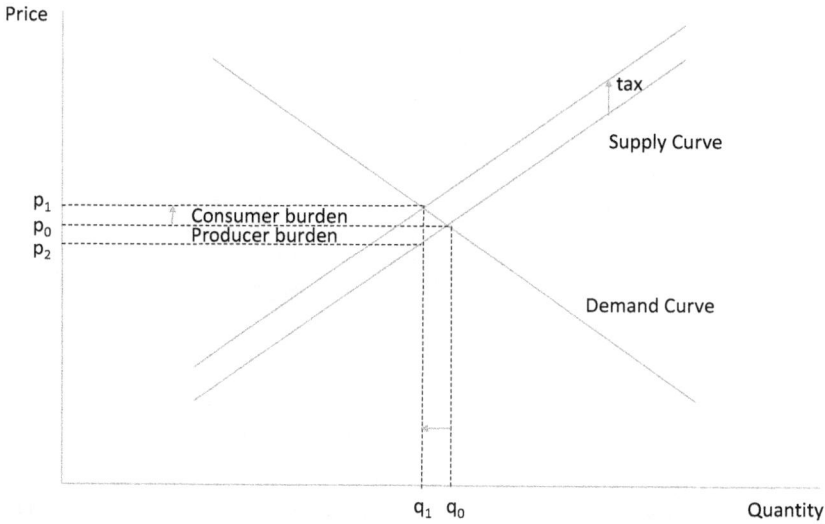

FIGURE 16.6   Effect of tax on price, consumer, and producer burden.

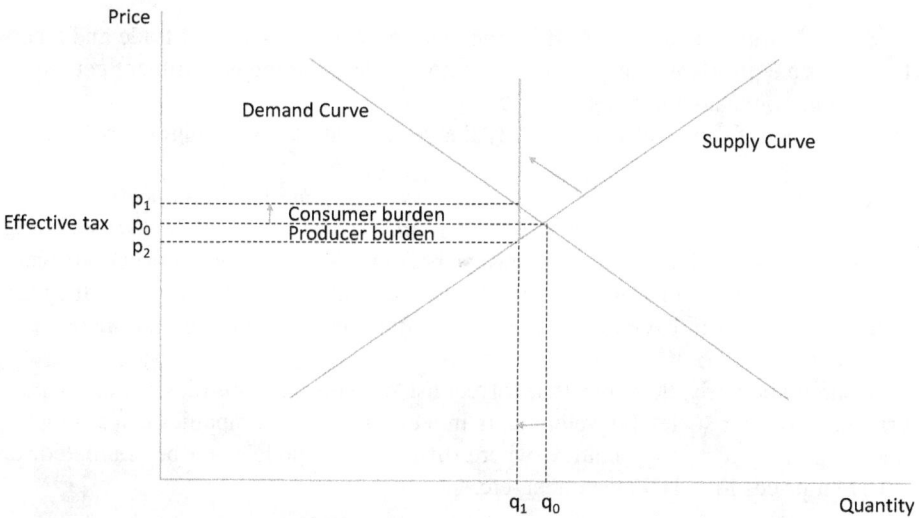

FIGURE 16.7   Effect of cap-and-trade on demand and supply curves.

But if output is capped at q_1, that difference is pure profit: a permit to produce one unit of output allows its owner to collect a rent equal to the difference between the selling price and the cost of production. If permits are traded, their price will be bid up so that their price will be equal to T. So where that money goes depends on how the permits are allocated in the first place. If the permits are simply given to existing emitters, then those profits are pocketed by the firms. If the permits are auctioned off, the price will be bid up to T, and the government gets the money (Gordon, 2012).

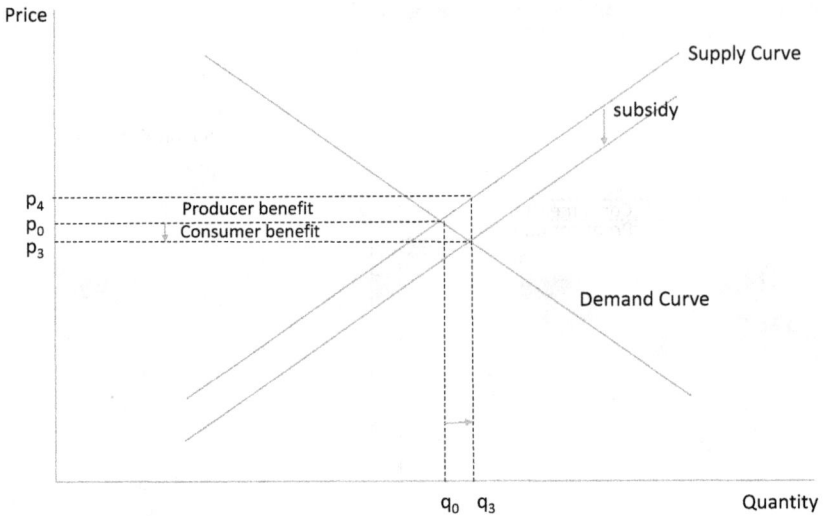

FIGURE 16.8   Effect of subsidies on pricing and quantities.

So if permits are auctioned off by the government, then cap-and-trade and a carbon tax are equivalent: same quantities, same prices, and the government gets revenues equal to the area in the green rectangle in the graphs.

The effect of subsidies is essentially that of a negative tax, as Figure 16.8 clearly shows.

The subsidies are generally paid to the producers, and both producers and consumers see a benefit. Naturally this means that the government is now spending to ensure that certain products/processes become favorable, but subsidies in general cannot be continued indefinitely and will have a limited life. However, they are extremely helpful in overcoming the initial adoption of technology that habits can indeed change.

It should be fairly clear now that correct use of taxes and subsidies can allow governments to steer societal developments in a certain way. Companies can anticipate and plan using a strategy matrix, where different tax models can be simulated, or different prices for GHG abatement, etc.

REFERENCE

1. Stephen Gordon, September 17, 2012, Econ 101: What you need to know about carbon taxes and cap-and-trade—Ottawa's climate change kabuki theater, https://macleans.ca/economy/business/why-the-difference-between-carbon-taxes-and-cap-and-trade-isnt-as-important-as-you-think/.

17 Low Carbon

17.1 INTRODUCTION

Current world consumption of energy is approximately 450 ExaJoules [6], of which the bulk (>75%) comes from fossil energy sources [1]. The associated CO_2 emissions are approximately 40 Gt. Clearly, these energy sources do not replenish at timescales relevant to the usage timescale, and contribute to the emissions of CO_2. Clearly, solutions which reduce the carbon footprint for energy generation must be explored in an effort to be either carbon neutral or negative.

17.2 HIERARCHY OF SOLUTIONS

A well-accepted framework in the energy transition is as follows:

1. Avoid: Don't make CO_2 and do not leak other greenhouse gases.
2. Reduce: If you cannot avoid making CO_2, be more efficient.
3. Capture CO_2 at source
4. Capture CO_2 from atmosphere

17.3 AVOID CO_2

At present, a large proportion of the energy system relies on combustion of hydrocarbons to generate heat, which then is transformed into power. Similarly, chemical plants depend on a delicate steam balance, where hydrocarbons are transformed into heat and steam at various places, to drive equipment or provide heating duties.

A key driver in today's energy transition is electrification, which in this context has two distinct meanings:

1. Replace steam driven pumps and compressors by electrically driven equipment, and replace fossil fuel based heating (i.e. combustion) by electrical heaters
2. Replace the conventional chemical processes by electrochemical processes, or drastically re-think the process routes

A basic premise is that abundant clean electricity is available, with low or no carbon footprint.

Due to the development of high-speed induction motors and voltage source inverters, standalone electric drivers are today an alternative to the traditional train

DOI: 10.1201/9781003304388-21

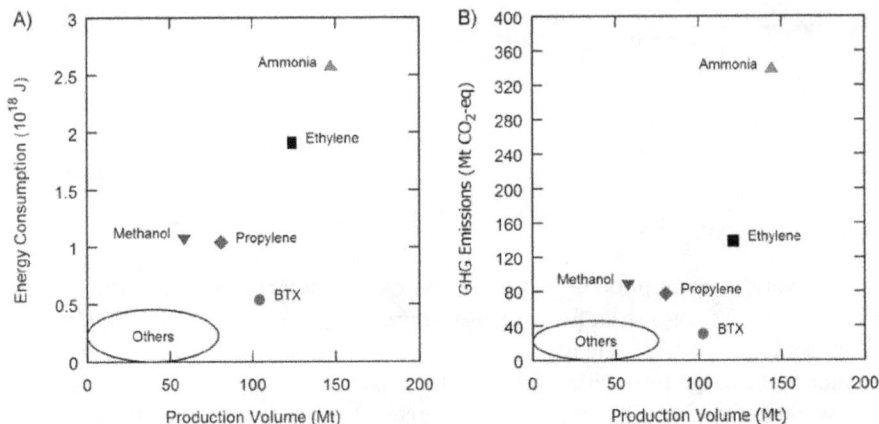

FIGURE 17.1 Energy consumption and carbon footprint of high-volume commodity chemicals. Comparison of (A) energy consumption and (B) greenhouse gas (GHG) emissions versus production volume for top chemicals by volume in 2010. GHG emissions are expressed as megatons of carbon dioxide equivalent (MtCO$_2$-eq). The five high-volume commodity chemicals labeled on the plots have the largest energy requirements, which correspond to large GHG emissions. The region labeled "Others" shows approximately where the next 13 largest contributors fall. (Figure adapted from IEA, ICCA, and DECHEMA data (Schiffer, 2017)).

driven by steam and gas turbines when regulating the operating speed of compressor, improving the system efficiency and reducing significantly the emission of greenhouse gases as requested by the new European regulations [3]. These have been successfully implemented in retrofits, but also technology providers such as AirLiquide are offering processes which use electrically driven equipment.

Figure 17.1 shows the energy intensity and associated CO$_2$ footprint of major chemical processes.

A key contributor to the energy intensity as well as the carbon footprint is the furnace, where heat is generated to drive the endothermic reactions by combustion of natural gas. The furnace can therefore be electrified, by replacing by a suitable technology that provides heat by electrical means, providing the steam balance and energy balance of the plant are appropriately redesigned, *and* electrical heaters are available/can be developed. This is an active area of engineering and research among large petrochemical producers [10] [8] [9]. At present, commercial implementations do not exist, however, they are expected soon.

On a longer time horizon, is the replacement of the chemistry by electrochemical processes in a different paradigm (Figure 17.2; Schiffer, 2017).

17.4 REDUCE CO$_2$

One of the major technology trends today is that of digitalization. Digitalization is that process which builds on digitization (i.e. capturing the information in digital

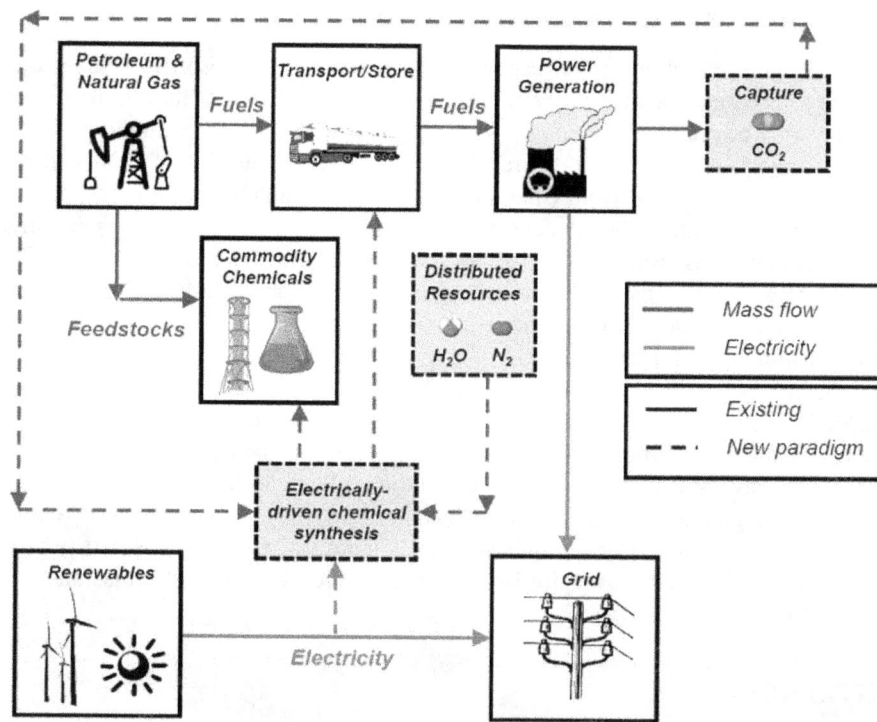

FIGURE 17.2 A new paradigm for chemical and energy industries (With permission from [7]).

format), automation of process steps, but introduces machine learning and artificial intelligence. In 2020, a survey by Ernst and Young [4] revealed that most chemical industry executives expected digital transformation to play a significant role in their companies. Indeed, artificial intelligence has the potential to drastically reduce unplanned shutdowns (with associated CO_2 and other GHG emissions), find optimal operating conditions on the go, and be prescriptive in recommendations, either in open or closed loop format.

The applications are not limited to just process, but also in product development, where the vast availability of computing power, computer memory, data transmission, and real time sensors (internet of things), computer modeling, phenomenological and AI based, can be used to design and synthesize either new products or process routes. This adoption of digital initiatives was only accelerated due to the Covid 19 global pandemic.

17.5 CAPTURE CO_2 AT SOURCE

Third in the hierarchy is the capture of CO_2 at source, which falls into the category of capture technology, and is described in Chapter 19. Figure 17.3 shows the emissions by sector, which allows for a strategic plan as to where to capture.

17.6 CAPTURE CO_2 FROM ATMOSPHERE

There is much talk about direct capture of CO_2 from the atmosphere. This should truly be viewed as a last resort, as the energetics and scale are truly mindboggling, and may only be practical under very specific circumstances. The technology used is similar to the capture of CO_2 at source. However, the concentration of CO_2 is around 400 ppm, and as recalled from Chapter 10, the minimum amount of energy goes as the natural logarithm of the concentration. The practical amount of energy required to capture one ton is around 10 GJ [2]. As stated earlier, the annual emissions of CO_2 due to energy generation and chemicals production are around 40 Gt. This would mean that around 400 ExaJoules are necessary to remove these emissions from the air (i.e. using direct air capture technology, rather than capture at source). Now, the world is producing and consuming around 450 ExaJoules of energy annually. If 400 ExaJoules of clean energy capacity is available it would make far more sense to use this to *displace* the carbon intensive energy generation. The estimate is meant as a back-of-the-envelope number to indicate that the scale is enormous, and would likely not work on a global scale. However, in certain cases, the energetics may be reasonable, for stranded power generation sources. Capture of CO_2 may indeed perhaps be best handled by nature inspired solutions, and the avoidance of emissions by a changing energy mix is critical.

REFERENCES

1. BP. *Statistical Review of World Energy*, 2022. www.bp.com/en/global/corporate/energy-economics/statistical-review-of-world-energy.html
2. Collins, L. The amount of energy required by direct air carbon capture proves it is an exercise in futility. *Recharge*, September 2021.
3. Durantay, L. V. A success story of steam turbine replacement by high speed electric system driven compressor. *PCIC* (p. Paper No. PCIC Europe EUR21_08), 2021.
4. Ernst and Young. *DigiSurvEY 2020*, 2021.
5. IEA, 2021. www.iea.org/reports/key-world-energy-statistics-2021/final-consumption (Retrieved from Key World Energy Statistics 2021, IEA, Paris www.iea.org/reports/key-world-energy-statistics-2021: www.iea.org/reports/key-world-energy-statistics-2021/final-consumption).
6. Schiffer, Z.J. Electrification and decarbonization of. *Joule* 2017, *1*(1), 10–14.
7. Tullo, A.S. Dow test electricity-based cracking. *C&E News* 2022, *100*(23).
8. Tullo, A.H. BASF, SABIC and Linde to test electric cracker designs. *C&E News* 2022, *100*(32).
9. Voermans, L. 2021. www.brightlands.com/en/brightlands-chemelot-campus/nieuws/accelerating-electrification-cracker-future-consortium(RetrievedfromBrightlands:www.brightlands.com/en/brightlands-chemelot-campus/nieuws/accelerating-electrification-cracker-future-consortium).

18 A Changing Energy Mix

18.1 INTRODUCTION

Today's energy mix is shown in schematic form here (Figure 18.1):

As stated earlier, the bulk of the energy (>75%) comes from fossil energy sources [3], and is clearly shown in Figure 18.1. The energy mix of the future will undoubtedly change, but it is instructive to examine how this has happened in the past (Figure 18.2).

For about 1800 to 1850, the dominant energy source was traditional biomass. In 1800 the population of our planet was approximately 1 billion, and energy consumption was 5653 TWh or 20 ExaJoule. The energy intensity per capita therefore was ~20 GJ/capita. The dominant source of energy (excluding animal and human energy) was obtained by burning wood, cropwaste, with coal picking up at the advent of the industrial revolution.

In 1900, the population was 1.6 billion, and energy consumption was 12,131 TWh or 44 ExaJoule. The associated energy intensity per capita was 27.5 GJ/capita. As of 2021, the world population was 7.88 billion, with a consumption of 176,431 TWh or 635 ExaJoule, and associated energy intensity of 81 GJ/capita. It is important to note that this is an *average* energy intensity number, and large variations exist globally (Figure 18.3).

Today's Energy Mix

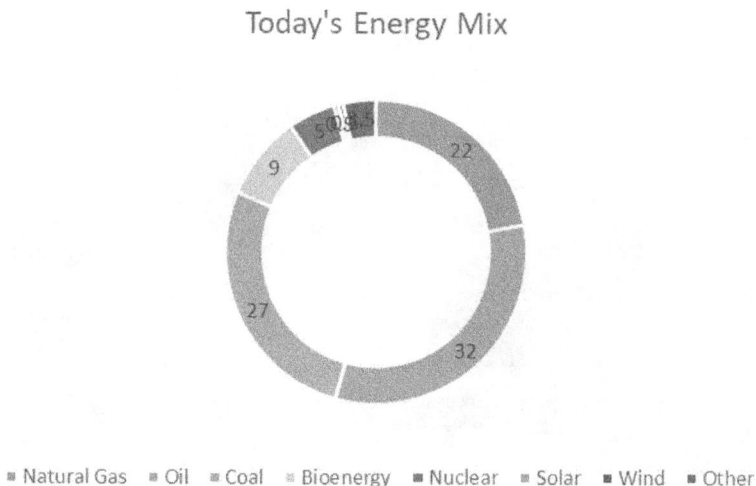

■ Natural Gas ■ Oil ■ Coal ■ Bioenergy ■ Nuclear ■ Solar ■ Wind ■ Other

FIGURE 18.1 Energy mix as of 2021 in terms of energy generation.

DOI: 10.1201/9781003304388-22

Global primary energy consumption by source

Primary energy is calculated based on the 'substitution method' which takes account of the inefficiencies in fossil fuel production by converting non-fossil energy into the energy inputs required if they had the same conversion losses as fossil fuels.

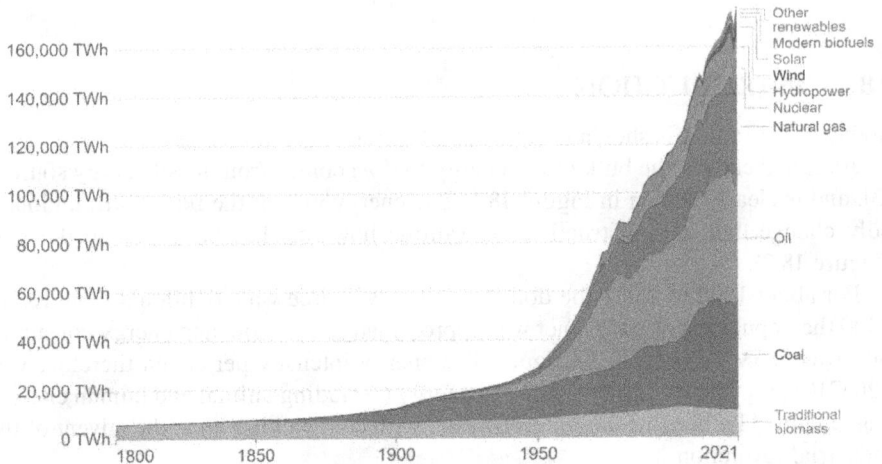

Source: Our World in Data based on Vaclav Smil (2017) and BP Statistical Review of World Energy   OurWorldInData.org/energy • CC BY

FIGURE 18.2 Changing energy mix from 1800 to 2021 in TWh. 1 TWh = 0.0036 ExaJoule.

GDP per capita vs. energy use, 2015

Annual energy use per capita, measured in kilowatt-hours per person vs. gross domestic product (GDP) per capita, measured as constant international-$.

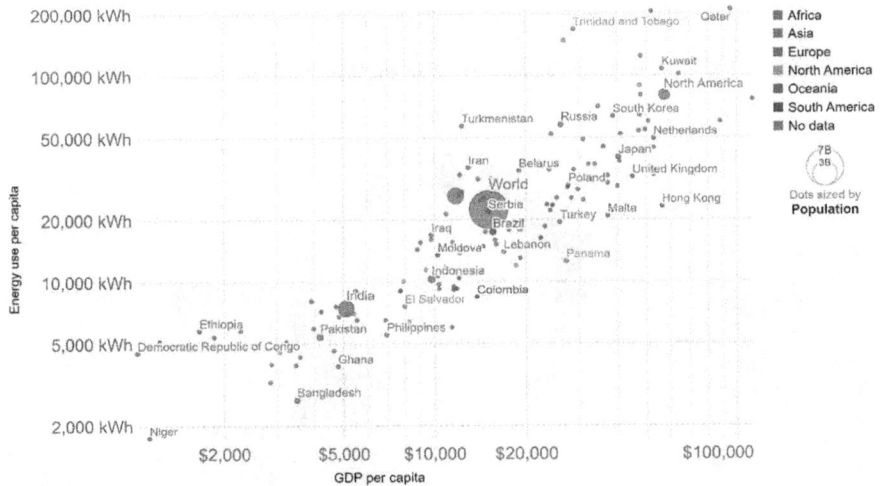

Source: Data compiled from multiple sources by World Bank   OurWorldInData.org/energy • CC BY

FIGURE 18.3 Energy intensity per capita as a function of GDP (With permission from [9]). 20,000 kWh = 72 GJ.

The first observation is that per capita GDP and energy consumption seem to be correlating well, but the second one is that undoubtedly, the average energy intensity globally will continue to increase, as countries develop and standards of living increase.

However, there are limits to growth as was highlighted in Chapter 13, and CO_2 cannot continue to grow—the energy mix needs to change.

18.2   FUTURE ENERGY MIX

There are many public commitments [11] [2] [10] [16] to be net zero by 2050 or various other target dates, but it may be instructive to outline what a potential 2050 energy system could be.

If we examine Figure 18.2 again, we readily notice that nuclear energy only entered by the 1960s, and that renewables such as solar and wind only became widespread, albeit at a low level, by the 1980s. What is particularly striking is that the transition and growth of the energy system took quite a while, and that transitions take time, and are generally evolutionary, and are based on need and technological success, as well as entrenched habits. Equally important to recognize is that energy security for a country is intimately tied to political security and is a key driver for prosperity. The goal now is to have a major overhaul of the energy systems in 25 years, which is a very ambitious goal.

Whereas we will only know in the future what the energy mix is, some educated projections can be made (Figures 18.4 and 18.5; Shell, 2022; [1]). What is common among these scenarios is that there is a decline in the use of fossil fuels, an uptick in the use of renewables, and nuclear, and use of carbon sequestration. It is interesting to first examine the projected pathways to reduced CO_2 emissions and netzero.

The scenarios all have in common that between 2025 and 2030, the CO_2 emissions seem to be plateauing, followed by a reduction.

Figure 18.6 suggests that to get to netzero goals, the total use must decrease, or equivalently, energy efficiency must increase. Hydrogen and electricity become main carriers (albeit from renewable sources).

The same broad conclusions can be drawn from the Shell analysis, Figure 18.7:

The Sky scenario shown next highlights how fossil will remain part of the energy mix, but with appropriate CO_2 capture.

What is clear is that very large investments need to be made to achieve this goal, but that these formidable challenges also provide enormous opportunities.

If we are successful, the world's energy system could look something along the lines predicted in the Shell scenario (Shell, 2022), Figure 18.9.

One thing that will have to have considerable investment is the energy grid, as grid stability could become an issue. Broadly speaking, at present, the demand and supply profiles are well known of electricity, and the supply can be ramped up or down accordingly to match the demand. Voltage and frequency are known, and transmission is designed to deal with the load. With renewables such as solar

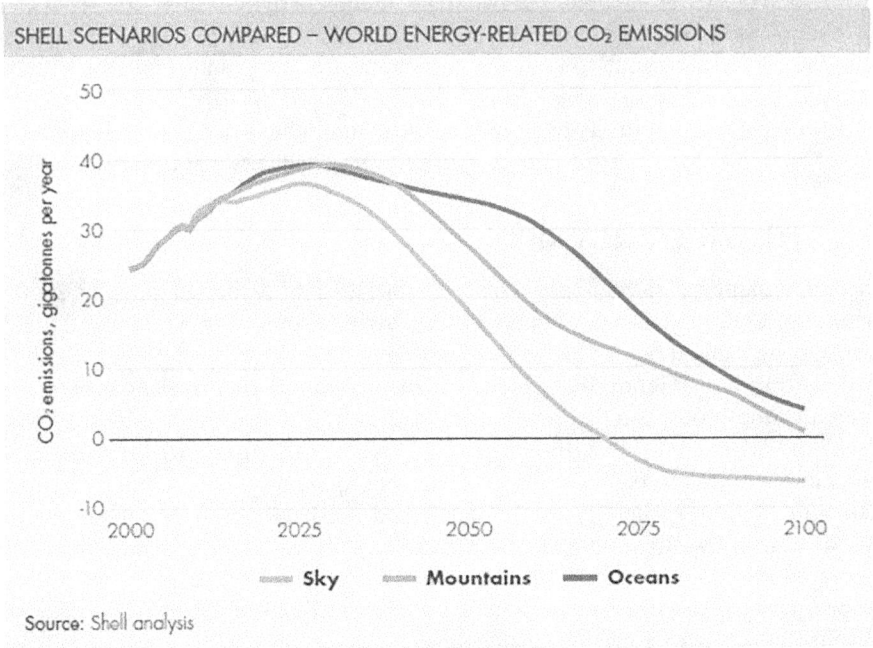

FIGURE 18.4 Shell scenarios compared. (Courtesy of Shell, 2022.)

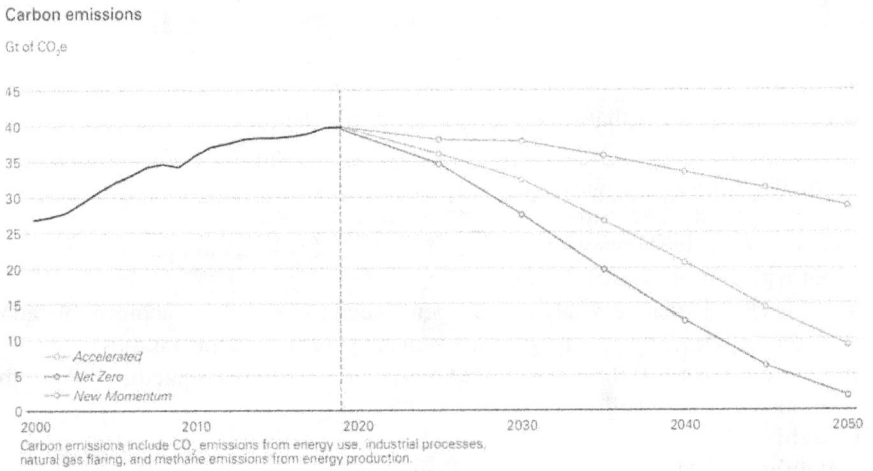

FIGURE 18.5 BP scenarios compared. (Courtesy of [1].)

Total final consumption

EJ

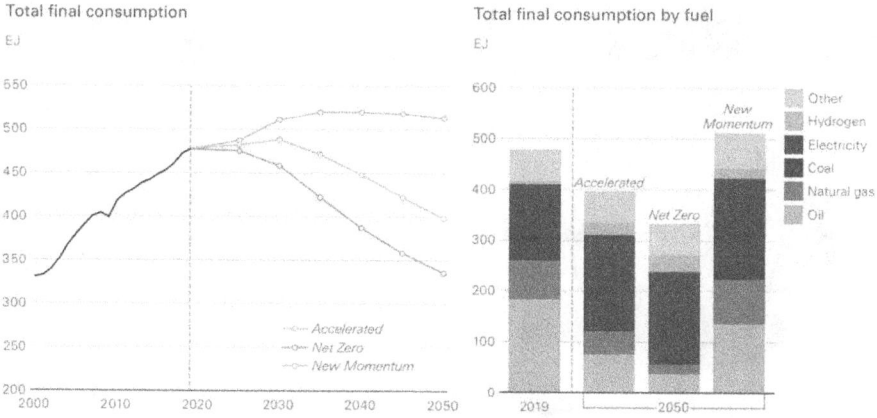

Total final consumption by fuel

EJ

FIGURE 18.6 Total energy consumption and total consumption by fuel. (Courtesy of [1].)

PRIMARY ENERGY BY SOURCE IN THE THREE SCENARIOS

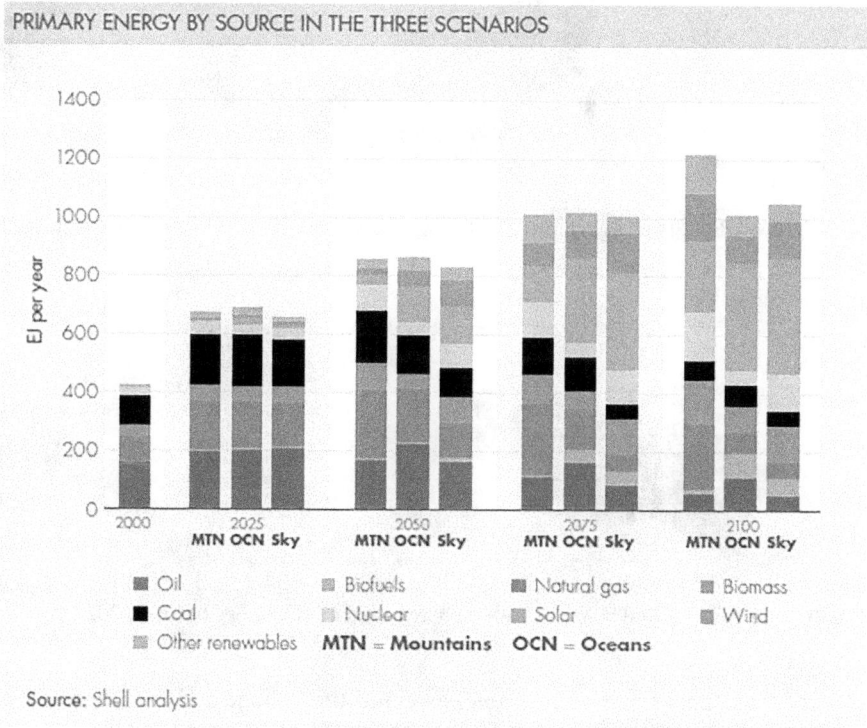

Source: Shell analysis

FIGURE 18.7 Comparison of three Shell scenarios in terms of energy mix.

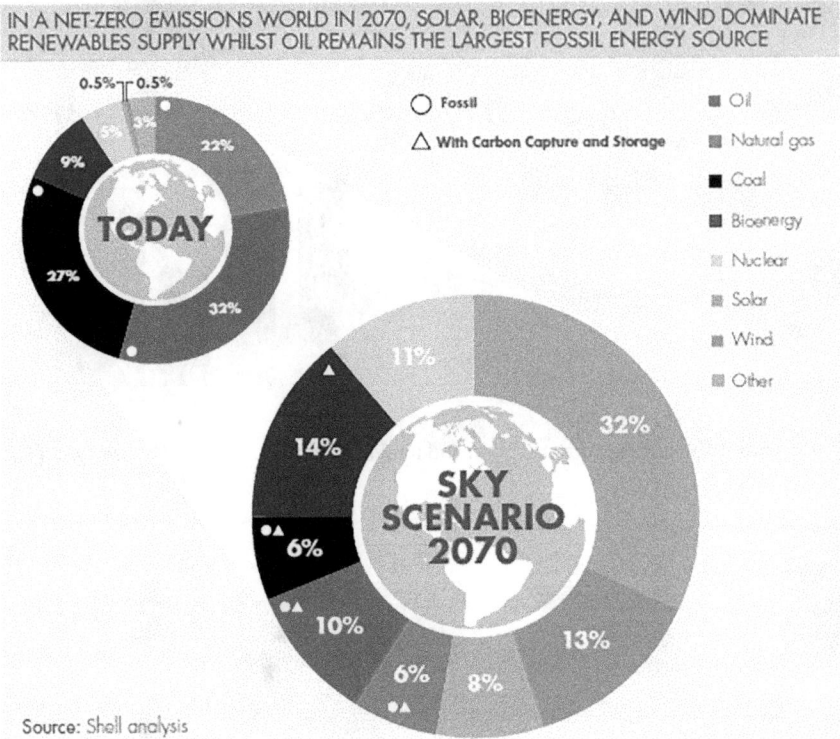

FIGURE 18.8 Energy mix according to the Sky scenario in 2070. (Courtesy of Shell, 2022.)

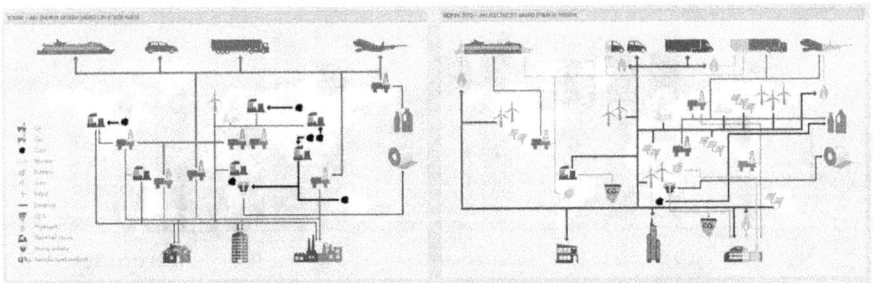

FIGURE 18.9 Today and potential future energy system. (Courtesy of Shell, 2022.)

and wind, the supply may be variable, as could the voltage and frequency. These therefore have the potential, as the proportion in the energy mix goes up, to disrupt the grid.

However, with appropriate storage of surge electricity this can be prevented, coupled with smart grid operation to arbitrate demand and supply across location.

THE EVOLVING ENERGY SYSTEM CO₂ BALANCE SHEET IN **SKY**

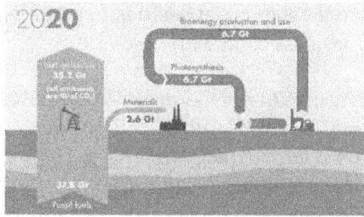

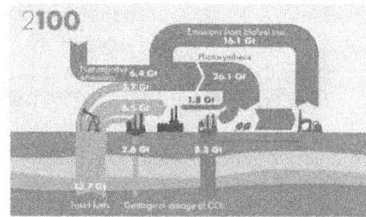

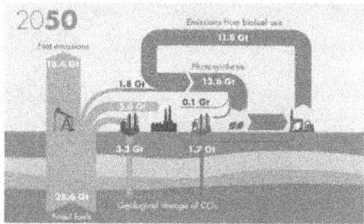

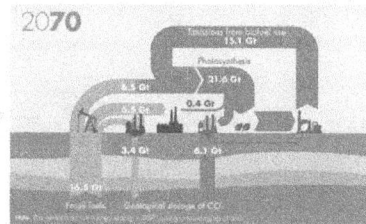

FIGURE 18.10   Evolving CO_2 as per Sky scenario. (Courtesy of [13].)

18.3   NET ZERO AND BEYOND

Concomitant with the transitioning energy mix, and reducing CO_2 and other GHGs, the CO_2 that is already in the atmosphere needs to be reduced to move away from climate tipping points. CO_2 needs to be removed from the atmosphere, and this can be achieved by nature based solutions, reforesting, preventing deforestation, and direct capture of CO_2, despite the challenging energetics. Figure 18.10 shows the evolving CO_2 system.

18.4   CONCLUSION

The energy mix of the future will be drastically different, and the convergence of many technologies is necessary, coupled with private and public investment, and behavioral changes. It is heartening to see the transition start.

REFERENCES

1. BP. *BP Energy Outlook 2023*, 2023.
2. BP. *Statistical Review of World Energy*, 2022. www.bp.com/en/global/corporate/energy-economics/statistical-review-of-world-energy.html
3. BP. *Statistical Review of World Energy*, 2022. www.bp.com/en/global/corporate/news-and-insights/reimagining-energy/net-zero-by-2050.html

4. *Our World in Data*, 2022. https://ourworldindata.org/; https://ourworldindata.org/grapher/energy-use-per-capita-vs-gdp-per-capita?yScale=log

5. *Reuters*, 2023. www.reuters.com/business/cop/saudi-sabic-targets-carbon-neutrality-by-2050-statement-2021-10-23/: www.reuters.com/business/cop/saudi-sabic-targets-carbon-neutrality-by-2050-statement-2021-10-23/

6. *Reuters*, February 11, 2021. www.reuters.com/article/us-shell-strategy-idUSKBN2A-B0LT: www.reuters.com/article/us-shell-strategy-idUSKBN2AB0LT

7. Shell. *Shell Scenario Sky*, 2022. The Hague: Shell International BV.

8. *UK COP 26*, 2021. https://ukcop26.org/cop26-goals/

19 CO_2 Capture and Sequestration

19.1 INTRODUCTION

The temperature of the earth has fluctuated in ancient and recent history, and according to data from the Climatic Research Unit [1] has been increasing steadily for the last couple of decades (Figure 19.1).

These data were based on actual temperature measurements. Using proxy data, that is inferred from other data, it is clear that the earth's temperature has been fluctuating and that now we are in a warm period (Figure 19.2) [2–4].

Depending on how we interpret the data, it seems that the last decades or so have been in a warming trend. The scientific conclusion reached, according to the World Meteorological Organization, is that the warming is real. CO_2 has been rising since the time of James Watt (1736–1819), inventor of the auto-controlled steam engine that helped jump-start the industrial revolution. Since then, coal, oil, and natural gas have powered our economies, as was alluded to in Chapter 9. Hydropower and nuclear power, albeit significant resources in Norway and France, are comparatively minor contributors to energy supply.

19.2 CO_2 EMISSIONS

Today, the annual amount of carbon dumped globally into the atmosphere corresponds, on average, to 1 ton per person on the planet. In the United States and China, carbon-based energy, in particular coal, is especially dominant (Figure 19.3) [5]. The average American per capita emission is 5 tons of carbon annually. In Sweden (with a similar standard of living as the United States), the carbon output is less than 2 tons of carbon per person per year. Figure 19.4 [6] shows the greenhouse gas emissions by sector in the United States. It is clear that there is not one sector which is the most dominant one. CO_2 is a greenhouse gas—it traps heat radiation that is attempting to escape from Earth. The physics of this process was established by the Irish physicist John Tyndall [7] (1820–1891), and the effect was calculated by Swedish chemist Svante Arrhenius (1859–1927).

The basic argument (i.e., that greenhouse gases keep the Earth comfortably warm) has never been challenged, and it follows that an increase in CO_2 in the atmosphere undoubtedly produces a rise in temperature at ground level.

First, we turn to the reconstruction of the rise of CO_2 since the time of James Watt. The early part of the CO_2 series is derived from extracting air in polar ice and measuring its CO_2 content. The later part is based on the measurements of Keeling

DOI: 10.1201/9781003304388-23

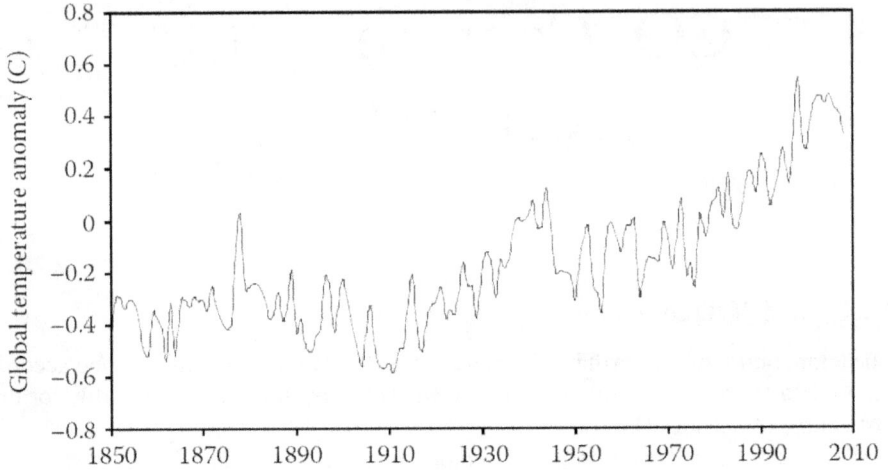

FIGURE 19.1   Global temperature anomaly.

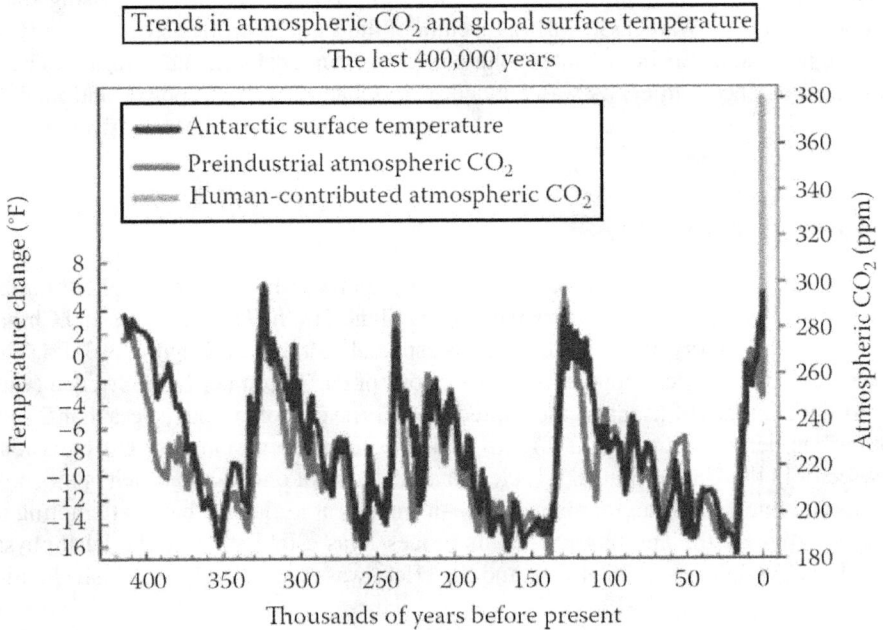

FIGURE 19.2   Inferred atmospheric CO_2 concentrations and Antarctic surface temperature. Atmospheric CO_2 prior to 3,000 years ago and Antarctic surface temperature prior to 100 years ago. (With permission from Petit et al., 1999.)

United States, 1949–2004

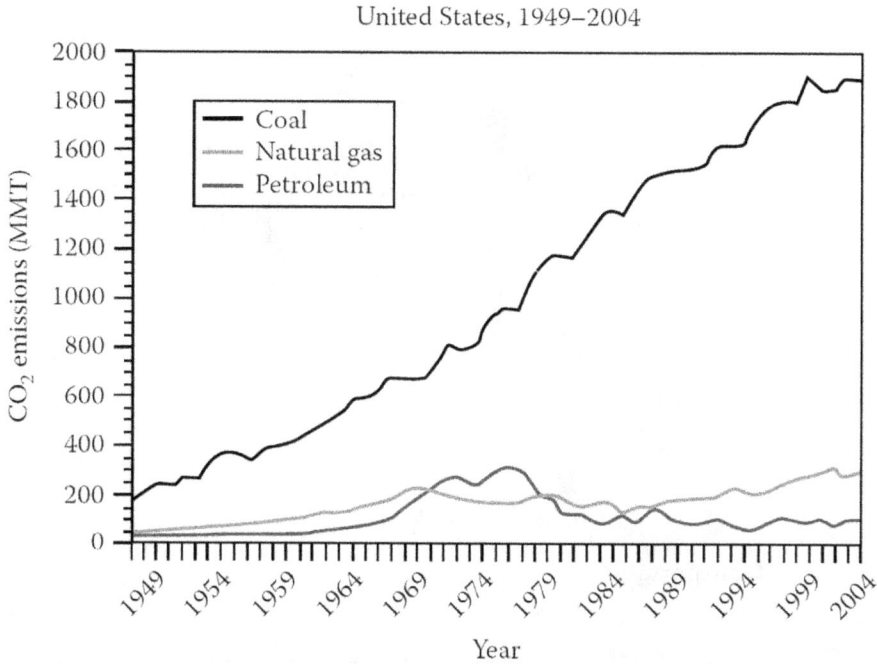

FIGURE 19.3 Trends in CO$_2$ emissions from the electric power sector. 1949–1959: Calculated from energy data in the Annual Energy Review. 1960–1989: Calculated from energy data in the State Energy Data Report. 1990–2004: Estimates documented in greenhouse gases in the United States 2004. (With permission from Report Emissions of Greenhouse gases in the United States 2004, DOE/EIA-573, 2004. Energy Information Administration, Washington DC.)

and Whorf, since 1957, on Mauna Loa [4]. The overall rise is from just below 280 ppm (the "preindustrial" value) to the present values above 360 ppm and the rise seems to be sustained (see Figures 19.2 and 19.5). According to various climate models, a doubling of CO$_2$ will result in an average temperature rise between 1.5°C and 5°C.

As of now, the anthropogenic contributions to CO$_2$ and global warming have been established. We will not examine this further, but focus on viewing CO$_2$ and carbon cycles through the lens of sustainability. With that in mind, it would seem reasonable to close the so-called carbon cycle.

19.3 THE CARBON CYCLE

A cartoon of the carbon cycle is given in Figure 19.6 [8]. A key reaction from photosynthesis uses the energy of the sun to convert CO$_2$ and water into oxygen and carbohydrate

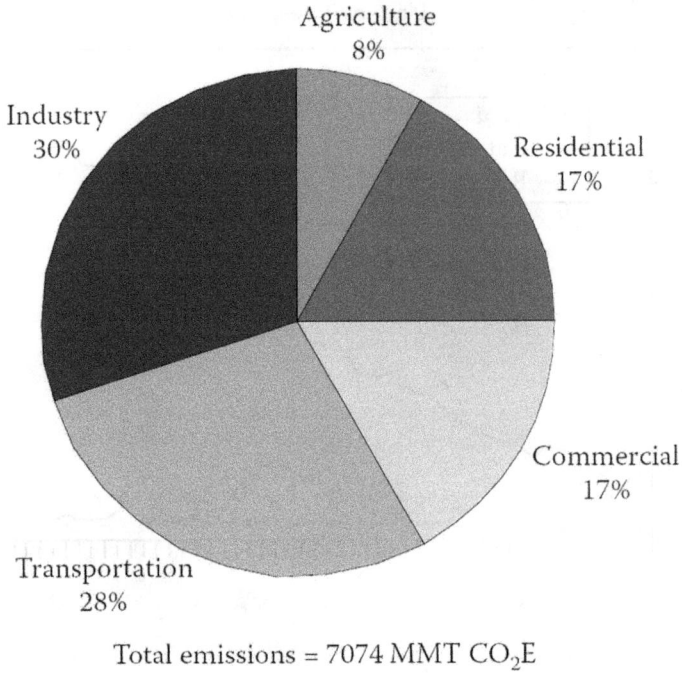

Total emissions = 7074 MMT CO_2E

FIGURE 19.4   Greenhouse gas emissions by sector in the United States.

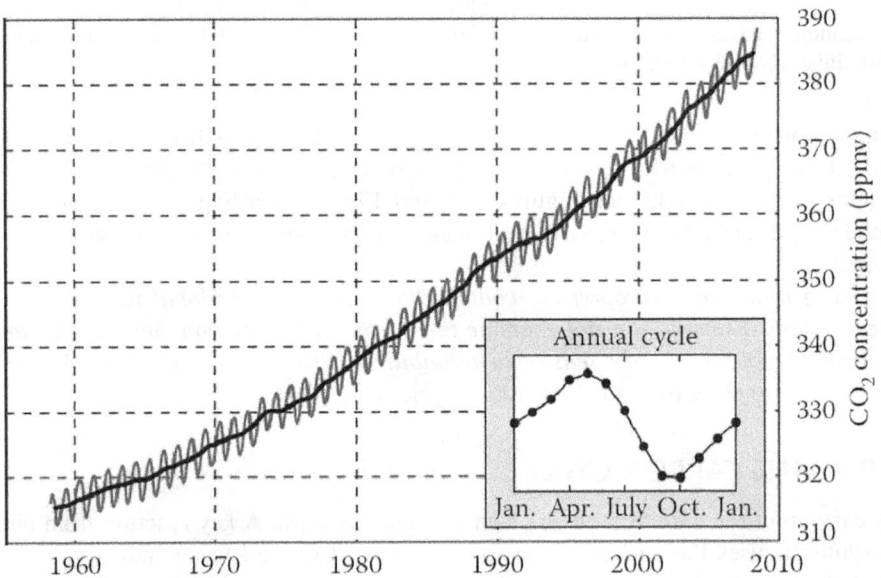

FIGURE 19.5   Atmospheric CO_2 concentrations measured at Mauna Loa, HI. (With permission from Inventory of U.S. Greenhouse Gas Emissions and Sinks: 1990–2007. www.epa.gov/climatechange/emissions/usgginventory.html.)

$$CO_2 + H_2O \rightarrow O_2 + \text{carbohydrate}$$

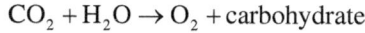

Both plants and animals undergo respiration in which

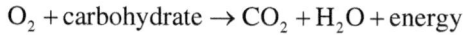

$$O_2 + \text{carbohydrate} \rightarrow CO_2 + H_2O + \text{energy}$$

Additionally, when plants and animals die, the dead organisms decay and give off CO_2.

According to some estimates [9], the present amount of CO_2 taken out of the atmosphere every year by plants is almost perfectly balanced by the amount of CO_2 put back into the atmosphere by respiration and decay. The CO_2 produced in this manner is part of a cycle in which new carbon does not enter the system, but rather keeps changing in form. In addition to the preceding, deforestation and the combustion of fossil fuels and traditional biomass release CO_2 into the atmosphere, and

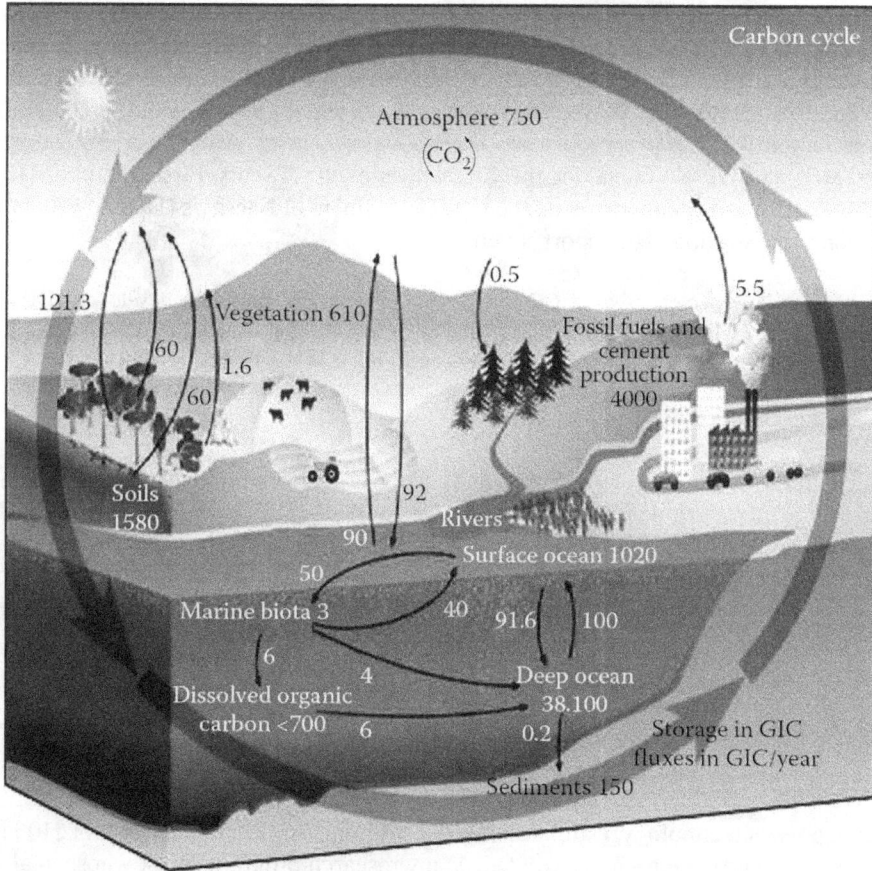

FIGURE 19.6   Artist's impression of carbon cycle.

the oceans absorb CO_2 in various forms, thus sequestering it. The carbon cycle is therefore large and complex. It is not entirely straightforward to predict how it will behave. Nonetheless, there is merit in closing the CO_2 cycles as much as possible, as this is consistent with sustainability.

19.4 CARBON SEQUESTRATION: SEPARATION AND STORAGE AND REUSE OF CO_2

As was stated in Chapter 9, fossil fuels will remain the mainstay of energy production well into the twenty-first century. The availability of these fuels to provide clean, affordable energy is essential for prosperity and economic endeavor. However, increased concentrations of CO_2 due to carbon emissions are expected unless energy systems reduce the carbon emissions to the atmosphere.

Roughly one-third of the United States' carbon emissions come from power plants and other large point sources (see Figures 19.3 and 19.4). To stabilize and ultimately reduce concentrations of this greenhouse gas, it will be necessary to employ carbon sequestration—carbon capture, separation, and storage or reuse. It is noteworthy that roughly another third comes from transportation, where carbon capture may prove much more difficult.

The President's Committee of Advisors on Science and Technology (PCAST) underscored the importance of carbon sequestration in its report "Federal Energy Research and Development for the Challenges of the Twenty First Century" [10]. PCAST recommended increasing the U.S. Department of Energy's (DOE's) R&D for carbon sequestration. The report stated:

> A much larger science-based CO_2 sequestration program should be developed. The aim should be to provide a science-based assessment of the prospects and costs of CO_2 sequestration. This is very high-risk, long-term R&D that will not be undertaken by industry alone without strong incentives or regulations, although industry experience and capabilities will be very useful.

The joint Office of Fossil Energy and Office of Science published a report titled *Carbon Sequestration: State of the Science* in 1999 [11], in which "key areas for research and development (R&D) that could lead to an understanding of the potential for future use of carbon sequestration as a major tool for managing carbon emissions" were formulated. To be successful, the techniques and practices to sequester carbon must meet the following requirements:

- Be effective and cost-competitive
- Provide stable, long-term storage
- Be environmentally benign

Using present technology, estimates of sequestration costs are in the range of $100 to $300/ton of carbon emissions avoided. Many research programs exist whose goal is to reduce the cost of carbon sequestration.

Further, achieving a stabilization scenario (e.g., 550 ppm CO$_2$) would not require wholesale introduction of zero-emission systems in the near term. This would allow time to develop cost effective technology over the next 10–15 years that could be deployed for new capacity and capital stock replacement capacity.

Modeling and assessments provide the capabilities to evaluate technology options in a total systems context (i.e., considering costs and impacts over the full product cycle). Further, the societal and environmental effects are analyzed to provide a basis for assessing trade-offs between local environmental impacts and global impacts.

In the mid-term, sequestration pilot testing will develop options for direct and indirect sequestration. The direct options involve the capture of CO$_2$ at the power plant before it enters the atmosphere coupled with "value-added" sequestration, such as using CO$_2$ in enhanced oil recovery (EOR) operation and in methane production from deep unmineable coal seams. "Indirect" sequestration involves research on means of integrating fossil fuel production and use with terrestrial sequestration and enhanced ocean storage of carbon.

In the long term, the technology products will be more revolutionary and rely less on site-specific or application-specific factors to ensure economic viability.

There are also research initiatives that aim to study the use of CO$_2$ as a raw material and synthesize other chemicals from this. These technologies are in the research stage and can be analyzed with the standard tools we have developed so far.

EXAMPLE 19.1

It has been proposed to convert CO$_2$ into useful products. This is known as *CO$_2$-chemistry*. We want to show with a compact simple thermodynamic analysis that here expectations can easily turn into disappointments.

Let us assume that, in theory, it is possible to convert CO$_2$ with ethylene into a useful polymer:

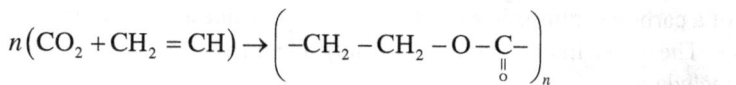

$$n\left(CO_2 + CH_2 = CH\right) \rightarrow \left(-CH_2 - CH_2 - O - \underset{\underset{O}{\|}}{C} - \right)_n$$

We know that the work available in CO$_2$, its exergy, is zero in the atmosphere. The value for pure CO$_2$ is about 20 kJ/mol. We estimate the work available in C$_2$H$_4$ to be about 1400 kJ/mol and assume that in the formation of the polymer no work is lost: one polymer unit is thus assumed to have the same exergy as the reactants together—1420 kJ/mol. In this way, CO$_2$ has been neatly stored in, we assume, a useful product and the work required is conserved in the product. There is one snag: the thermodynamic efficiency. Any real process requires more work than the theoretical amount. We assume for this reaction η to be equal to 0.5 or 50%. Then, the reaction requires double the amount of ethylene, one part for the polymer, the other part for the work lost in the process. This part of ethylene may well end up in CO$_2$ and so converting 1 mol CO$_2$ will produce 2 mol of CO$_2$ emission. For $\eta = 67\%$, which is rather high for a chemical process, the process plays even: 1 mol of CO_2 converted and captured in the polymer, produces 1 mol of CO_2 emission. The conclusion is that using hydrocarbons or hydrocarbon-derived reactants for CO_2-chemistry may produce an adverse result: an increase in CO_2 emission rather than a reduction. The story is of course different if the work lost in the process is compensated with work from a nuclear power station or, even better, from solar, wind, or geothermal, that is, renewable energy.

19.5 CARBON CAPTURE RESEARCH

Before CO_2 gas can be sequestered from power plants and other point sources, it must be captured as a relatively pure gas. On a mass basis, CO_2 is the 19th largest commodity chemical in the United States [12], and CO_2 is routinely separated and captured as a by-product from industrial processes such as synthetic ammonia production, H_2 production, and limestone calcination.

Existing capture technologies, however, are not cost-effective when considered in the context of sequestering CO_2 from power plants. Most power plants and other large point sources use air-fired combustors, a process that exhausts CO_2 diluted with nitrogen and excess air. Flue gas from coal-fired power plants contains 10%–12% CO_2 by volume, while flue gas from natural gas combined cycle plants contains only 3%–6% CO_2. For effective carbon sequestration, the CO_2 in these exhaust gases must be separated and concentrated.

CO_2 is currently recovered from combustion exhaust by using amine absorbers and cryogenic coolers. The cost of CO_2 capture using current technology, however, is on the order of $150/ton of carbon—much too high for carbon emissions reduction applications. Analysis performed by SFA Pacific, Inc. indicates that adding existing technologies for CO_2 capture to an electricity generation process could increase the cost of electricity by 2.5–4 ¢/kWh depending on the type of process [12].

Furthermore, CO_2 capture is generally estimated to represent 75% of the total cost of a carbon capture, storage, transport, and sequestration system.

The most likely options currently identifiable for CO_2 separation and capture include

- Absorption (chemical and physical)
- Adsorption (physical and chemical)
- Low-temperature distillation
- Gas separation membranes
- Mineralization and biomineralization

Opportunities for significant cost reductions exist since very little R&D has been devoted to CO_2 capture and separation technologies. Several innovative schemes have been proposed that could significantly reduce CO_2 capture costs, compared to conventional processes. "One box" concepts that combine CO_2 capture with the reduction of criteria pollutant emissions are being explored as well.

Feng et al. [13] showed that for a sample steam reforming process, using the tools of exergy analysis, the efficiency is around 70%. When complete capture of CO$_2$ was practiced, the efficiency dropped to about 60%.

19.6   GEOLOGIC SEQUESTRATION RESEARCH

CO$_2$ sequestration [14] in geologic formations includes oil and gas reservoirs, unmineable coal seams, and deep saline reservoirs. These are structures that have stored crude oil, natural gas, brine, and CO$_2$ over millions of years. Many power plants and other large emitters of CO$_2$ are located near geologic formations that are amenable to CO$_2$ storage. Further, in many cases, the injection of CO$_2$ into a geologic formation can enhance the recovery of hydrocarbons, providing value-added by-products that can offset the cost of CO$_2$ capture and sequestration.

The primary goal of the U.S. Department of Energy's sequestration research is to understand the behavior of CO$_2$ when stored in geologic formations. For example, studies are being done to determine the extent to which the CO$_2$ moves within the geologic formation, and what physical and chemical changes occur to the formation when CO$_2$ is injected. This information is key to ensure that sequestration will not impair the geologic integrity of an underground formation and that CO$_2$ storage is secure and environmentally acceptable.

19.6.1   OIL AND GAS RESERVOIRS

In some cases, production from an oil or natural gas reservoir can be enhanced by pumping CO$_2$ gas into the reservoir to push out the product, which is called Enhanced Oil Recovery (EOR). The United States is the world leader in EOR technology, using about 32 million tons of CO$_2$ per year for this purpose. From the perspective of the sequestration program, EOR represents an opportunity to sequester carbon at low net cost, due to the revenues from recovered oil/gas.

In an EOR application, the integrity of the CO$_2$ that remains in the reservoir is well understood and very high, as long as the original pressure of the reservoir is not exceeded. The scope of this EOR application is currently economically limited to point sources of CO$_2$ emissions that are near an oil or natural gas reservoir.

19.6.2   COAL BED METHANE

Coal beds typically contain large amounts of methane-rich gas that is adsorbed onto the surface of the coal. The current practice for recovering coal bed methane is to depressurize the bed, usually by pumping water out of the reservoir. An alternative approach is to inject CO$_2$ gas into the bed. Tests have shown that the adsorption rate for CO$_2$ is approximately twice that of methane, giving it the potential to efficiently displace methane and remain sequestered in the bed. CO$_2$ recovery of coal bed methane has been demonstrated in limited field tests, but much more work is necessary to understand and optimize the process, particularly the capacity of coal beds with respect to CO$_2$ and methane.

Similar to the by-product value gained from EOR, the recovered methane provides a value-added revenue stream to the carbon sequestration process, creating a low net cost option. The U.S. coal resources are estimated at 6 trillion tons, and 90% of it is currently unmineable due to seam thickness, depth, and structural integrity. Another promising aspect of CO_2 sequestration in coal beds is that many of the large unmineable coal seams are near electricity generating facilities that can be large point sources of CO_2 gas. Thus, limited pipeline transport of CO_2 gas would be required. The integration of coal bed methane with a coal-fired electricity-generating system can provide an option for additional power generation with low emissions.

19.6.3 SALINE FORMATIONS

The sequestration of CO_2 in deep saline formations does not produce value-added by-products, but it has other advantages. First, the estimated carbon storage capacity of saline formations in the United States is large, making them a viable long-term solution. It has been estimated that deep saline formations in the United States could potentially store up to 500 billion tons of CO_2.

Second, most existing large CO_2 point sources are within easy access to a saline formation injection point, and therefore sequestration in saline formations is compatible with a strategy of transforming large portions of the existing U.S. energy and industrial assets to near-zero carbon emissions via low-cost carbon sequestration retrofits.

Assuring the environmental acceptability and safety of CO_2 storage in saline formations is a key component of this program element. Determining that CO_2 will not escape from formations and either migrate up to the earth's surface or contaminate drinking water supplies is a key aspect of sequestration research. Although much work is needed to better understand and characterize the sequestration of CO_2 in deep saline formations, a significant baseline of information and experience exists. For example, as part of EOR operations, the oil industry routinely injects brines from the recovered oil into saline reservoirs, and the U.S. Environmental Protection Agency (EPA) has granted permits to some companies to inject into deep saline formations.

In addition, the Norwegian oil company, Statoil, is injecting approximately 1 million tons per year of recovered CO_2 into the Utsira Sand, a saline formation under the sea associated with the Sleipner West Heimdel gas reservoir. The amount being sequestered is equivalent to the output of a 150 MW coal-fired power plant.

19.6.4 CO_2 MINERALIZATION

Everywhere on earth, we can find large quantities of silicates that are able to convert CO_2 into carbonates. In fact, the erosion of silicate minerals and their transformation into carbonates is the most important process to control, on a very long timescale, the CO_2 balance on this earth. So it does not come as a surprise that scientists, aware of these enormous silicate reserves, have proposed to mine these reserves and spread them at selected sites over the earth's surface to capture CO_2 from the atmosphere.

Such large-scale projects may take a lot of time to materialize because they have global, transnational, and regional dimensions. Closer to realization seem projects in which industrial waste streams, such as the slags of blast furnaces, which contain substantial amounts of these silicates, are brought at certain conditions at which they have great reactivity toward CO$_2$. The waste stream becomes a feedstock for CO$_2$ capture and mineralization, and the product may be useful as a filler and/or whitener.

19.6.5 EFFICIENCY OF CO$_2$ CAPTURE AND SEQUESTRATION

At the moment, there are many different process options to capture CO$_2$ and sequester it. Various pilot studies are also being conducted, such as the one by Shell CO$_2$ Storage, that are examining the feasibility of underground storage of CO$_2$ in spent gas-fields, as currently practiced off-shore [15]. With all these options, it is not possible to make any statement as to which process is the most efficient from a thermodynamic perspective, and which will become the most widespread. By now, the reader must recognize that since all of these involve processes, once adequate models or predictions can be made regarding the conversions and yields, the tools developed in this book are applicable, and rigorous statements can be made as to their efficiencies.

19.7 CARBON TAX AND CAP-AND-TRADE

There is talk about carbon tax and cap-and-trade schemes. Essentially, a carbon tax is a tax on the carbon content of fuels—effectively a tax on the CO$_2$ emissions from burning fossil fuels. Thus, *carbon tax* is shorthand for *carbon dioxide tax* or *CO$_2$ tax.*

Carbon is present in every fossil fuel—coal, oil, and gas—as is hydrogen. Essentially all carbon atoms are converted to CO$_2$ when the fuel is burned. In contrast, non-combustion energy sources—wind, sunlight, falling water, and atomic fission—do not convert carbon to carbon dioxide. Accordingly, a carbon tax (or CO$_2$ tax) is conceptually a tax on the use of fossil fuels, and only fossil fuels.

In *cap-and-trade*, each large-scale emitter, or company, will have a limit on the amount of greenhouse gas that it can emit. The firm must have an "emissions permit" for every ton of CO$_2$ it releases into the atmosphere. These permits set an enforceable limit, or cap, on the amount of greenhouse gas pollution that the company is allowed to emit. Over time, the limits become stricter, allowing less and less pollution, until the ultimate reduction goal is met. This is similar to the cap-and-trade program enacted by the Clean Air Act of 1990, which reduced the sulfur emissions that cause acid rain, and it met the goals at a much lower cost than industry or government predicted.

It will be relatively cheaper or easier for some companies to reduce their emissions below their required limit than others. These more efficient companies, who emit less than their allowance, can sell their extra permits to companies that are not able to make reductions as easily. This creates a system that guarantees a set level of overall reductions, while rewarding the most efficient companies and ensuring that the cap can be met, it is thought, at the lowest possible cost to the economy.

Neither carbon tax nor cap-and-trade are practiced as of today, and obviously practical details will have to be worked out. Both these approaches put a price on carbon, and will affect the economics of processes.

19.8 CONCLUDING REMARKS

There are political and sustainability drivers to capture and sequester CO_2. Various options exist, but none of these have reached technological maturity. Legislation and other schemes may accelerate the development of these technologies. As with any technology or process, one can apply the principles of sustainability and exergy analysis to make a judgment as to which process is most attractive from a technological perspective. The reader must bear in mind, though, that it is not always the best technological option, but the best societal and economic option that typically prevails.

REFERENCES

1. www.cru.uea.ac.uk/
2. Petit, J.R.; Jouzel, J. et al. Climate and atmospheric history of the past 420,000 years from the Vostok ice core, Antarctica. *Nature* 1999, *399*, 429–436.
3. Indermuhle A.; Stocker, T.F. et al. Holocene carbon-cycle dynamics based on CO_2 trapped in ice at Taylor dome, Antarctica. *Nature* 1999, *398*, 121–126.
4. Keeling, C.D.; Whorf, T.P. Atmospheric CO_2 records from sites in the SIO air sampling network. In *Trends: A Compendium of Data on Global Change*, Carbon Dioxide Information Analysis Center, Oak Ridge National Laboratory, U.S. Department of Energy: Oak Ridge, TN, 2005.
5. *Inventory of U.S. Greenhouse Gas Emissions and Sinks: 1990–2007*. www.epa.gov/climatechange/emissions/usginventory.html
6. Report Emissions of Greenhouse gases in the United States 2004. *DOE/EIA-573, 2004*, Energy Information Administration, Washington, DC, 2004.
7. Tyndall, J. On the absorption and radiation of heat by gases and vapours, and on the physical connection of radiation, absorption, and conduction. *The Philosophical Magazine* 1861, *4*(22), 169–194, 273–285.
8. http://earthobservatory.nasa.gov/Library/CarbonCycle/carbon_cycle4.html
9. Earth Gases—Carbon Dioxide. *BBC Weather*. MET Office: Devon and Exeter, 2006. http:/www.bbc.co.uk/weather/features/gases_carbondioxide.shtml.
10. Report to the President Federal Energy Research and Development for the Challenges of the Twenty First Century, Presidents Committee of Advisors on Science and Technology, Panel on Energy Research and Development, 1997. www.ne.doe.gov/pdfFiles/pcast.pdf
11. Office of Fossil Energy and Office of Science. *Carbon Sequestration: State of the Science*, 1999. www.netl.doe.gov/publications/press/1999/seqrpt.pdf
12. www.fossil.energy.gov/programs/sequestration/capture/
13. Feng, W.; Ji, P.; Tan, T. Efficiency penalty analysis for pure H_2 production processes with CO_2 capture. *The AIChE Journal* 2007, *53*(1), 249–261.
14. http://fossil.energy.gov/programs/sequestration/index.html
15. www.shell.nl/home/content/nld/responsible_energy/co2_storage/-; www.shell.com/home/content/responsible_energy/shell_world_ stories/2009/barendrecht/

20 Sense and Nonsense of Green Chemistry and Biofuels

"Green" is a very popular word these days. It expresses the best of intentions to create a clean, healthy, and safe environment. Green wedding, green universities, even green coal and green GDP are examples we can regularly hear and read about in the media. However good the intentions are, the meaning of the word green is usually vague and ill defined.

In process engineering, we need to be much more precise. What do we mean by green chemistry, green plastics, or green fuels? In this chapter, we present a scientifically based method to make a proper assessment to what extent a process is "green," that is, sustainable and efficient. Some examples will be given for illustration.

20.1 INTRODUCTION

20.1.1 WHAT IS GREEN?

Green is the name of a beautiful and friendly color. We associate it with plants, flowers, and fields, and it may give us feelings of refreshment. But these days, the word is also used for things beyond their color. Hotels claim to be green if they encourage their guests to be economical with using water, electricity, and towels. Several universities claim that they have turned "green," in the United States one may be invited to "my green wedding," France's President Nicolas Sarkozy wants France to be "a green country," and *The New York Times* [1] claims that China's economy may not need a white or a black but rather a "green" cat. On a national level, the concept of "Green GDP" has been introduced, which means that the total output of a country's production and services needs to be corrected for the cost of depleting natural resources and of repairing the degradation of the environment. For sure, the use of the word green is no reference to the color, but to a concept that goes beyond the color and of which we understand more or less the meaning and its sensibility. It must make sense when hotel guests use less water, electricity, and towels. But how sensible is it when people speak about "green coal" or about "green plastics" when plastics are grown in the field together with crops for food? What actually do we mean by "green"? Usually, "green" expresses our intentions and efforts to come close to operating efficiently, sustainably, healthily, and safely, but in our profession we need to be more precise and define the concept of "green" in a more quantitative sense. In Section 20.2, we will adopt the definition for *green chemistry* as has been given by the American Chemical Society.

DOI: 10.1201/9781003304388-24

20.1.2 What Is Biomass?

Processes involving *biomass* as the raw material are frequently touted as green. The natural question now becomes, what is biomass?

Biomass is among the oldest resources known to man and an important contributor to the world economy [2]. Biomass comprises any organic matter—wood, plants, crops, and animal waste are good examples. It is both an energy resource and a raw material. The burning of wood for purposes of heat and light has been commonplace for millennia, and early fabrics were comprised entirely from biomass. Indeed, even today, cotton and wool are still popular materials in the clothing industry. Recently, a true revolution in the life sciences has begun. This revolution has the potential to radically change the green plants and products we obtain from them. Green plants developed to produce desired products and energy could be possible in the future.

Accordingly, biomass has become increasingly popular again. The reasons are simple. Biomass is, per definition, renewable and sustainable if the amount utilized equals the amount that is naturally replenished, for instance, by replanting in the case of wood utilization (Figure 20.1). Broadly speaking, biomass can be utilized as (1) a source of renewable chemicals and materials and (2) an energy source.

Now, biomass is a "substance," if you may, which consists of primarily carbon, hydrogen, nitrogen, and oxygen and has a complicated chemical structure. It can be a source of chemicals or materials by a minor modification of the biomass (e.g., cotton, wool). Substantial modification of the biomass yields products that are sources of carbon and hydrogen. This can be achieved by gasification, yielding carbon monoxide and hydrogen, also referred to as synthesis gas; by hydrothermal upgrading yielding an oily substance that has been termed bio-crude oil; or by other technologies. The synthesis gas or bio-crude can then be subjected to processing steps similar to those used in the traditional petrochemical industry to yield valuable products. When viewing biomass as an energy carrier, one can convert the biomass

Nonrenewable

$$\frac{\text{Rate (generation)}}{\text{Rate (consumption)}} < 1 \qquad \text{e.g., oil}$$

Renewable

$$\frac{\text{Rate (generation)}}{\text{Rate (consumption)}} > 1 \qquad \text{e.g., solar}$$

FIGURE 20.1 Renewability of resources. Rates are worldwide per unit time.

into a fuel by a number of conversion technologies such as hydrothermal upgrading toward bio-crude, bacterial decay yielding methane, or one may elect for direct combustion.

20.1.3 BIOMASS AS A RESOURCE

It is not hard to understand why biomass is of interest as a source of both chemicals and energy. Biomass is a renewable resource, and the amount therefore constitutes an effectively infinite source of raw material if the rate of consumption is equal to the rate of biomass regeneration or growth. The use of biomass has certain restrictions, of which three important ones are (1) biomass should be used in such a way that biodiversity is preserved, (2) the quality of biomass varies from source to source, and (3) the use of land for biomass cultivation competes with the use of the land for food cultivation. Furthermore, when viewing biomass as an energy source, any CO_2 emission should be counterbalanced by the amount of CO_2 used during growth. Biomass combustion is seen as a means of closing the carbon cycle, as, in effect, solar energy is converted to chemical energy via photosynthesis and ultimately thermal energy.

It is natural to wonder why the combustion of biomass is an example of sustainable technology, as combustion releases CO_2 in the atmosphere and therefore does not seem to close the material cycles. While the combustion does release CO_2 into the atmosphere, the growth of biomass consumes CO_2 (see Figure 20.2). Therefore, the final result is that there is no net release or consumption of CO_2 if *the biomass is replenished*. For example, consider that a certain unit of active biomass absorbs one unit of CO_2 per unit time while growing. If one unit of this biomass is combusted, thus releasing one unit of CO_2, the result is that there is no net increase or decrease in CO_2 emission. The analogous argument holds for the other materials used during the growth of the biomass such as minerals. Note that the minerals are transformed to ash during combustion, which is a form that cannot be readily assimilated during growth. The whole process is sustainable provided the use of one unit of biomass is counterbalanced by the replenishment of biomass by growth. However, it is important to note that with the current state of technology, additional use of fossil fuels in the biomass conversion process is inevitable in various process steps and the complete process may not be entirely renewable; a closer look at these processes is necessary to determine their renewability. Finally, we note that the depletion of forestlands is not an example of sustainable technology since there is no reforestation to balance the consumption. The recycling of minerals, for example, N, S, P, and K, is also important to replenish the soil.

In this chapter, we will focus on the sustainability and efficiency of processes, and show that what seems or is meant to be green doesn't always turn out to be green, sometimes against our intuition. As is common in this book, thermodynamics and some of its most relevant concepts will be used to analyze claims on the achievements of a process or route. The claim will be made that a product or process can only be called "green" after a proper assessment has been made. Often, such an assessment has necessarily a multidisciplinary character with contributions from disciplines with which we may be less familiar.

Solar energy

Growth of biomass

CO_2, N-compounds, etc.   Biomass

Use of biomass

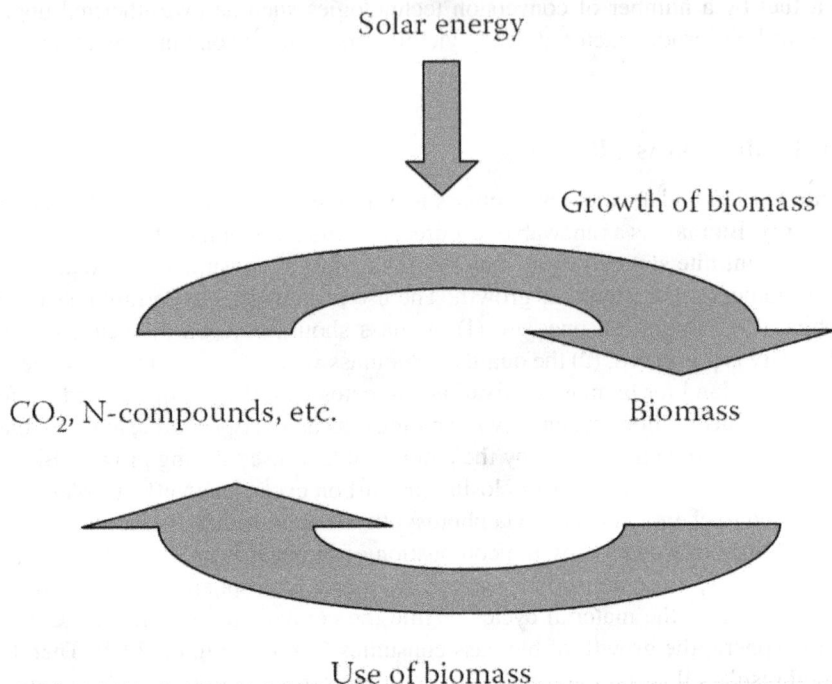

FIGURE 20.2   Schematic of generation and use of biomass illustrating closed elemental cycles.

20.1.4 STRUCTURE OF THIS CHAPTER

We examine some common principles of green chemistry in Section 20.2. Raw materials including recycling are discussed in Sections 20.3, and 20.4 reviews biomass conversion technologies. Sections 20.5 through 20.7 discuss the sense and nonsense of green chemistry and biofuels from various perspectives.

20.2 PRINCIPLES OF GREEN CHEMISTRY

First, we turn to the American Chemical Society [3] for the following definition:

> Green chemistry is the design of chemical products and processes that reduce or eliminate the use and generation of hazardous substances. Green chemistry is environmentally benign, linking the design of chemical products and processes with their impacts on human health and the environment.
>
> [p. 30]

Green chemistry and associated technology is not necessarily new chemistry or a new discipline. It is the same discipline that requires a new "state of mind." The focus of attention may be on new chemical reactions, but in "the back of our mind," we should be thinking of the health and safety of people, the quality of the environment, and the future of our offspring. Of course, this should not keep us from thinking freely about new chemistry, but in the end, we should apply above "boundary conditions."

When former President Bill Clinton was still in the White House, his staff published a document on industrial ecology. By industrial ecology is understood the interaction between industry and its environment. The document was called "The New Paradigm in Environmental Protection" [4]. These days, one often reads about "paradigms" and "a shift in paradigm." Looking this up in the dictionary [5], one finds that a *paradigm* is "a very clear example or model of something" and thus *a shift in paradigm* is observed "when the usual way of practicing it or thinking about it (chemistry, in our case) is changed."

20.3   RAW MATERIALS

Most organic chemicals in use today are derived from nonrenewable petroleum and natural gas, with a small proportion still made from coal. After use, the products ultimately end up as CO_2, the main greenhouse gas causing global warming. For a sustainable future, increasing CO_2 levels is not desirable, and the closing of elemental cycles should be pursued. In other words, renewables from fields and forests should be utilized where possible. Before the advent of the petroleum era, many products were made using renewables, and it would be instructive to examine the viability of these old techniques.

The largest volume chemical produced in 1995 in the United States [6] was ethylene (46.97 billion lb), followed by ammonia (35.60 billion lb) and propylene (25.69 billion lb). These compounds are frequently used to form polymers such as polyethylene (PE) (41% of the world's total) and polypropylene (21% of the world's total) [7].

Before the advent of petroleum, many natural polymers were being used or being studied for use in plastics. Since then, however, synthetic polymers have dominated the marketplace. It should be possible to make plastics that we need from cellulose, starch, lignin, or other natural products.

Cellulose is found in wood, along with hemicellulose and lignin, and in cotton, linen, hemp, and other similar products. It is a polymer of glucose. It can also be made from sucrose by bacterial means [8]. This bacterial cellulose has high mechanical strength and may become an important material if the cost can be brought down.

Regenerated cellulose fibers are known as rayon and regenerated films as cellophane. They are made by the xanthate process, which uses sodium hydroxide and carbon disulfide:

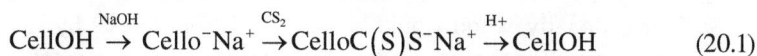

$$CellOH \xrightarrow{\text{NaOH}} Cello^-Na^+ \xrightarrow{\text{CS}_2} CelloC(S)S^-Na^+ \xrightarrow{\text{H+}} CellOH \qquad (20.1)$$

The xanthate solution is converted to fiber or film by passage into a bath of acid. This relatively polluting process can be, and is being, replaced by a process in which a solution of cellulose in N-methylmorpholin-N-oxide is passed into water [9–13]. The solvent is recovered and recycled. This method produces stronger materials, because the resulting molecular weight is higher.

A possible way to lower the costs of fibers and films of regenerated cellulose would be to run cellulose through a twin-screw ultrasonic extruder with a minimum of solvent and pass the extrudate through a stream of hot air to recover the solvent for reuse. This stronger cellophane could be used in place of many plastic films used today. A great number of derivates of cellulose have been made. Methyl, ethyl, carboxymethyl, hydroxyethyl, and hydroxypropyl ethers are made commercially today. These are used as water-soluble polymers, except for ethylcellulose, which is a tough plastic used in screwdriver handles and such.

The hemicellulose from the pulping of trees is an underused resource. A small amount is currently being hydrolyzed to xylose for hydrogenation to the sweetener xylitol. A good use could be as a substrate for fermentation.

Many examples of materials for a sustainable economy can be found by lateral thinking and good chemistry. A comprehensive list is given in [14].

20.3.1 BIOMASS

As mentioned earlier, any organic matter can be considered as biomass. Biomass acts as a CO_2-sequestering agent by using the CO_2 during photosynthesis. In this process, the energy of the sun is stored in chemical form (see Figure 20.2). In a sense, therefore, the sun is the ultimate source of energy, and the biomass is the intermediate. All plants are examples of biomass. Plant and yard cuttings are also biomass, as is wood waste such as sawdust. Agricultural residues such as bagasse from sugarcane, corn fiber, rice straw and hulls, and nutshells are also considered biomass. Now, paper trash is also biomass since it ultimately was made from wood, and a lot of municipal waste is also biomass [15]. It is convenient to further classify the biomass based on function. We choose to consider biomass that is cultivated to serve as a biomass source as *dedicated* or *primary* source, and consider biomass residue such as wood waste a *secondary* source or biowaste. Primary biomass is cultivated with a specific purpose in mind, and the conversion technologies that use this biomass are dedicated conversion plants. For example, sugarcane is grown for the purpose of producing sugar and is therefore a primary source. The residues of the sugarcane processing are biowaste and can be either processed separately or mixed up with other biowaste and therefore deemed a secondary source.

When classifying biomass for purposes of generating energy, it is common to do this based on its source. The following classification is generally used [16]:

1. *Woody biomass*: As mentioned earlier, wood is the oldest biofuel. Conventionally, wood would be harvested from forests by simple logging. This conventional method does not replenish the harvested wood and is therefore *not* sustainable. The current paradigm is to grow wood specifically for purposes of harvesting by means of short-rotation forestry.

2. *Energy crops*: Energy crops are simply crops grown for this specific purpose and are continually replanted once harvested. Woody short-rotation crops are examples of these crops. Trees such as the poplar, the eucalyptus, the willow, and the conifers are expected to be useful [16–22], as are agricultural crops such as the sweet sorghum and the algae.
3. *Residues*: Residues from the forest such as twigs, from the agricultural industry, from wood processing (wood dust), as well as from animal farming (manure) are good examples of residue biomass.
4. *Municipal waste*: Municipal waste contains large amounts of biomass. The use of biowaste as a source transforms a negative-value substance, namely, the waste, into salable products such as energy and/or compost [17]. It is therefore useful in reducing waste, and providing fully renewable energy, since biological waste will be available. The United States alone generated around 208 million tons of municipal solid waste in 1995 [18]. The waste consisted of 66.7% biomass (paper and paperboard, yard trimmings, food, and wood), the remainder being made up by non-biomass items such as glass, plastics, and metals. The breakdown in other countries seems to be similar [19].

As mentioned earlier under the heading of residue, manure is a good biomass source, and there is certainly no shortage of that. The estimated annual U.S. manure generation as a by-product of farming in 1997 in dry weight units is as follows [23]: cattle, 118,424,288 tons; poultry, 17,859,625 tons; and swine, 9,341,288 tons. Others estimate the manure production by cows alone to be approximately 1 billion tons [24]. We can estimate the exergetic value of the manure by examining the caloric values of the dry manure, which we consider to be fuel (see Chapter 9). The caloric value of manure depends strongly on the type of cow, and ranges between 16,771 kJ/kg for fresh beeflot manure to 10,607 kJ/kg for dairy cow manure [25–28]. For purposes of obtaining an order-of-magnitude estimate, we will simply use the mean of these two numbers: 13,689 kJ/kg. The production of cattle manure is between 0.1 and 1 billion tons, or $0.1 \times 2,000/2.2 = 91 \times 10^9$ kg and $1 \times 2,000/2.2 = 910 \times 10^9$ kg. We note, however, that based on the estimates of manure generation in the Netherlands (shown later), the larger number is more likely to be accurate. The approximate exergetic value of cattle manure is therefore between $91 \times 10^9 \times 13,689 \times 10^3 = 1.2 \times 10^{18}$ J and 1.2×10^{19} J. To put this figure in perspective, we compare this to the electricity demand in the United States in 2001, which was 1.3×10^{19} J and the exergetic value of the oil consumption, which was 19.7 million barrels a day, or $19.7 \times 10^6 \times 365 \times 6105.6 \times 10^6$ J/bbl $= 4.4 \times 10^{19}$ J [29]. The numbers suggest that the exergetic value of the manure, which is a waste product, is comparable in order of magnitude to the electricity demand and the exergetic value of the oil consumption. A similar calculation can be made for the Netherlands. In the Netherlands, in 1986, 10^{11} kg of manure was produced [30]. Using the same method described earlier, this is equivalent to $10^{11} \times 13,689 \times 10^3 = 1.4 \times 10^{18}$ J. Compare this to the total energy consumption of the Netherlands in 2001, which was 4.1×10^{18} J, and it is clear that these numbers are of similar order of magnitude [31]. Clearly, the use of biomass as a source for energy production is not hampered by the lack of it!

The use of crops for the purpose of energy generation has been described in the literature [16–22,32]. The use of energy crops reduces the net CO_2 emissions by a significant amount. More information on CO_2 emissions can be found in the literature [33]. Typically, crops used for these fuel purposes include short-rotation woody crops and short-rotation herbaceous crops such as switch grass [34].

20.3.2 RECYCLING

The minimization of waste is an important issue. Recycling is a good option but should only be considered if *reuse* is not possible. For a sustainable future, it is necessary to recycle as much as possible. The amount of waste varies from country to country, with the United States leading the list with 0.88 ton per person per year, followed by Australia (0.74 ton per person per year), and Canada (0.5 ton per person per year) [35,36]. Only 27% of the municipal solid waste generated in the United States in 1995 was recycled. Materials typically recycled included paper, plastic, wood, steel, aluminum, and glass.

Recycling is simplified if the material is free of contaminants. This requires as much separation as possible at the source. However, this process is complicated because many consumer items consist of more than one material. Furthermore, materials such as paper, steel, aluminum, glass, and plastic come in different grades and compositions. For example, some paper may be dyed or waxed. Window glass has a different composition than that commonly found in containers and must be kept separate.

Some curbside recycling programs allow the commingling of materials that are easily separated to save work for the consumer. For example, steel can easily be separated from aluminum by use of a magnet.

The reuse or recycling of paper has advantages on incineration to recover energy from it [36–40], despite contrary claims by some authors. Recycling saves more energy than that obtained by incineration, because making new paper requires energy to harvest new trees, and so on. Recycling creates three times as many jobs as incineration. In England, recycling helps the balance of payments and avoids landfill costs. People who recycle paper are working toward a goal of 100% recycled fiber content, with zero wastewater discharge from the plant. In a few instances, this is possible today [14]. As the techniques of recycling gradually improve, the postconsumer content of recycled paper is rising.

Some paper can be recycled without de-inking. The resulting sheet will be somewhat gray but can be reprinted in a legible fashion. For some uses, such as toilet tissue, there should be no need for de-inking or bleaching. However, in most cases, de-inking is required, and the degree of de-inking depends on the ultimate use.

The removal of toners and inks used in photocopiers can offer problems. Toners are low-melting resins containing carbon black. Various techniques are employed to facilitate their removal. One way is to melt them during the de-inking process, so they can conglomerate, then cool the mixture to harden them, and remove them by screens. Ultrasound can be used to break up the toner particles for easier removal as well.

Mixed waste paper can also be utilized for uses other than paper. The uses include (1) molding of paper fibers mixed with PE or polypropylene to form door panels, trunk liners, and plastic lumber; (2) nonwoven mats of up to 90% paper fibers held together by synthetic fibers; and (3) composites of wood, paper fiber, and gypsum.

In short, a lot can be done using recycling. The reduction of the amount of waste is also a good step toward a sustainable society. For example, the use of a drinking glass eliminates the use of a disposable polystyrene cup. Bringing shopping bags (as is common in the Netherlands) eliminates the use of disposable plastic bags. Also, if less material can be used to make more of the product (by improving, e.g., the mechanical strength of the plastic so the bottle can be thinner), less waste will be generated.

20.4 CONVERSION TECHNOLOGIES

The following conversion technologies are recognized to convert biomass to energy:

1. Combustion
2. Pyrolysis
3. Gasification
4. Upgrading by chemical or biochemical means

Combustion is by far the oldest method of utilizing biomass and has traditionally involved the burning of wood, dung, and other materials. The other methods are, comparatively speaking, more recent. In this section, we will outline these methods.

20.4.1 COMBUSTION

Conceptually speaking, the technology for combustors is similar to that used for coal combustion (see Chapter 9) if solid biomass fuels are used. Typically used configurations for biomass combustion include [41] (1) pile-burned, (2) stoker-fired, (3) suspension-fired, and (4) fluidized-bed combustors.

As the name suggests, in pile burners, the biomass is arranged in a pile-like fashion in a furnace and burned. The combustion air comes from above and below the pile. Advantages of this technology include the flexibility in the choice of fuel and the simple design. However, boiler efficiencies are generally lower than those of the other technologies, and with poor control over the combustion process, these are drawbacks of this technology. A characteristic of the stoker-fired boilers is that there is a fuel feeding system that puts a thin and evenly distributed layer of fuel on a grate, which can be sloping, traveling, or vibrating.

The suspension-fired boilers are similar to the pulverized coal firing technology and involve combusting the fuel in the form of small particles as they are fed into the boiler. A great deal of pretreatment is required of the fuel, which is a potential disadvantage. However, the higher boiler efficiency is an advantage.

Fluidized-bed systems combust the fuel that has been fluidized by high-velocity air. In general, fluidized-bed systems are flexible in terms of fuel requirements. As a result, they are quite suitable for the simultaneous combustion of biomass and other fuels.

Biomass differs from conventional fossil fuels in the variability of fuel characteristics, higher moisture contents, and low nitrogen and sulfur contents of biomass fuels. The moisture content of biomass has a large influence on the combustion process and on the resulting efficiencies due to the lower combustion temperatures. It has been estimated that the adiabatic flame temperature of green wood is approximately 1000°C, while it is 1350°C for dry wood [41]. The chemical exergies for wood depend heavily on the type of wood used, but certain estimates can be obtained in the literature [42]. The thermodynamic efficiency of wood combustors can then be computed using the methods described in Chapter 9.

Now, while the combustion of biomass typically requires modifications to coal combustors, the simultaneous combustion of coal and biomass, also referred to as co-firing, in coal combustors requires very little, if any, modification to the combustors [43]. Biomass co-firing within the existing infrastructure of pulverized coal utility boilers is viewed as a practical means of encouraging the use of renewable energy while minimizing capital cost requirements and maintaining the high efficiencies of pulverized coal boilers. The wide dispersion of pulverized coal boilers (in number and capacity) translates into significant potential opportunities for biomass utilization [44].

We note in passing that countries with large deposits of coal are unlikely to switch to biomass-only combustion and are likely to opt for co-firing instead. The operating temperature is partly determined by the composition of the ash-forming compounds present in the biomass.

20.4.2 Pyrolysis

Pyrolysis is the degradation of macromolecular materials with heat alone in the absence of oxygen [45]. The development of pyrolysis processes for the production of liquids has gained much attention in the last decade because they offer a convenient way to convert low-value woody residues into liquid fuels and value-added products. Biomass pyrolysis is of growing interest as the liquid product can be stored and easily transported [46]. Pyrolysis processes yield a mixture of gas, liquid, and solid products. If pyrolysis is practiced alone, that is, without a subsequent gasification step (see Section 20.4.3), the process conditions are usually chosen to maximize liquid product yields.

20.4.3 Gasification

Biomass gasification involves the transformation of biomass into a mixture of carbon monoxide and hydrogen, also referred to as synthesis gas, and CO_2. These chemicals can then be transformed to a variety of chemicals using Fischer-Tropsch synthesis or biochemical means to fuels such as ethanol [47], or used as feed for fuel cells, turbines, and so on to generate electricity. The technology is therefore very

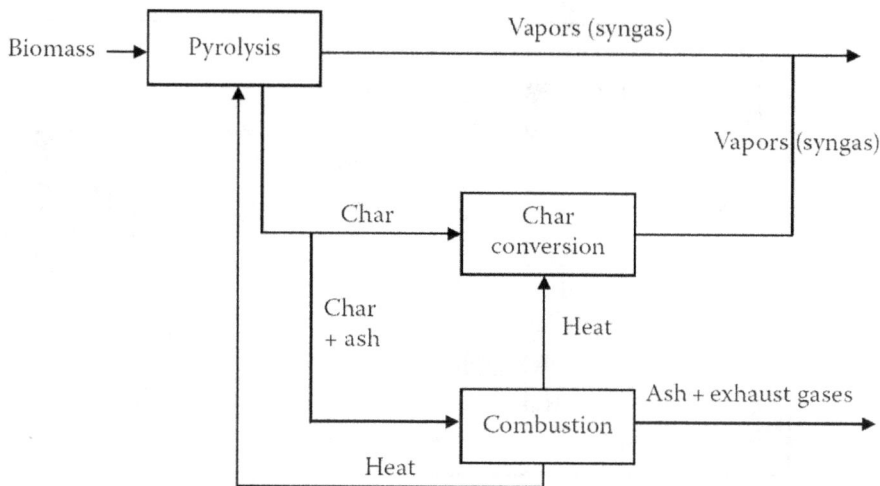

FIGURE 20.3   Schematic representation of biomass gasification.

attractive given the flexibility in the use of the products. Gasification provides a fuel that can easily be incorporated into the existing infrastructure, and it allows for the easy removal of components that cause problems for downstream power generation. Moreover, the technology allows the processing of many biomass wastes that cannot be combusted.

Gasification consists of two endothermic steps, namely, (1) pyrolysis, where volatile components in the biomass are vaporized, and (2) char conversion, which involves reacting the char with steam. The technology is analogous to that used for coal. The pyrolysis step is more important in the case of biomass than it is for coal, since the content of volatiles is higher (70%-86% on a dry basis compared to the 30% of coal). The heat requirements are generally supplied by combusting part of the char. A simple schematic is given in Figure 20.3. A noteworthy example shown in the literature discusses the exergetic efficiency of biomass gasification (Figure 20.4) in which Ptasinski et al. [48] rigorously analyzed the gasification of various sources of biomass to synthesis gas.

These results are important since they highlight that once biomass has been gasified to synthesis gas, the usual processing technologies can be used for chemical synthesis, which are well understood by the industry. As the efficiency of the gasification step is fairly competitive with coal, this may make the gasification of biomass a fairly viable option, but the other metrics developed in this book must be applied as well.

20.4.4   UPGRADING BIOMASS

The upgrading of biomass either by fermentation or by direct liquefaction has been the topic of research for many years. It has been known for decades that landfills produce gas by bacterial decay (anaerobic fermentation) of the biological constituents of

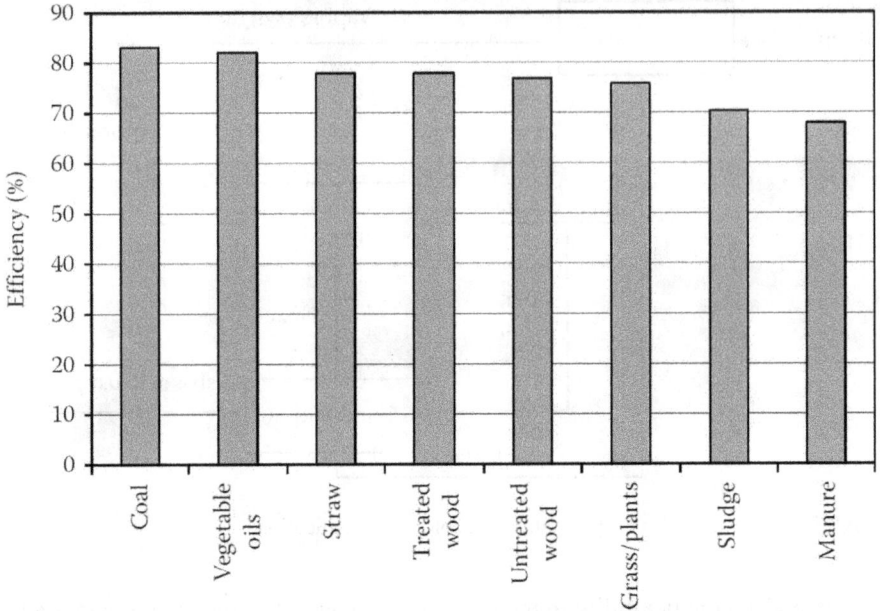

FIGURE 20.4 Comparison of gasification efficiency for various fuels. (With permission from Ptasinski, K.J. et al., *Energy*, 32, 568, 2007.)

the waste. This landfill gas consists mainly of methane and CO_2. In the late 1970s, landfill gas was first used as a fuel in the United States, and since then the technology to collect and use it has improved gradually. This method of producing energy is now regarded as one of the most mature and successful in the field of renewable power. It is well known that methane is a more potent greenhouse gas than CO_2, and its utilization as an energy source is therefore beneficial. The gas is typically around 50% CO_2 and 50% methane, though variation is possible depending on the conditions.

Numerous landfills throughout the world produce gas used to generate electricity [49,50], and additional increases in the landfill gas utilization capacity are expected [50,51]. As noted earlier, landfill gas technology is well established. The gas is taken from the landfill through a series of wells and piped to a processing station, where it is combusted to generate electricity. A typical landfill has the potential to produce gas for 50–100 years [51,52]. The production may only be economically feasible with current technology for about 15 years. The yield of biogas can be increased by pretreating the waste by steam pressure disruption [53] and can yield a net increase of power. The size of the generators at the landfill sites can be several megawatts. For example, a 5 MW plant would produce approximately 42 million kilowatt-hours per year—enough to supply about 3,200 homes in the United States [52].

In Europe [53], approximately two-thirds of the waste goes to landfills, and therefore the waste constitutes a potential source of gas. However, bacterial decay, as it occurs in the landfills, has also been applied to waste generated in farms [49]. The use of household and agricultural waste to provide energy to distant rural communities is therefore believed to have potential (e.g., [54,55]). In certain rural communities in China, for example, bacterial decay of biomass and sewage provides compost and biogas that are subsequently used for soil quality improvement and heating and cooking purposes [56,57]. The anaerobic fermentation of agricultural waste products such as manure can yield valuable amounts of methane and compost.

The aerobic fermentation of pretreated biomass, on the other hand, can yield alcohols, which can be used as liquid fuels. This bioethanol is a potentially useful fossil fuel substitute [58]. It is believed that the addition of ethanol to gasoline reduces carbon monoxide emissions [58] and increases the octane rating in internal combustion engines. With current internal combustion engine technology, however, the mileage does reduce somewhat, but efforts are underway to use the characteristics of the fuel to obtain higher mileage [58].

Bioethanol is suitable for internal combustion engines that run on gasoline. Similarly, biodiesel is designed for diesel engines. Biodiesel is a fuel manufactured from various oils and fats. These acids are chemically transformed to fatty acid methyl esters. By blending the fatty acid methyl or ethyl esters in the right proportions, the properties of the fuel can be influenced [59] and potentially mimic the properties of petrochemically derived diesel. Biofuel efficiency generally is the same as for fossil-derived diesel fuel [59].

A large body of literature exists describing liquefaction processes (see for a review, e.g., [60]).

20.5   HOW GREEN ARE GREEN PLASTICS?

20.5.1   OPTIMISM IN THE UNITED STATES

It is a good thing to read the finest newspapers and magazines of the world and at the same time to read and question them with all the objectivity that one is able to generate. *The New York Times* 2008 article "Harnessing Biology, and Avoiding Oil, for Chemical Goods" [61] points to the enormous dependency of society on oil. Not only do we need it as a transport fuel for the car, but look inside the car. Most of the materials used have been produced from oil as the base chemical. More than 10% of oil is spent on making chemicals, oil is not only the basic resource for transport fuel. This, the article continues, is of course an excellent motivation for switching to renewable sources like corn or switch grass. Microorganisms play a crucial role in such a switch toward another feedstock. And, the article points out, this new technology may not only lead to *biochemicals* but even to *biofuels*, and the traditional oil refineries may be replaced by *biorefineries*. The article then sums up how large companies like Dupont and Cargill are more and more involved in converting sugar from corn or vegetable oils in all sorts of interesting industrial chemicals and even biofuels like bioethanol and biodiesel. In all this, the potential of genetic engineering in "tailoring" microorganisms so that they will do what we want them to do, seems unlimited.

20.5.2 INITIATIVES IN EUROPE

In Europe, a similar development takes place and is called *white biotechnology*, a popular term for industrial biotechnology. Searching with *Google* for this term leads to a "Commentary" in the famous American scientific journal *Science*. In this article, three factors are mentioned for man's great interest in exploring the switch to plant-based biomass for energy production and as a chemical feedstock: the rapidly increasing CO_2 level in the atmosphere, the dwindling fossil fuel reserves, and their rising costs. The authors of this article make a plea for a substantial increase in political and financial investment to move toward a sustainable production of energy and materials. Carbon fixation with the help of solar radiation as in photosynthesis can lead to capturing about 1% of incident light over a year, for sure a significant resource reservoir. The article mentions Brazil and the United States where the emphasis is on ethanol production from sugarcane and corn, respectively, the polylactic acid (PLA) production in the United States from plant-based lactic acid for biodegradable packaging, the biodiesel production from rapeseed oil in Europe and the potential of palm oil in South-East Asia for the same transport fuel. It also mentions the scientific challenge of converting the lignocellulosic cell wall components into components more readily suitable for conversion into chemicals and fuels. This would open the way toward the use of biowaste and dedicated plants rich in such cell wall components, as switch grass, in this bio-industry. A range of technologies such as gasification, steam reforming, and Fischer-Tropsch synthesis is available to complete the conversion from these "inaccessible" plant components to products ready for the market. Finally, this article points to the potential of this industry to "reinvigorate" rural economies: a bio-based economy that goes far beyond food would give rural communities an impulse without precedent to participate in the economy, from local to even global economies, a marvelous prospect for industry, the scientific and the agricultural world and well worth strong cooperation.

In Europe, the bio-industries are associated in EuropaBio. This organization submitted in 2003 six studies within "White Biotechnology" to a number of independent peer reviewers. All studies of processes making use of microorganisms and enzymes in producing industrial antibiotics, vitamins, detergents, bioplastics, and fibers (Figure 20.5) seemed to save on the use of water, energy, and raw materials with environmental benefits as reducing CO_2 emissions.

Five case studies scored high on economic value. These results appeared to be in line with earlier results in similar studies made for the OECD, the Organization for Economic Cooperation and Development. In a joint initiative of EuropaBio and the European Chemical Industry Council, a Technological Platform on Sustainable Chemistry has been set up and a roadmap designed for a "Strategic Research Agenda." Some companies declared White Biotechnology to be an "emerging business area."

20.5.3 FROM A HYDROCARBON TO A CARBOHYDRATE ECONOMY?

Finally, in an analysis made in a report from the European Molecular Biology Organization, the author reminds us that for tens of thousands of years, man relied completely on nature to make his life comfortable: food, clothes, housing materials,

FIGURE 20.5 Bioplastics.

and so on. Then, petroleum (oil) and organic chemistry turned this world upside down in the twentieth century and now biology seems to take revenge for all the reasons we have mentioned before: Not only chemicals and fuels seem to become bio-based in the future but also medicine and agriculture may well benefit from this development as they already do with vitamins as food additive for humans and animals alike. The article mentions that "the carbohydrate economy" may well replace "the hydrocarbon economy" but then it needs some financial assistance as a reward for the former in reducing CO_2 emissions or as a penalty to the latter for not doing so. The United States seems to become more active in this development, more so than Europe, because the United States seems to feel less comfortable with its dependency on oil from other parts of the world.

20.5.4 Feelings of Discomfort

In studying all these articles, in prime newspapers or scientific journals, one cannot help but feel uncomfortable about what is said about the economics of such processes and their energy consumption. Often statements are conflicting, or are ignored, and one feels the need for "a proper assessment" of the many claims that are made. How such a proper assessment may lead to a completely unexpected outcome will be shown in the following case study.

20.5.4.1 Case Study 20.1: Green Plastics

In 1999, two American biochemical engineers with experience in microbial enzymology and genetics, both in the academic and the industrial world, Gerngross and Slater

[62], were delighted with an executive order issued by former President Clinton. The order insisted that researchers should work toward replacing fossil resources with plant material both as raw material and fuel. Both men had a lot of experience in *growing plastic* in plants. This sounds like fiction, but it is not. Their dreams seemed to come true because they assumed that this plant-based plastic would be "green" in two ways: Use had been made of renewable resources and upon disposal they would break down and leave no waste.

These are two of the many principles that have been defined for *green chemistry*. We mentioned some other principles earlier: health, safety, energy efficiency, and so on. Such technology can best be quantitatively analyzed with a thorough thermodynamic analysis in which the concepts we presented in Part II, and which are closely related to energy, its quality and its losses, are most prominent. But let us follow these researchers as they reported their remarkable achievements and conclusions in the *Scientific American* [62].

Three main approaches exist to replace conventional plastics with plant-derived materials. Some major industrial enterprises are in this field: Cargill, Dow Chemical, Monsanto, and ICI. Two "green" plastics have drawn most attention: PLA and polyhydroxyalkanoate (PHA). Three different routes can be distinguished. The first is the conversion of plant sugars via microorganisms into lactic acid (LA), with a subsequent chemical step for the synthesis of PLA, which has properties similar to those of polyethylene terephthalate (PET), the plastic well known for its use in soda bottles and clothing fibers. The second route is a nice example of "the microbial factory" where a specific microorganism converts the sugar directly into PHA granules that can build up to 90% of a single cell's mass. The third route is the most remarkable. The plastic grain, PHA, is grown directly *in the plant*, corn, which by proper genetic engineering ensures not only the growth of plastic grains but also that this growth does not interfere with the production of food grains and will take place in the leaves and in the stem. So in addition, there is no competition for land, and one product, grains of plastic, can be harvested after the other, grains of corn.

This route sounds too fantastic to be true, but the route is real and exists. However, the aforementioned researchers were, quite unexpectedly, in for some unpleasant surprises. We will only mention what caused them the greatest concern. This concern was so great that they questioned whether it was worth continuing the process development. They calculated all the energy and raw materials required to arrive at the final product and, in their own words, "discovered that this approach would consume even more fossil resources than most petrochemical manufacturing routes" [62]. Table 20.1 shows that the conventional fossil-based petrochemical process to manufacture PE requires a total of 2.2 kg fossil fuel per kg product, but 1 kg ends up *in* the product, as available work, exergy, and 1.2 kg fossil fuel is spent and lost in the process, lost work. In thermodynamic terms: Of 2.2 units of work made available to the process, 1 unit is conserved and ends up as the exergy of the product, and 1.2 units are spent in the process and dissipated. In contrast, the plant route to produce PHA requires 2.65 kg fossil fuel per kg product, of which *nothing* ends up in the product and *all* is spent, dissipated in the process. The exergy in the plastic grain stems from the solar energy source, not from the fossil fuel. How to calculate the

TABLE 20.1

Fossil Fuel and Solar Energy Contributions to the Production of Polyethylene (PE), Polylactic acid (PLA), and Polyhydroxyalkanoate (PHA) in kg Fuel/kg Product.

	Fossil Fuel	Solar Energy[a]	In Product	Process
PE	2.2	—	1.0	1.2
PLA[b]	1.2–2.0	1.0	1.0	1.2–2.0
PHA[b]	2.39	1.0	1.0	2.39
PHA	2.65	1.0	1.0	2.65

[a] Inefficiency of solar energy recovery not taken into account.
[b] Via fermentation.

thermodynamic efficiency: All fossil fuel is lost in the process, the captured solar energy is conserved in the products such as corn and plastic grains. Things are somewhat more favorable for microbial-made PHA and still better for the PLA process. For this last process, it is expected that exergy dissipation, the work lost, will come close to that of PE manufacturing.

Yet there is another point of concern. If exergy dissipation for 1 kg of PLA and 1 kg of PE is comparable, PLA seems to be preferred because its exergy content is of solar origin. But the fossil fuel for the petrochemical industry is oil, whereas the fossil fuel for the agricultural industry, at least in many American states, is coal. From Chapter 4, it is clear that for the same amount of exergy required, the CO_2 emission is at least twice that of oil. So plant-based plastic creates a CO_2 emission that is twice that for petrochemical plastics, indeed a serious point of concern. In the meantime, it has become obvious that the focus of attention should not be the exergy source for the exergy contained in the plastic, but the source of exergy for the exergy dissipated in the process. The solution then seems to be that exergy from a renewable exergy source should pay for the dissipation, for the work lost. One of the surprising conclusions in this field of activities is that both emissions and depletion of fossil resources can be abated by continuing to make plastics from oil (i.e., to transfer exergy from oil to plastics), while paying for the process energy cost, the lost work in the process, should be done with renewable biomass as the fuel. Indeed, Monsanto calculated that the power that could be generated from the corn waste after harvesting corn and PHA plastic grains was more than sufficient to run the process. Other points of attention should be that the renewable PLA and PHA are biodegradable but conventional fossil fuel-based plastics are not. On the other hand, the biodegradation of renewable plastics may generate undesired emissions such as CH_4. PLA and PHA produced via fermentation compete with other needs for land; PHA grown in corn does not. The researchers conclude that there is no single strategy that can overcome all environmental, technical, and economic limitations of the various process alternatives at the same time. Biodegradability, CO_2 and other emissions, and competition for land are factors to consider.

20.5.5   Short Memory: Ignorance or Not Welcome?

It is interesting to observe that in all later publications, *no reference* is made to this study with its important and cautioning conclusions. It may be that scientific results are often ignored, or, if not ignored, are not considered welcome, or in conflict with political interests. A similar observation can be made with respect to the development of biofuels, see Case Study 20.2, although here awareness seems to grow. In Europe, governments seem, as from 2008, to temper some of the expectations that were raised by their own earlier decisions to introduce more biofuels in the transportation market. These examples show how dependent governments are of proper, selected scientific information and how politics, economics, and science and technology may get in conflict with each other. The great challenge for leadership is to align them.

20.6   BIOFUELS: REALITY OR ILLUSION?

For many problems in our world, it is not enough to depend on one discipline. Instead, more often than not, such problems require a multidisciplinary approach. A specific competence and education may be most useful but not enough, and contributions from quite different disciplines may also be required.

20.6.1   Multidisciplinarity

We can hardly think of a better example to illustrate this than the food crisis[1] that emerged in 2008 and perhaps even earlier. Food prices started to rise to such an extent that for the poorest among our world population, for whom the cost of food is a very significant part of their income, buying food and feeding their livestock became a serious problem. The crisis was ascribed to at least three factors:

1. The increase in living standard of some large developing countries like India and China, allowing them to eat more and to eat more meat
2. Speculation on the prices of wheat, rice, and soy at the world's largest commodities markets
3. The emergence of biofuels from corn, wheat, soy, rapeseed, and palm oil, taking land and food away from regular food supply

Let us focus on one factor where we can, to some extent and based on our scientific and engineering background, apply some well-founded judgment: biofuels.

20.6.1.1   Case Study 20.2: Bioethanol from Corn

In 2001, three Canadian researchers published a full thermodynamic analysis on the renewability of bioethanol from corn, produced at conditions prevailing in the Canadian province of Quebec. The authors [63] define a resource to be fully renewable if "regeneration mechanisms exist for the resources which maintain it intact without disturbing the environment." They recall the essence of a thermodynamic cycle, in particular the thermomechanical cycle (Figure 20.6), where

Heat from source

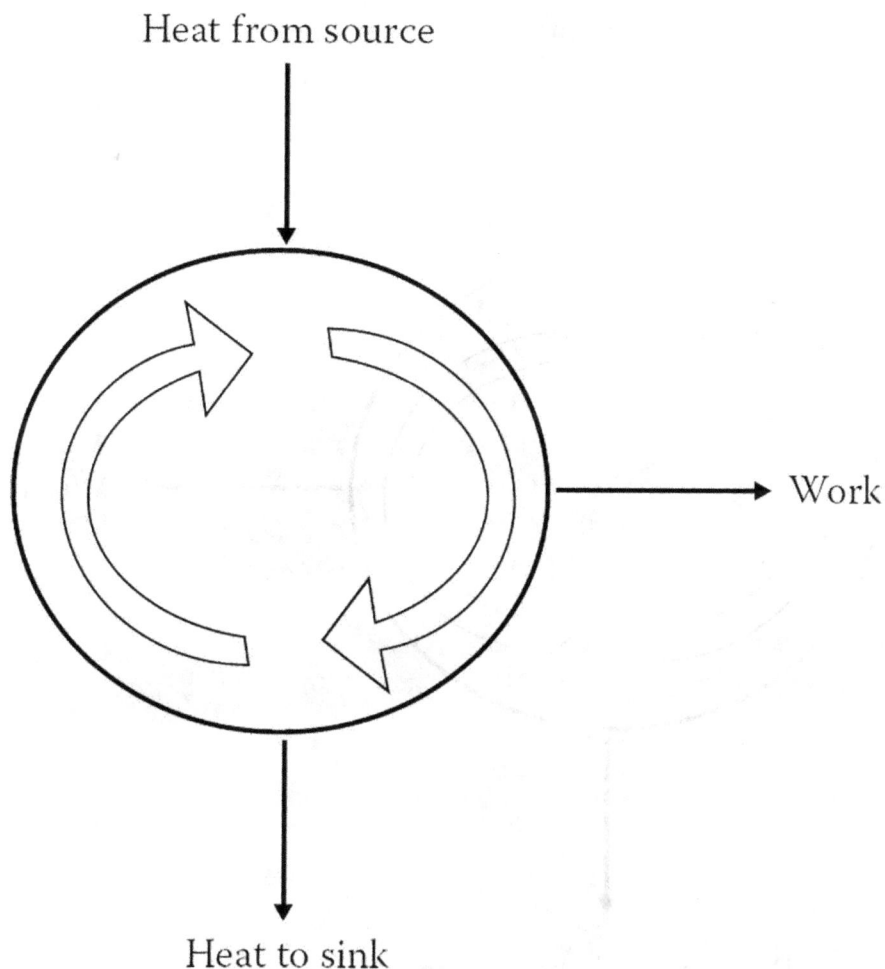

Work

Heat to sink

FIGURE 20.6   Essentials of the thermomechanical cycle.

heat from a high-temperature reservoir, the *source*, and with the help of a work-ing fluid, produces work, rejecting the remaining heat to a low-temperature reser-voir, the *sink*. Usually the working fluid does not change its chemical composition, although, as in biochemical cycles, this is possible. An ecosystem can be consid-ered as a natural thermodynamic cycle, where the sun acts as the source, space acts as the sink, and compounds in the hydrosphere, the atmosphere, and the lithosphere act as the working fluid. This cycle generates the work to sustain life (Figure 20.7) and, over geological times, may even accumulate exergy such as pres-ent in fossil fuels, which are considered nonrenewable because they are consumed much faster in human utilization than they are produced by nature. The ideal ther-modynamic cycle that the authors have focused on is the natural growth of corn, the

Solar radiation

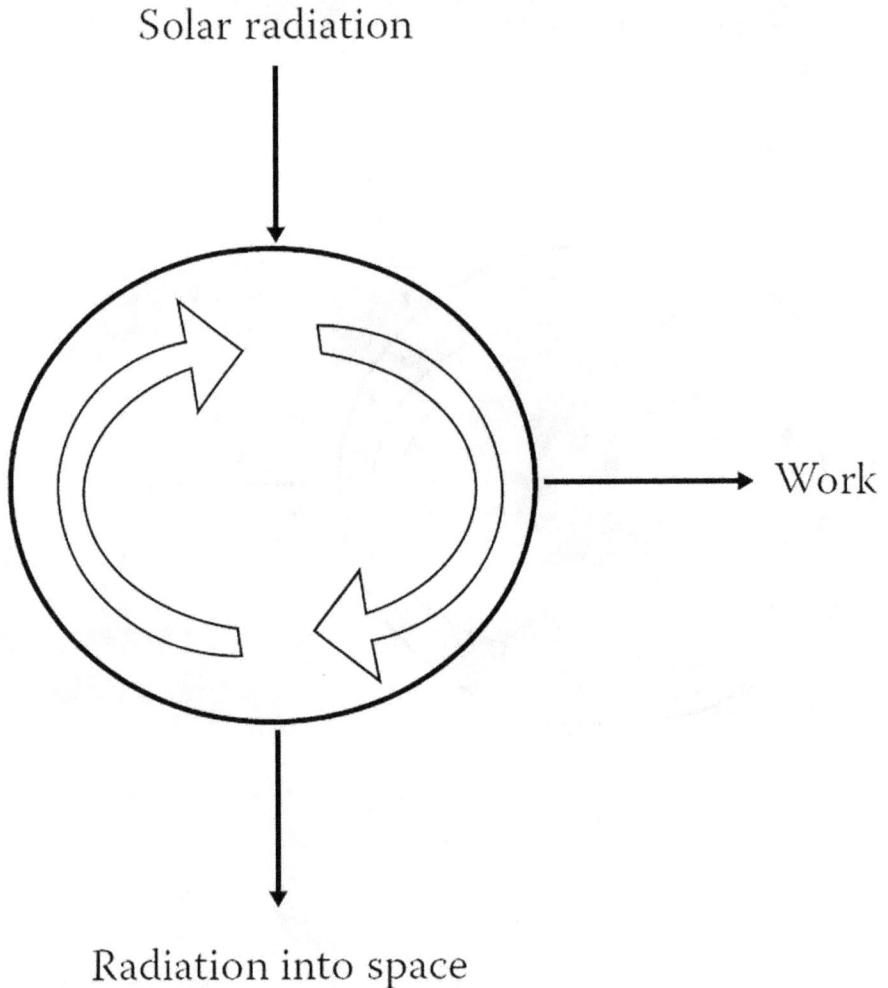

Work

Radiation into space

FIGURE 20.7   The ecosystem as a thermodynamic cycle.

human-assisted transformation of the corn's glucose into ethanol, and the subsequent combustion of ethanol back to the natural components CO_2 and H_2O for the production of work as in the engine of a car.

In their analysis, the authors soon realized that in practice there is a departure from ideal behavior in the sense that nonrenewable resources such as fossil fuel are unavoidable in carrying out this cycle (Figure 20.8). Large amounts of nonrenewable resources slip into the process for both resource processing as well as waste treatment; for example, fossil fuels such as diesel for transportation, propane for heating and drying, fertilizers, pesticides, lime, and electricity for the wastewater treatment installations. Although in Quebec the electricity is from hydropower, which is a renewable resource, the associated consumption of nonrenewable resources for the

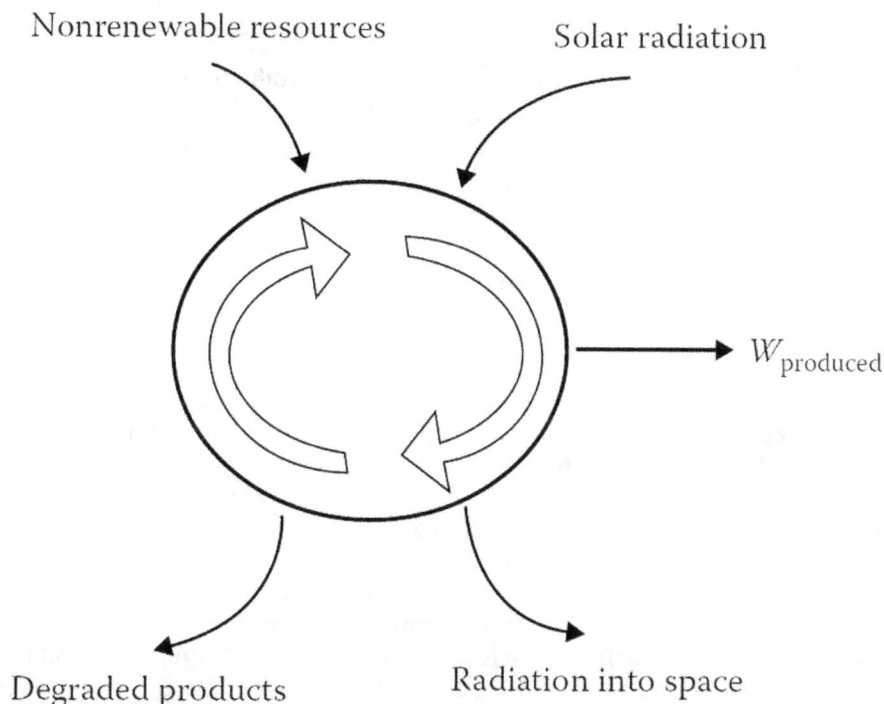

FIGURE 20.8 Departure from the cycle of nature.

construction of the dam is not insignificant and has to be included. Instead of a conventional exergetic efficiency for electricity production of 25%, this efficiency for electricity in Quebec is close to 83%, assuming that 95% of the electricity is generated from hydroelectric sources and 5% from fossil fuel combustion.

Next, the authors remind us of the concept of *cumulative exergy consumption*. For example, in our example of PE manufacturing in Case study 20.1, we mention that the production of 1 kg of PE requires 2.2 kg fossil fuel. Assuming as an approximation that both product and fuel consist of ethylene ($-CH_2CH_2-$) segments, the product's own exergy or available work, requires an additional 1.2 times this work for processing, resulting in "lost work," making the total work requirements, the cumulative exergy consumption, 2.2 times the product's exergy. In this example of ethanol production from solar exergy, CO_2 and H_2O via corn, the largest fraction of the work from nonrenewable resources ends up outside the product as the exergy contained in the product originates in the solar radiation. Table 20.2 summarizes the total work required in MJ, not originating from captured sunlight, to cover the total work required and dissipated for the production of ethanol from 1 ha of corn, which represents, 56,000 MJ of chemical energy produced from solar radiation. In other words, in order to produce one unit of work available in ethanol, more than one unit of work from other sources is required. Things get worse if hydroelectricity is not available and if, for example, coal or natural gas is used, as is not uncommon in

TABLE 20.2

Total Work Required in MJ for the Production of 56,000 MJ Ethanol from 1 ha of Corn.

Production from corn	27,788
Conversion into ethanol	
Diesel	49,507
Electricity	8,764
Water treatment	
Electricity	25,619
Total	111,678

American agricultural states. The electricity consumption in terms of nonrenewable resources will then sharply increase, with a factor of five or more, and the same holds for the CO_2 emission. Based on these findings, *the authors conclude that bioethanol from corn is not renewable.* So far for this case study.

In 2004, the U.S. Department of Agriculture (USDA) [64] published a report that pictures a much more favorable outcome for the production of ethanol from corn than the Canadian publication in 2001 that we have just discussed. The USDA claimed that the bioethanol output from corn is 1.67 units for each unit of input of fossil fuel (heat, electricity, chemicals, fertilizers, and insecticides) and 1.06 against 1 *without* allocating energy input to the by-products. These results seem to contradict the criticism of other prominent researchers, David Pimentel and Tad Patzek from Cornell University and the University of California in Berkeley, respectively, to whom we will refer again later.

20.6.2 SECOND-GENERATION BIOFUELS

In 2005, Shell reported on the activities of the Canadian company Iogen Energy, part of a producer of enzymes, in which Shell has bought a share in 1997. This company has developed an enzyme that is able to catalyze the conversion of the nonfood "difficult to digest" building materials of the cell (cellulose, hemicellulose, and lignocellulosic materials), into carbohydrates that can then be readily converted into ethanol with yeast (Figure 20.9). This route has been claimed to have the advantage that it does not compete with food, does not require subsidies, and makes use of biowaste or nonfood crops (such as switch grass, a herbaceous plant). A demonstration plant was built, and since 2004, this plant runs on straw. Its energy requirements are claimed to be covered by the waste of the process (Table 20.3).

Ethanol from waste is an example of what is called "second-generation" biofuels, products that are not competing with food. Another example is the production of biofuels of variable specification and end-use from gasification and the subsequent liquefaction of biomass. The German company CHOREN [65] showed in 2003 to be able to gasify biomass as wood chips and straw into a synthetic gas. Shell has a strong position in the so-called gas to liquid (GTL) technology making use of the Fischer-Tropsch technology and so it is not surprising that in 2005, Shell bought a minority share in this German company for the joint development of "second-generation" biofuels.

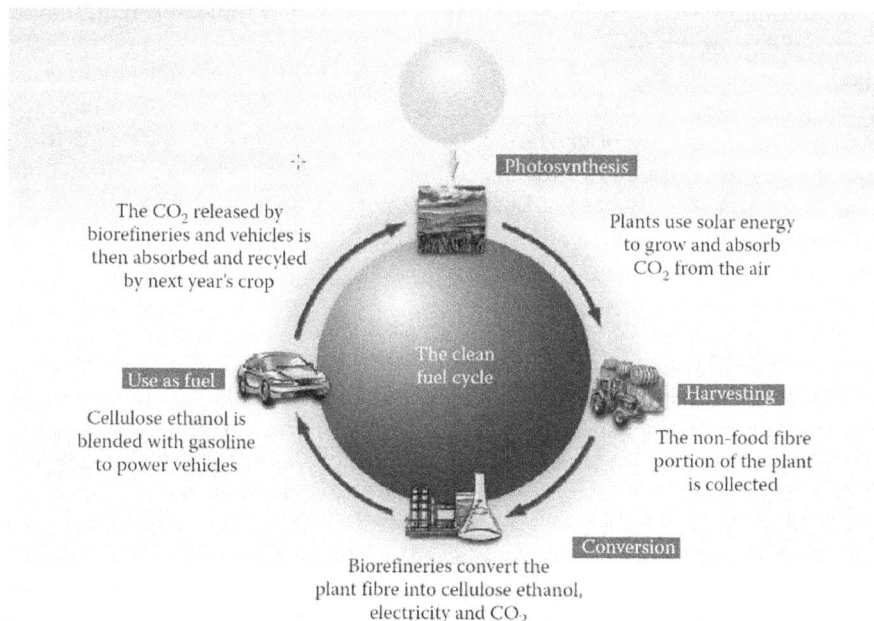

FIGURE 20.9 Second-generation biofuels. (With permission from http://www.geni.org/globalenergy/library/technicalarticles/generation/future-fuels/iogen/advanced-renewable-biofuel-that-can-be-based-intodays-cars/index.shtml.)

TABLE 20.3
Second-Generation Biofuels.

Shell and EcoEthanol		EcoEthanol Plant
Feedstock:	Plant- and agricultural waste, not corn, wheat, etc.	1.1 million barrels/year $250 million investment Requires an area with a radius of 70 km of which 25%–40% is for wheat production
Claims:	No competition with food; no subsidies required; no net production of CO_2	→ 700,000 tons waste/year (straw)
Process:	Conversion of waste into sugar by super enzyme; fermentation (yeast); process waste products for process energy	

Since late 2007, the Energy Biosciences Institute in Berkeley has been the center for cooperation between scientists from the University of California and the Agricultural Department of the University of Illinois for the production of fuels from so-called energy crops like switch grass. In this second-generation biofuel project that is financed over a 10-year period with $500 million by oil company BP, biomass is converted with the help of synthetic catalysts, for example, organometallic compounds, in a special solvent medium, better known as ionic liquids, into hydrocarbons with properties close to automotive fuels.

The development of second-generation biofuels may however require more patience. Some of the most popular biocrop feedstocks seem to belong to what are called "invasive species," better known as "weeds," with a high tendency to escape biofuel plantations and overrun adjacent farms and natural land [66]. One expert stated that investors have often started these new ventures in the expectation to produce biofuels in return and in the not-too-far future. It is understandable that they do not like negative assessments. But clearly this is another example where a proper assessment has to be made and patience needs to be applied.

20.6.3 The Fossil Load Factor

In 2002, Wassenaar [67] made an analysis of the fossil fuel requirements of the Dutch potato industry and its main products. The products are of course, in terms of their "energy" that is, available work or exergy, of solar origin. Therefore, he introduced the concept of "fossil load factor," defined as the number of fossil fuel exergy units per exergy unit of potato-based useful products. Of course, this ratio varied with the product: the fossil load factor for frozen fries (0.80) is of course larger than for fresh fries (0.64) and much larger than for selected potatoes (0.12).

This fossil load factor seems to be an important issue in the debate on biofuels. When we speak of the fossil load, we must think of fertilizers, fuels, herbicides and insecticides, electricity for wastewater treatment, heat for separation, and so on. If, as the Canadian researchers earlier claimed after a thorough thermodynamic analysis, this factor is larger or even much larger than 1, what is the sense of making biofuels? It is then more advantageous to use the fossil fuel right away without the exergy-costly bio-detour. And even if the USDA data are correct, this factor is still 0.6 meaning that each volume unit of bioethanol requires around half a volume unit of fossil fuel, which is still high for building an industry upon this concept. Meanwhile, it is clear that one objective has been reached, in particular in the United States: reinvigorating agriculture and the world of farming. At the same time, this world and the political world get confused because more and more articles appear in prominent newspapers and specific journals putting question marks behind this "biofuel movement." Not only the claims on reduced CO_2 emissions are brought in doubt but, even more seriously, caution is expressed with regard to a possible food crisis where the poorest people on this Earth, spending most of their income on food, cannot afford the rising food prices due to the "pull" of the biofuel industry. So a new question looms: If the fossil load is unavoidable, should it be spent on food for human and animal beings or on fueling cars? Considering the enormous volumes of fuel that transportation requires in comparison with food consumption and due to the rate of consumption, the question does not seem difficult to answer.

20.6.4 Sustainability and Efficiency

The fossil load factor is an important issue and its origin so evident and often unavoidable that we asked ourselves the question what the consequences are when this factor is reduced to zero. Whenever, in a biomass conversion process, a fossil fuel contribution was spotted, we replaced this contribution by one from biomass origin. For example, the process may require electricity, which is supplied by a nearby coal-fed power station. Then this amount of electricity was thought to be generated by a power station fed by biomass. Or the process may require heat or chemicals and again biomass is the raw material from which these requirements were met. Dr. Feng Wei[2] made such an analysis for a process where a diesel-type product was obtained from wood chips as a feedstock. His work has been discussed as an example at the end of Chapter 13.

20.6.5   ALGAE

A very recent development in the world of biofuels is the production of biodiesel from *algae*. Biodiesel is known to be produced from soybeans in the United States, from rapeseed in Europe, and from palm oil in Asia. But these routes have the disadvantage that their fossil load factor may be prohibitive, their production capacity is limited, they require areal land and thus compete with food, and they need fresh water. Shell, in late 2007, has set up a pilot facility in Hawaii to grow marine algae and produce vegetable oil for conversion into biofuel. It is hoped that this is a feasible, sustainable route starting from nonfood raw materials. Algae grow rapidly, can be harvested far more frequently than agricultural crops, have productivity per hectare that may be more than 50 times that of soybeans, are rich in oil, and can be cultivated in ponds of seawater, minimizing the use of fertile land and fresh water. Little is known yet about the fossil fuel requirements, but the first indications are that again the fossil load factor may be a handicap.

20.6.6   THE FUTURE

Looking even further into the future as was done at a meeting of the European Science Foundation in 2007 and in line with the conclusions of the Inter Academy Council (IAC), the international association of the world's most prominent scientific academies, scientists agreed that the future of electricity, hydrogen, and other fuels or chemicals is to be found in the abundance of solar energy. The challenge is to "harness" the work available in the solar radiation that permanently reaches our Earth. Scientists trust that they will be inspired by biological systems and learn to design "The Artificial Leaf" in which important principles of photosynthesis can be mimicked and the biological intermediate, the plant, can be circumvented as well as the synthesis of carbohydrates. Instead, the conversion should make use of CO_2, H_2 from splitting water and go straight for the fuels of interest for fulfilling our energy needs (*Note:* These conversions may capture much more solar energy than photosynthesis in living systems).

The most drastic proposal for transportation "fuels" can be found in a presentation at the headquarters of the OECD (Organization for Economic Cooperation and Development) in Paris in late 2007 by Professor Tad Patzek, then from the University of California in Berkeley [68]. His paper was based on his close cooperation with Professor David Pimentel from Cornell University, thus covering expertise from both the engineering and the plant sciences. Their work is firmly based on the laws of thermodynamics making extensive use of the concepts of exergy and cumulative exergy consumption. In his paper "How can we outlive our way of life?" Professor Patzek concludes that free private transportation should become limited by good public transportation. He argues and explains with conclusive calculations that fossil fuels and biofuels (suppose they have indeed entered the market) should be replaced by electricity while at the same time the internal combustion engine should be replaced by the all-electric engine with a thermodynamic efficiency of nearly 2.5 that of the internal combustion engine. Initially, the batteries will be charged with electricity from conventional power stations but eventually from photovoltaic cells. Even mediocre solar PV-cells with a lifetime of 30 years, taking one-third of their operational life to cover for the exergetic cost of their production, will be at least 70 and in most cases more than 100 times more efficient than biofuels. Table 20.4 shows the ratios of land area needed to power cars using different power sources, including the exergetic cost of production. The unit is the motive power produced by one square meter of a medium quality oil field producing its crude for 30 years at a constant rate. These numbers seem to indicate that any effort to produce transportation fuels via biomass is without prospect. It seems far more efficient to skip living systems and even skip material fuels like carbohydrates or hydrocarbons. The table suggests that the most efficient road from solar power to motive power is by the transportation of electrons. This table may be very upsetting for many believing strongly in the future of biofuels and therefore it needs to be scrutinized and subjected to proper assessment.

20.6.7   Sense or Nonsense? Discussion

We cannot blame the reader who is somewhat confused after digesting the material in this chapter. This reader may have sympathy and understanding for the principles of green chemistry. These principles make sense: replace nonrenewable by renewable resources, prevent or repair damage to the environment, reduce waste, increase efficiency, and so on. And so, if initiatives are taken to grow plastics from plants instead of producing them from fossil fuels, this seems to make sense too and the term "green plastics" seems justified. Until scientists in one of the most famous scientific journals, the *Scientific American*,[3] report, to their great disappointment, that this route consumes more fossil fuel than when produced from fossil fuels in a petrochemical plant [69]. Here, confusion sets in: "Green plastics" are not "green"? What seemed to be a sustainable process appears to be non-sustainable. And although this was reported in 2000, with all the reasons why, the era of "white biotechnology" in Europe heralded an era of close cooperation between government, industry, and scientists, without ever referring to the American article. And *The New York Times* reported in 2008 [70] on the research efforts to "harness biology" in efforts to replace all fossil fuel needed for producing chemicals from renewable sources like corn and switch grass . . . without any reference to the 2008 article [71]. Is this ignorance, optimism, entrepreneurial spirit, or less innocent, the wish to make short-term profits? The least one can say is that the scientists back in 2000 have helped to create awareness and prudence with regards to claims and expectations in "greening" the world.

TABLE 20.4

Ratios of Land Areas to Power Cars Using Different Motive Power Sources and Including the Area for Their Production.

Technology	Ratio
Oil field	1
PV cell	3
Wind turbine	37
Acacia + electricity	125
Sugarcane ethanol	214
Acacia Fischer-Tropsch (FT) fuel	324
Acacia cellulosic ethanol	410
Eucalypt electricity	593
Corn grain ethanol	620
Corn stover cellulosic ethanol	1299
Eucalypt FT fuel	1342
Eucalypt cellulosic ethanol	1917

Things really become confusing, overwhelming, and even frightening if one enters the world of biofuels. Confusing because of the many conflicting statements and claims, overwhelming because what looked like something of regional or national importance, quite suddenly takes global dimensions, frightening because what may have started as an impulse for agriculture and a stimulus for farmers suddenly is connected to more hunger among the millions of the poorest people on this Earth. From stimulating agriculture and increasing energy security (United States, Brazil), we are suddenly faced with increased greenhouse gas emissions (CO_2 and N_2O), competition for land between food and energy, and rising food prices hitting the poorest of this world. Again, back in 2001, Canadian researchers cautioned us in a solid applied thermodynamics journal against biofuels, pointing out that the fossil load factor per unit of biofuel far exceeded 1. Another moment of growing awareness that seems to have escaped the attention of many governments who "jumped on the bandwagon," undoubtedly often with the best of intentions.

A well-known Dutch professor in agricultural science and at the same time senator in the Dutch parliament, who has studied this development for 30 years, states that food security in this world can be attained, feed security is a problem (if more and more people start eating meat), but bioenergy will show to be an illusion. Land requirements relate roughly as 1:10:100 for somebody who *does not* eat meat, for somebody who *does* eat meat, and for somebody who fills his or her tank with biofuel. Biofuels are often subsidized, but this may soon stop. *The New York Times* calls, in 2008, biofuels a "terrible mistake," a senior authority of the United Nations calls Europe's measures to add biofuels to regular transportation fuels "a crime against humanity," and the secretary general of the United Nations states that "enough is enough" when he sees the food prices rise so fast [72]. Ironically, some bioethanol producers have started complaining about the rising prices for raw materials.

Slowly but surely, the world seems to discover that first-generation biofuels (from corn, soy, rapeseed, palm oil, i.e., crops that produce food ingredients) should be abandoned (Figure 20.10). Second-generation biofuels (from waste of food crops, or nonfood crops) look still promising, but the first reservations have already been expressed. Biofuels from algae have not been enough explored to be conclusive. On the longer term, solar fuels, where fuels are obtained directly from solar radiation, skipping biomass as an intermediate, seem promising and some claim that even carbon can be eliminated and solar radiation via photovoltaics should produce electricity, to be stored in a car's battery, driving the car's electric engine. A proper assessment with a thorough thermodynamic analysis will always be required, but this will only be *one* contribution in a total effort of experts from many disciplines. It is difficult to think of a better example of interdisciplinary activity than the production of food, feed, energy, and materials from solar radiation. And we have not even touched yet the problems of consumerism, in which 15% of the world's population consumes at any moment 85% of the world's resources [73], of the incompatibility of the world's economic growth with measures against climate change [74], and of how a growth-economy should be transformed into a steady-state economy that does not aim for growth but for sustainability [75]. We like to quote Professor Herman Daly in Figure 20.11.

(a)

(b)

Green dreams

(c)

(d)

FIGURE 20.10 Visions of a green fuels in the future: (a) biofuels from corn, (b) biofuels from rapeseed, (c) biofuels from sugarcane, and (d) biofuels from algae.

"The larger system is the biosphere and the subsystem is the economy. The economy is geared for growth... whereas the parent system doesn't grow. It remains the same size. So as the economy grows... it encroaches upon the biosphere and this is the fundamental cost..."

HERMAN DALY
Professor, School of Public Policy, University of Maryland.
Former Senior Economist, World Bank.

FIGURE 20.11   Professor Herman Daly's argument for a steady-state economy.

20.7   CONCLUDING REMARKS

From this chapter, we may conclude that "green" is a complex concept. A number of principles of "green chemistry" make sense: increase a process' efficiency, reduce or eliminate the damage done to the environment, reduce waste, make more and better use of renewable resources, and so on. But "green" may seem obvious when it is not. Then "green" does *not* make sense. Green is associated with efficiency and sustainability. Efficiency may be high without sustainability. Sustainability may have been accomplished without efficiency. Biofuels and bioplastics illustrate how far the consequences of such products reach. A proper, complete assessment needs to be made for all processes and products, including a thorough thermodynamic analysis. Such an analysis may be ignored in the beginning, but it may give off important signals. The total assessment may require experts from many disciplines, from chemistry, molecular biology, all the way to ecology and agroscience. Political considerations may blur the issues, but in the end, hard scientific evidence will prevail. Good leadership will insist on being informed by the best scientists such as those from the national academies and international academic associations such as the IAC.

More specifically, we wish to conclude that green chemistry often challenges the question of why things are done as they are done and invites innovation. It is not so much new chemistry that is addressed, but rather a new state of mind. And as for biomass, there seems to be neither shortage of supply, nor shortage of conversion routes to materials and fuels. But caution is required; is the rate of conversion indeed compensated by the rate of replenishment? Is there no competition for food? And are conversion processes truly sustainable or is too much "fossil slipping into the process"? A full thorough assessment and analysis, including second-law analysis, should give the answer.

NOTES

1 It did not help that this crisis coincided with another crisis: the initially local American financial crisis leading to a global crisis.
2 Dr. Feng Wei is a professor at Beijing University of Chemical Technology. He performed this work during an extended research period at the Delft University of Technology as guest of one of the former authors (JdSA).

3 At the occasion of the "ice-breaking" visit to China in 1972 of the late former President of the United States, Richard Nixon, the American government offered the Chinese people all editions of the *Scientific American* as a gift.

REFERENCES

1. Friedman, T.L. (Op-Ed). Bring in the green cat. *New York Times*, November 15, 2006. http://select.nytimes.com/2006/11/15/opinion/15friedman.html?_r=1
2. Chum, H.L.; Overend, R.P. Biomass and renewable fuels. *Fuel Processing Technology* 2001, *71*, 187–195.
3. *The American Chemical Society.* www.acs.org/education/greenchemistry (ACS uses definition based on Anastas, P.T.; Warner, J.C. *Green Chemistry: Theory and Practice*, Oxford University Press: New York, 1998, p. 30, By permission of Oxford University Press).
4. White House Report. *The New Paradigm in Environmental Protection*, 1991.
5. *Compact Oxford Dictionary*, Oxford University Press: Oxford, 2008.
6. Kirschner, E.M. Growth of top 50 chemicals slowed in 1995 from very high 1994 rate. *Chemical & Engineering News* 1996, *17*, 8.
7. Reich, M.S. *Chemical & Engineering News* 1997, *14*, 26.
8. Tajima, K.; Fujiwara, M.; Takai, M. Biological control of cellulose. *Macromolecular Symposia* 1995, *99*, 149.
9. Mortimer, S.A.; Peguy, A.A.J. The influence of air-gap conditions on the structure formation of lycocell fibers. *The Journal of Applied Polymer Science* 1996, *60*(305), 1747.
10. O'Driscoll, C. Spinning a stronger yarn. *Chemistry Bromine* 1996, *32*(12), 27.
11. Dobson, S. Man-made fiber markets recent agitation and change. *Chemistry and Industry (London)* 1995, 870.
12. Parkinson, G. Borax's new image: Beyond the forty mule team. *Chemical Engineering* 1996, *103*(10), 50.
13. Hirami, M.J. New developments in the Rayon industry. *Macromolecular Science, Part A: Pure and Applied Chemistry* 1996, *A33*(12), 1825.
14. Matlack, A.S. *Green Chemistry*, Marcel Dekker, Inc.: New York, 2001.
15. National Renewable Energy Laboratory. *Learning about Renewable Energy.* www.nrel.gov/learning/re_biomass.html
16. LIOR International NV. *Biomass Combustion.* www.lior-int.com/biomass_home.htm
17. Nonhebel, S. Energy yields in intensive and extensive biomass production systems. *Biomass & Bioenergy* 2002, *22*, 159–167.
18. Hallam, A.; Anderson, I.C.; Buxton, D.R. Comparative economic analysis of perennial and annual intercrops for biomass production. *Biomass & Bioenergy* 2001, *21*, 407–424.
19. van den Broek, R.; Teeuwisse, S.; Healion, K.; Kent, T.; van Wijk, A.; Faaij, A.; Turkenburg, W. Potentials for electricity production from wood in Ireland. *Energy* 2001, *26*, 991–1013.
20. Malik, R.K.; Green, T.H.; Brown, G.F.; Beyl, CA.; Sistani, K.R.; Mays, D.A. Biomass production of short-rotation bioenergy hardwood plantations affected by cover crops. *Biomass & Bioenergy* 2001, *21*, 21–33.
21. Zan, C.S.; Fyles, J.W.; Girouard, P.; Samson, R.A. Carbon sequestration in perennial bioenergy, annual corn and uncultivated systems in southern Quebec. *Agriculture, Ecosystems & Environment* 2001, *86*, 135–144.
22. Tuskan, G.A.; Walsh, M.E. Short-rotation woody crop systems, atmospheric carbon dioxide and carbon management: A US case study. *The Forestry Chronicle* 2001, *77*, 259–264.

23. Kellogg, R.L.; Lander, CH.; Moffit, D.C; Gollehon, N. Manure nutrients relative to the capacity of cropland and pastureland to assimilate nutrients: Spatial and temporal trends for the United States. *Publication number NPS-00–0579*, Washington, DC: US Department of Agriculture, National Resource Conservation Center, Economic Research Service.

24. How America's diet style is destroying its ground water, *Farm Sanctuary News*, 1998. http://www.jivdaya.org/farm_animal_waste_and_the_clean_water_dilemma.html

25. Kitani, O.; Hall, C.W. *Biomass Handbook*, Gordon & Breach Science Publishers: New York, 1989.

26. De Rijk, P.J.; Zegwaard, M.J. *Literature survey quality organic domestic waste components (Literatuuronderzoek kwaliteit GFT-componenten)*, Publicatie reeks afvalstoffen VROM, Delft, 1993/7, 1993.

27. Sweeten, J.M.; Korenberg, K.; LePori, W.A.; Annamalai, K. Combustion of cattle feedlot manure for energy production. *Energy in Agriculture* 1986, *5*, 55–72.

28. *Phyllis, Database for Biomass and Waste.* www.ecn.nl/phyllis/

29. U.S. Energy Information Administration. U.S. household electricity uses: A/C, heating, appliances. www.eia.doe.gov/emeu/reps/enduse/er01_us.html

30. Landbouw Ekonomisch Instituut. *Jaarcijfers*, Landbouw Ekonomisch Instituut: The Hague, 1986.

31. U.S. Energy Information Administration. *EIA International Energy Outlook 2009* (Electricity). www.eia.doe.gov/oiaf/ieo/electricity.html

32. Graham, R.L.; Allison, L.J.; Becker, D.A. ORECCL—Oak Ridge Energy Crop County Level Database. *Bioenergy* 1996 (Partnerships to develop and apply biomass technologies. In *Proceedings of the Seventh National Bioenergy Conference*, Nashville, TN, September 5–20, 1996). *Southeastern Regional Biomass Energy Program* 1996, *1*, 552–529 (NICH Report No. 24419).

33. www.epa.gov/globalwarming/emissions/national/co2.html

34. Tillman, D.A. Biomass coking: The technology, the experience, the combustion consequences. *Biomass & Bioenergy* 2000, *19*, 365–384.

35. Levy, G.M. (ed.). *Packaging in the Environment*, Blackie Academic and Professional: London, 1993.

36. Matos, G.; Wagner, L. Consumption of materials in the United States 1900–1995. *The Annual Review of Environment and Resources* 1998, *23*, 107.

37. Berman, F. *Trash to Cash*, St. Lucie Press: Delray Beach, FL, 1996.

38. Thompson, C.G. *Recycled Papers—The Essential Guide*, MIT Press: Cambridge, MA, 1992.

39. Bateman, B. *Pap-Tech Engineers & Associates* 1996, *37*, 15 (as cited in Matlack, A.S. *Green Chemistry*, Marcel Dekker, Inc.: New York, 2001).

40. Karna, A.; Engstrom, J.; Kutinlahti, T. *Pulp and Paper Industry* 2001, 38 (as cited in Matlack, A.S. *Green Chemistry*, Marcel Dekker, Inc.: New York).

41. Broek, R.; van den, A.; Faaij, A.; van Wijk. Biomass combustion power generation technologies, Study performed within the framework of the extended JOULE-IIA Program of CEC DGXII, Energy from biomass: An assessment of two promising systems for energy production (Project). *Report no. 95029*, Department of Science, Technology and Society, Utrecht University, Utrecht, 1995.

42. Szargut, J.; Morris, D.R.; Steward, F.R. *Exergy Analysis of Thermal, Chemical and Metallurgical Processes*, Hemisphere Publishers: New York, 1988.

43. *Advanced Coal-Based Power and Environmental Systems '98 Conference*, session 4.2. www.netl.doe.gov/publications/proceedings/98/98ps/ps4-2.pdf

44. *Advanced Coal-Based Power and Environmental Systems '98 Conference*, session 4.3. www.netl.doe.gov/publications/proceedings/98/98ps/ps4-3.pdf

45. D. Meier.; O. Faix; State of the art of applied fast pyrolysis of lignocellulosic materials—A review. *Bioresource Technology* 1999, *68*, 71–77.

46. Bridgwater, A.V.; Meier, D.; Radlein, D. An overview of fast pyrolysis of biomass. *Organic Geochemistry* 1999, *30*, 1479–1493.

47. Klasson, K.T.; Elmore, B.B.; Vega, J.L.; Ackerson, M.D.; Clausen, E.C.; Gaddy, J.L. Biological production of liquid and gaseous fuels from synthesis gas. *Applied Biochemistry and Biotechnology* 1990, *24/25*, 857–873.

48. Ptasinski, K.J.; Prins, M.J.; Pierik, A., Exergetic evaluation of biomass gasification. *Energy* 2007, *32*, 568–574.

49. www.ad-nett.org/html/news.html, DOE/EIA-0603(96) Distribution Category UC-950, Renewable Energy Annual 1996, Energy Information Administration, Office of Coal, Nuclear, Electric and Alternate Fuels, U.S. Department of Energy, Washington, DC. http://www.eia.doe.gov/cneaf/solar. renewables/renewable.energy.annual/contents.html, April 1997.

50. Weiland, P. Anaerobic waste digestion in Germany—Status and recent developments. *Biodegradation* 2000, *11*, 415–421.

51. *Energy from Waste, Irish Energy Centre*, 2002 (Archive 94). http://www.irish-energy.ie/publications/index.html

52. *TVA Landfill Gas Energy.* www.tva.gov/greenpowerswitch/landfill.htm

53. Liu, H.W.; Walter, H.K.; Vogt, G.M.; Vogt, H.S.; Holbein, B.E. Steam pressure disruption of municipal solid waste enhances anaerobic digestion kinetics and biogas yield. *Biotechnology and Bioengineering* 2002, *77*, 121–130.

54. Purohit, P.; Kumar, A.; Rana, S.; Kandpal, T.C. Using renewable energy technologies for domestic cooking in India: A methodology for potential estimation. *Renewable Energy* 2002, *26*, 235–246.

55. Misi, S.N.; Forster, C.F. Batch co-digestion of two-component mixtures of agrowastes. *Process Safety and Environmental Protection* 2001, *79*(B6), 365–371.

56. Spath, P.L.; Dayton, D.C. *Preliminary Screening—Technical and Economic Assessment of Synthesis Gas to Fuels and Chemicals with Emphasis on the Potential for Biomass-Derived Syngas*, National Renewable Energy Laboratory: Golden, CO, 2003. www.nrel.gov/docs/fy04osti/34929.pdf

57. *Clean Energy Resource Teams.* www.cleanenergyresourceteams.org/technology/biogas-digesters

58. *Ethanol: Separating Fact from Fiction*, 4/99. http://www.ott.doe.gov/biofuels/publications.html# bioethanol.

59. *Biodiesel Handling and Use Guidelines.* http://www.ott.doe.gov/biofuels/publications.html bioethanol.

60. Elliot, D.C.; Beckman, D.; Bridgewater, A.V.; Diebold, J.P.; Gevert, S.B.; Solantausta, Y. Developments in direct thermochemical liquefaction of biomass: 1983–1990. *Energy & Fuels* 1991, *5*, 399–410.

61. Bhattacharjee, Y. Harnessing biology, and avoiding oil, for chemical goods. *New York Times*, April 9, 2008. www.nytimes.com/2008/04/09/technology/techspecial/09chem.html

62. Gerngross, T.U.; Slater, S.C. How green are green plastics. *Scientific American*, August 2000.

63. Berthiaume, R.; Bouchard, C.; Rosen, M.A. Exergetic evaluation of the renew-ability of a biofuel. *International Journal of Exergy* 2001, *1*(4), 256–268.

64. Shapouri, H.; Duffield, J.A.; Wang, M. *The 2001 Net Energy Balance of Corn Ethanol*, USDA Office of the Chief Economist: Washington, DC, 2004.

65. Choren Industries. *Electricity and Heat From Biomass, The Carbo-V process*, Choren Industries: Freiburg, Sachsen, 2003.

66. Rosenthal, E. New trends in biofuels has new risks. *The New York Times*, May 21, 2008.

67. Wassenaar, J. *Sustainability of the Potato Production Chain; A Thermodynamic Approach.* Master thesis, University of Technology, Delft, 2002.

68. Patzek, T.W. How can we outlive our way of life, *Paper prepared for the 20th Round Table on Sustainable, Development of Biofuels: Is the Cure Worse than the Disease?* OECD Headquarters, Château de la Muette, Paris, September 11–12, 2007. http://petroleum.berkeley.edu/papers/Biofuels/OECDSept102007TWPatzek. pdf

69. Slater, S.C.; Gerngross, T.U. How green are green plastics, *Scientific American*, August, 2000, pp. 37–42.

70. Chang, K.; Revkin, A.C. At a sleek bioenergy lab, a lens on a cabinet pick. *New York Times*, December 22, 2008.

71. Zeller, T. Scientists on cellulosic fuel: Avoid mistakes before they happen, *New York Times*, October 1, 2008.

72. Krugman, P. Grains gone wild, *New York Times*, April 7, 2008.

73. Diamond, J. What is your consumption factor? *The New York Times*, January 2, 2008.

74. Sachs, J.D. Keys to climate protection. *Scientific American*, March 18, 2008.

75. Daly, H. *Towards a Steady-State Economy*. www.theoildrum.com/node/3941

21 Solar Energy Conversion

In this chapter, we will focus on solar energy and its conversion to other forms of energy. Solar energy is renewable and as an immaterial source, radiation, its emission is equally immaterial. In nature, it induces closed material cycles with overall outputs that closely resemble those of industry: mechanical, chemical, and electrical energy, and chemical products. These properties (i.e., renewability, no emissions, and closed material cycles) make solar energy an excellent candidate for energy supply to a sustainable society in the making.

After the introduction, we will briefly discuss the main characteristics of solar radiation. Consistent with the scientific principles on which this book is founded, a rigorous thermodynamic analysis will then follow of the creation of wind energy, of photothermal and photovoltaic energy conversion, and of photosynthesis. Most of this chapter has been based on the monograph *Thermodynamics of Solar Energy Conversion* by De Vos [1].

21.1 INTRODUCTION: "LIGHTING THE WAY"

In October 2007, the Inter Academy Council (IAC) published a report on the road the world needs to take to attain a sustainable energy future [2]. The IAC represents the 150 science and engineering academies of this world and the report was written upon the request of Brazil and China (www.interacademycouncil.net). The co-chairman for this project was the 1997 Nobel laureate for physics, and current, as of 2009, Secretary of State, Steven Chu from Stanford University.

The title of the report is very suggestive: "Lighting the Way" and is a nice play of words [2]. But using sunlight was not the only recommendation. The recommendations were as follows:

1. Secure power to the world's poorest people
2. Increase the efficiency of the use of fossil fuel
3. Capture CO_2 from coal and other fossil fuel utilization
4. Extract energy from the sun and other renewable resources
5. Invite the United Nations to examine the use of nuclear power on security and waste

The first recommendation applies to the global society as a whole. Every citizen of this world should feel and act as a global citizen and bear responsibility for the energy needs of those who are worst off. Of course this is a personal choice but most likely the choice of most of our readers. The second recommendation refers to the low thermodynamic efficiency of many industrial processes: "the best new power station is the one you do not have to build because that does not need building if the existing power stations have been made more efficient." In Chapters 3 and 4, we have

DOI: 10.1201/9781003304388-25

made the same observation. Recommendation 3 is also obvious and stems from our worries about climate change. Recommendation 5 originates in the conviction of many countries, like China, that so much time is still required to reach a sustainable energy situation, and that growing needs of the developing world, like China and India, make the competition for fossil fuels so fierce, that the use of nuclear fuel, albeit temporarily, is unavoidable.

Recommendation 4 is obvious but not simple to fulfill. Much research and development has to be undertaken before solar energy becomes competitive. But it has the advantage of being an immaterial source of energy, devoid of emissions, except for radiation. And it is abundant. The incident solar radiation is about 5×10^6 EJ/year (1 EJ = 1 ExaJoule = 10^{18} J). The biological energy flux is 7×10^3 EJ/year, and the world energy consumption is in the range from 350 EJ/year in 1995 up to an estimated 900 EJ/year in 2050. So there is all possible reason to transform our fossil fuel-based society into a solar-fuel-based society (Figures 21.1 and 21.2). The power stations of the future may look somewhat different: as the photovoltaic power station of Figure 21.3, where incident radiation is converted into electricity in numerous solar cells or as the photothermal power station of Figure 21.4, where a battery of numerous mirrors reflect and focus incident radiation onto an area where this heat is captured by a hot fluid that acts as the familiar cycling agent of a heat engine.

The road to a sustainable energy future will not be easy but Chu reminds us that when the United States decided to put a man on the moon, it was successful [2]. If the world decides to have a similar ambition with sustainable energy, success will be assured when comparable funds are made available.

Chu was not the first Nobel laureate to make this observation. A few years before, in 2004, Smalley [3], who received the Nobel Prize for chemistry in 1996, expressed in a speech his hope that President George W. Bush, at the beginning of his second term, would inspire the next generation of U.S. scientists and engineers: the new "Sputnik Generation." This generation should develop replacements for fossil fuel resources and find solutions to the problems associated with climate change. Smalley

Extraction   Emission

FIGURE 21.1   The fate of fossil fuels in the globe's engine.

Radiation in   Radiation out

FIGURE 21.2   A solar-fueled globe.

FIGURE 21.3   A photovoltaic power station in which solar radiation is converted into electricity.

mentioned that the resources available to meet these challenges, solar, wind, and geothermal energy, are now hardly exploited. He pointed out that most of the hydro-electric options have already been tapped, that biomass is a very interesting energy resource if it were not for its competition for land with food and its water requirements. Hydrogen was neither a good option as it is not a source for energy but rather

FIGURE 21.4 A photothermal power station in which sunlight is reflected by a large collection of mirrors and focused on "the hot spot" where heat is input to the cycle of a heat engine.

a medium for storage and transportation in which respect hydrocarbons do a much better job (see Chapter 18).

Smalley then referred to the calculations of the research group of Lewis at the California Institute of Technology [4]. The group focuses on the energy of the biggest, most proven and sustainable nuclear reactor: the sun. This reactor sends about 165,000 TW (1 TW is 10^{12}W) to the Earth. Six strategically chosen sites (Figure 21.5) on our Earth, each of 100 x 100 km^2, should together capture 200 TW of solar radiation and convert it into electricity in photovoltaic cells with a thermodynamic efficiency of 10%. The total electricity produced would correspond with that from 60 TW fossil fuel converted in conventional power stations with a thermodynamic efficiency of about 30%.

The resulting electric power corresponds to the energy consumption of 2000 W by each of 10 billion people, that is, allowing the future world population a power consumption more or less corresponding to that of somebody living in the developed world now. This electricity should be transported and distributed in giant grids with intermittent small local storage units for supply to industries, the house, the car, and so on. Such storage units do not exist yet and this should, when Smalley wrote his article (2004), be the focus of research for the years to come, together with efforts to increase the thermodynamic efficiency of photovoltaic cells. Smalley is excited about this prospect of electricity flowing instead of oil or gas (see Chapter 18), is convinced that it can be realized, and hopes at the same time that the world will produce the leaders who can inspire the people with the skills to accomplish such challenging tasks.

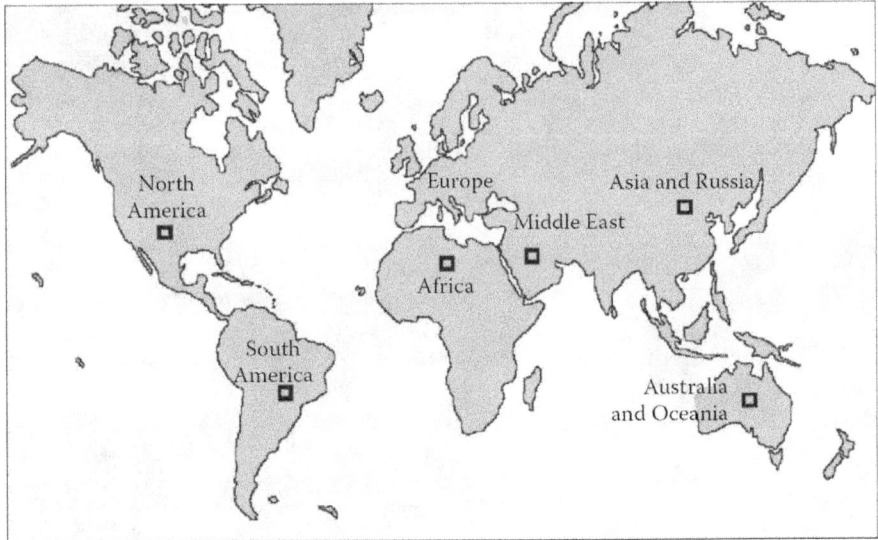

FIGURE 21.5 Strategically chosen sites to capture 200 TW of solar radiation. (From Smalley, R.E., *MRS Bull*, 30, 412, 2005.)

21.2　CHARACTERISTICS

Solar radiation consists of photons of different energies E. Of particular interest is the spectral distribution $\dot{n}(E)$, which describes how the photons are distributed over the different energy values. The quantity $\dot{n}(E)$ indicates the number of photons of specific energy E per unit surface area per unit energy per unit time. From this distribution we define the total photon flux as

$$\dot{N} = \int_0^\infty \dot{n}(E)\,dE \qquad (21.1)$$

and the associated energy flux as

$$\dot{E} = \int_0^\infty \dot{n}(E)\,E\,dE \qquad (21.2)$$

Planck's law expresses the spectral distribution $\dot{n}(E)$ as a function of E according to

$$\dot{n}(E) = \varepsilon \cdot \frac{2\pi}{c^2 h^3} \cdot \frac{E^2}{\exp(E/kT) - 1} \qquad (21.3)$$

in which

    ε denotes the emissivity of the photon emission reservoir
    c is the velocity of light
    h is the Planck's constant
    k is the Boltzmann constant

Figure 21.6 gives n as a function of E with E in units kT and $\dot{n}$ in units k^2T^2/c^2h^3, that is, $m^{-2}\,s^{-1}\,J^{-1}$. This curve holds for an emissivity ε = 1 when the spectrum is called a black-body spectrum. Note that the curve displays a maximum for an intermediate value of E, which is commonly referred to as Wien's energy. Wien's displacement law expresses that the coordinates of this maximum change with temperature T. Figure 21.7 shows the Planck spectrum for three different temperatures $T = 5762$ K, the temperature of the sun; $T = 288$ K, the average temperature of our planet; and $T = 2.7$ K, the temperature of the cosmic background. The double-logarithmic scale allows for the clear visualization of how the three spectra are distributed over the most common ranges of light. From Planck's law, Equation 21.3, we can now proceed to Equations 21.1 and 21.2 and obtain the corresponding expressions for the photon flux $\dot{N}$ and energy flux $\dot{E}$ and find

$$\dot{N} = \varepsilon\sigma'T^3 \tag{21.4}$$

and

$$\dot{E} = \varepsilon\sigma T^4 \tag{21.5}$$

Both constants σ' and σ are expressions of and can be calculated from the natural constants k, h, and c, which were discussed earlier. Equation 21.5, which relates the

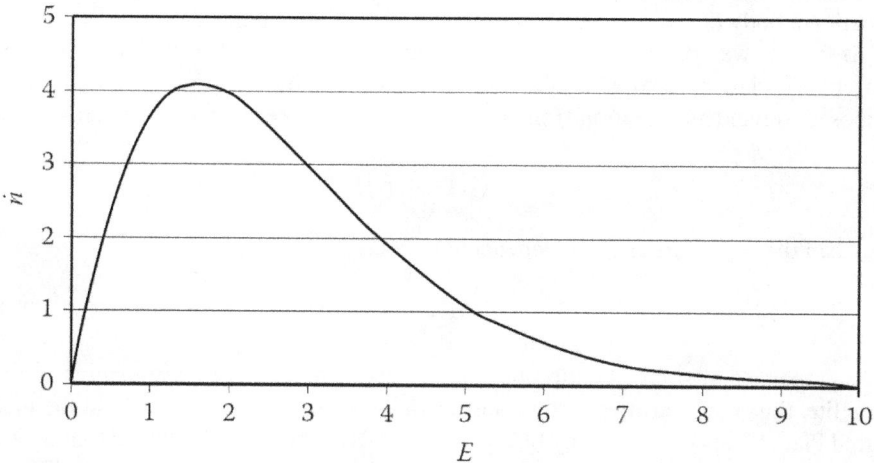

FIGURE 21.6 The Planck spectrum with linear scales. E in units kT; $\dot{n}$ in units k^2T^2/c^2h^3.

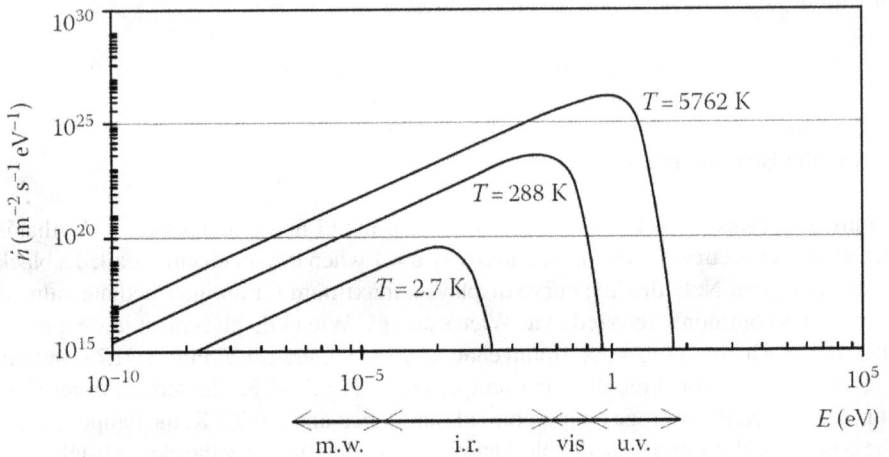

FIGURE 21.7 The Planck spectrum with logarithmic scales (m.w.: microwave; i.r.: infrared; vis: visible light; u.v.: ultraviolet radiation).

energy flux to the absolute temperature, is known as the Stefan-Boltzmann equation. The coefficient σ is known as the Stefan-Boltzmann constant. The average photon energy can be calculated from

$$\overline{E} = \frac{\dot{E}}{\dot{N}} = \frac{\sigma T}{\sigma'} \tag{21.6}$$

and is found to be 2.70 kT.

Finally, we should mention Kirchhoff's law. The emissivity ε expresses which fraction a body of temperature T emits to bodies of lower temperature.

If $\varepsilon = 1$, we speak of black-body radiation, otherwise of gray-body radiation. Kirchhoff's law compares the emissivity α with the absorptivity a of a body when exposed to incident radiation from a body with a higher temperature and states that

$$\alpha(E) = \varepsilon(E) \tag{21.7}$$

or when these properties are independent of energy

$$\alpha = \varepsilon \tag{21.8}$$

The properties of solar radiation have been established from measurements from a satellite, thus eliminating all influences from the earth's atmosphere. From the measured Planck's spectrum and by fitting Equation 21.3, it can be concluded that the sun is a black body [5], that its emissivity $\varepsilon = 1$, and that its temperature is $T_s = 5762$ K. The sun emits a photon energy flux $\dot{E}_s$, and therefore, the amount of energy emitted per unit time for 1 m^2 of the sun's surface equals $1 \times \dot{E}_s$. The total energy emitted is obtained by multiplying the total external surface area with the flux: $4\pi R_s^2$ emits a photon flux $\dot{E}_s$; then the total energy flow rate emitted by the sun is $4\pi R_s^2 \dot{E}_s$. However, when this energy reaches earth, it is distributed over a much larger area, thus reducing the energy flow rate. With R_0 the radius of the earth's orbit around the sun, 1 m of the earth's surface is irradiated by a flux $\dot{E}_e$ equal to

$$\dot{E}_e = f\dot{E}_s \tag{21.9}$$

where the dilution factor f can be calculated from

$$f = \frac{4\pi R_s^2}{4\pi R_0^2} = \frac{R_s^2}{R_0^2} \tag{21.10}$$

and yields the numerical value of 2.6×10^{-5}. From Equations 21.5 and 21.9 it follows that

$$\dot{E}_s = f\sigma T_s^4 \tag{21.11}$$

which can be calculated to be 1353 W/m². This radiation intensity from the sun upon the earth's upper atmosphere is called the *solar constant*, S. The corresponding photon flux is

$$\dot{N}_e = f\sigma' T_s^3 \tag{21.12}$$

The average energy of one photon is

$$\bar{E} = \frac{\dot{E}_e}{\dot{N}_e} \tag{21.13}$$

and is calculated to be $2.70\, kT_s = 1.34\text{eV}$.

These results allow us to calculate the temperature of the respective planets of our solar system by assuming that these planets prevail in thermal balance, with the incoming energy of photons from the sun balancing the energy of photons emitted by the planet. For the earth, the incoming energy of radiation is $\dot{E}_{in} = \pi R_e^2 \cdot \dot{E}_e = \pi R_e^2 \cdot f\sigma T_s^4$ with R_e being the radius of the earth and πR_e^2 the surface area of the projection of our planet on a plane perpendicular to the sun's radiation. The energy emitted by our planet is $\dot{E}_{out} = 4\pi R_e^2 \cdot \sigma T_e^4$. By setting $\dot{E}_{in} = \dot{E}_{out}$, we can calculate the average temperature of our planet from

$$\pi R_e^2 \cdot f\sigma T_s^4 = 4\pi R_e^2 \cdot \sigma T_e^4 \tag{21.14}$$

The result is $T_e = 278$ K, which corresponds reasonably well with the temperature found from experiment, $T_e^{\text{exp}} = 288\,K$.

Repeating these calculations for the other planets of the solar system gives a remarkably good correspondence between experimental and calculated temperatures for the respective planets, save one notorious exception, namely, Venus! The experimental temperature is 733 K, whereas the calculated temperature is only 327 K. This difference is caused by the pronounced *greenhouse effect* of its dense CO_2 atmosphere. Introducing the number ρ as the albedo of the planet and the number γ as the greenhouse coefficient, Equation 21.14 can be corrected for the fact that not all the sun's incident energy is absorbed, nor all the planet's energy emitted:

$$(1-\rho)\cdot \pi R_p^{\,2}\cdot f\sigma T_s^{\,4}=(1-\gamma)\cdot 4\pi R_p^{\,2}\cdot \sigma T_p^{\,4} \qquad (21.15)$$

with T_p being the average temperature of the planet concerned. With ρ = 0.7 and γ = 0.99 for Venus, the experimental temperature for this planet, $T^{\exp}$ = 733 K, agrees quite well with the calculated value of 765 K. Correcting by means of σ and γ for the earth results in a calculated temperature $T_e^{\text{calc}}=289\,K$, which is only 1 K above the experimental temperature of $T_e^{\exp}=288\,K$.

This should have summed up the major characteristics of solar radiation that are relevant for the context of this book. For further details the interested reader is referred to the monograph by De Vos [1] on which, as has been mentioned in the beginning, most of the material presented earlier has been based.

21.3   THE CREATION OF WIND ENERGY

When one side of a planet is irradiated by the sun, the other side is not and a temperature difference is created. With the planet's atmosphere as the working fluid operating between the planet's two extreme temperatures, macroscopic cycles can be performed that act as the origin of winds. Gordon and Zarmi [6] generated the first model for this wind creation, which was subsequently refined by De Vos and Flater [7]. Both models and their outcome will be briefly discussed in this section.

The sunny side of the planet is assumed to be a heat reservoir at temperature T_3, while the dark side is a reservoir at temperature T_4. The illuminated side is receiving energy from the sun, but is also emitting energy. The dark side is emitting energy only. The net heat flow rate in sustaining the heat reservoir at T_3 is

$$\dot{Q}_{\text{in}}=\pi R_p^{\,2}\cdot(1-\rho_p)\,f\sigma T_s^4-2\pi R_p^{\,2}\cdot(1-\gamma_p)\sigma T_3^4 \qquad (21.16)$$

in which R_p, ρ_p, and γ_p represent the planet's radius, albedo, and greenhouse coefficient, respectively.

Of particular interest is $\dot{Q}_{\text{absorbed}}$, the total amount of solar energy absorbed by the planet per unit time:

$$\dot{Q}_{\text{absorbed}}=\pi R_p^{\,2}\cdot(1-\rho_p)\cdot f\sigma T_s^4 \qquad (21.17)$$

which is the first term in Equation 21.16.

At the cold side, with temperature T_4, the planet emits energy with a rate

$$\dot{Q}_{out} = 2\pi R_p^{\ 2} \cdot (1 - \gamma_p) \cdot \sigma T_4^4 \tag{21.18}$$

The Carnot engine operating between these temperatures T_3 and T_4 with heat absorbed at a rate $\dot{Q}_{in}$ and rejected at a rate $\dot{Q}_{out}$ will have an efficiency

$$\eta \equiv \frac{\dot{Q}_{in} - \dot{Q}_{out}}{\dot{Q}_{in}} = 1 - \frac{T_4}{T_3} \tag{21.19}$$

Equations 21.16 and 21.18 can be rewritten in the form

$$\dot{Q}_{in} = g_1 \left(T_1^4 - T_3^4 \right) \tag{21.20}$$

and

$$\dot{Q}_{out} = g_2 \left(T_4^4 - T_2^4 \right) \tag{21.21}$$

with

$$T_1 = \left[\frac{(1 - \rho_p) \cdot f}{2(1 - \gamma_p)} \right]^{1/4} T_s \tag{21.22}$$

with T_2 the temperature of the cosmic background radiation, which is assumed to be 0K. Combining Equation 21.22 with Equation 21.15, which relates the average planet temperature T_p to the temperature of the sun, we find that

$$T_1 = 2^{1/4} \cdot T_p = 1.19 T_p \tag{21.23}$$

and

$$g_1 = 2\pi R_p^{\ 2} \sigma (1 - \gamma_p) \tag{21.24}$$

with radiation conductancies

$$g_1 = g_2 \tag{21.25}$$

We have now arrived at the so-called Stefan-Boltzmann engine. A black surface at temperature T_1 emits energy at a rate $\dot{Q}_{in}$ to a black surface at T_3 that absorbs it. This is an irreversible process described by Equation 21.20. Next, a reversible engine operates between T_3 and T_4. Heat at a rate $\dot{Q}_{out}$ is emitted by the black body at T_4, according to the irreversible process described by Equation 21.21, where the cosmic background radiation has been neglected and thus $T_2 = 0$ K. This is the setup of an endoreversible engine for which all irreversibilities are concentrated within the interaction of the engine with its environment, that is, with heat transfer, with no irreversibilities taking place within the engine. We have seen this before in Section 5.2, the only difference being that in the latter case, heat was exchanged by conduction and convection rather than radiation. The characteristics of this Stefan-Boltzmann engine are shown graphically in Figure 21.8. As described earlier in Chapter 5, this engine also has maximum power performance. Given $T_1 = 2^{1/4}.T_p = 1.19\ T_p$ and $T_2 = 0$ K, numerical calculations show that $T_3 = 1.107\ T_p$ and thus the efficiency η is given by

$$\eta = 1 - \frac{T_4}{T_3} = 0.307 \qquad (21.26)$$

This efficiency refers to the engine and relates the net power output $\dot{W} = \dot{Q}_{in} - \dot{Q}_{out}$ to the heat input rate $\dot{Q}_{in}$:

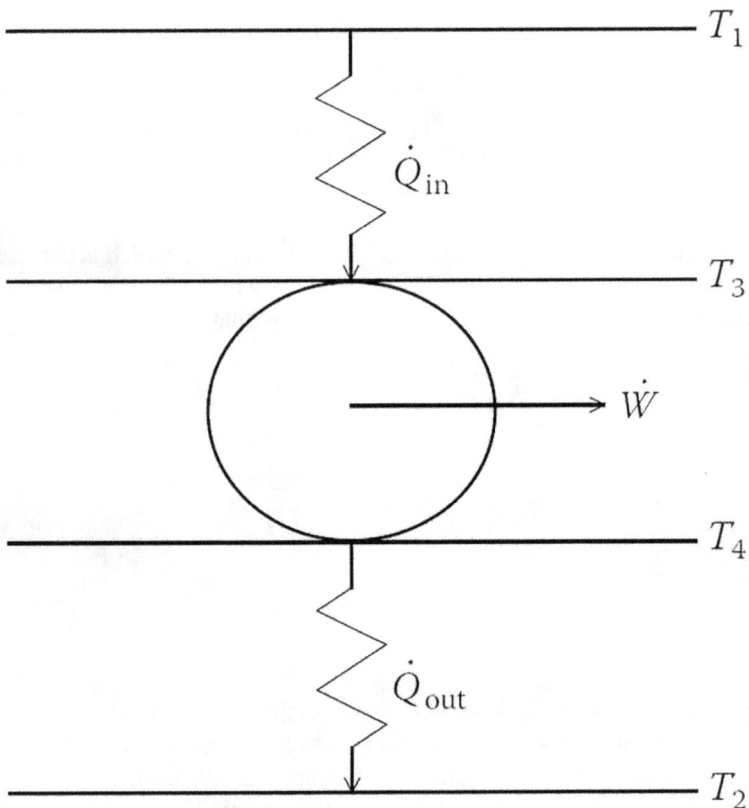

FIGURE 21.8   The Stefan-Boltzmann engine.

$$\eta = \frac{\dot{W}}{\dot{Q}_{in}} \tag{21.27}$$

In contrast to this engine efficiency, one defines the *solar energy efficiency w* as

$$w = \frac{\dot{W}}{\dot{Q}_{abs}} \tag{21.28}$$

The quantity $\dot{Q}_{abs} = \pi R_p^2 (1-\rho) \cdot f\sigma T_s^4$ is the total solar energy absorbed by the planet per unit time. Calculations show that at maximum power, where $\eta = 0.307$, the efficiency $w = 0.0767$, or 7.67%. This value is independent of f, ρ, and γ and thus is the same for all planets, even those in other solar systems since w is also independent of T_s.

Figure 21.9 shows a refinement on the model as proposed by De Vos and Flater [7]. They consider the ultimate fate of the wind power output $\dot{W}$ and assume that an

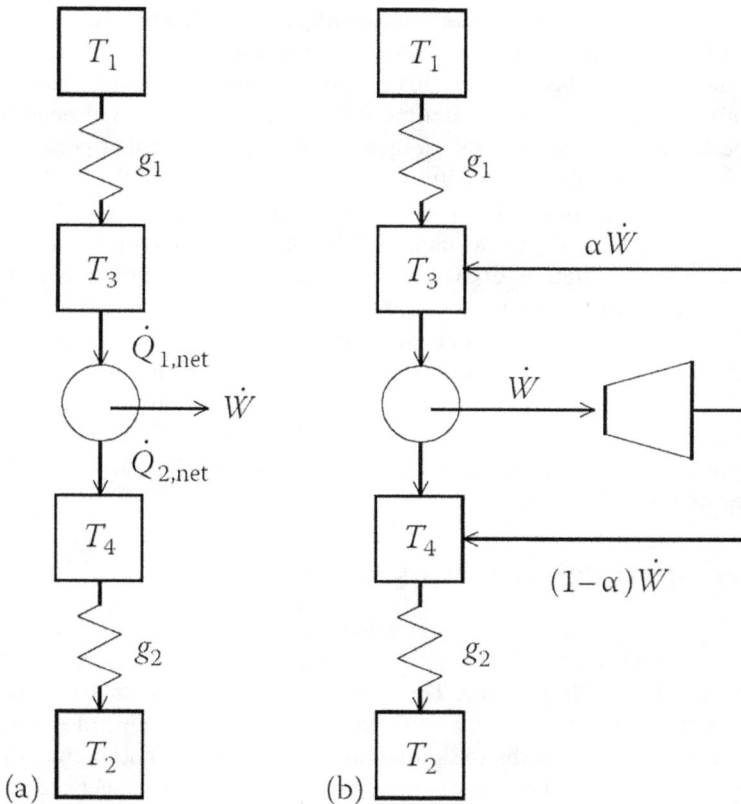

FIGURE 21.9 Two models for the conversion of solar energy into wind energy, (a) preliminary model and (b) extended model.

amount of heat equal to $\dot{W}$ is absorbed within the planet's atmosphere. Defining α as the fraction of this heat that is absorbed at the illuminated side of the planet, they find that for

$$\alpha = 0 \quad w = 6.01\%$$

$$\alpha = 0.5 \quad w = 6.90\%$$

$$\alpha = 1.0 \quad w = 8.30\%$$

It is interesting to note that in an experimental study of the limits to wind power utilization, Gustavson [8] found $w = 3\%$.

This section has discussed the creation of wind energy from solar radiation and has established that at maximum power (see Chapter 5) its efficiency is limited to some 6%–8%. In particular, we are of course equally, and perhaps even more, interested in the fraction of this energy that can be captured as electricity in a windmill park.

We should realize, however, that the building and operation of a windmill park requires a great exergetic investment, for example, into building its material components and assembly. Most likely, this exergy has been supplied by fossil energy. A windmill park with a certain lifetime (t_1) and power output will need to run a certain period (t_2) to earn back its exergetic investment. After that period the park supplies purely renewable electricity, typically at an efficiency between 10% and 20%. Its fossil load factor (see Chapter 16) will then be given by the ratio of t_2 and t_1.

A Japanese estimate (Yoda, as cited in Chapter 13) for a 1000 kW photovoltaic power station with a lifetime of 30 years, is that it takes 10 years to earn back the exergetic expenditures, which are assumed to be of fossil origin. The fossil load factor would then be 1/3: 30 years of exergy output for 10 years of exergy input. The multiplication factor is 3 and compares unfavorably with that for nuclear power generation, 55, and fossil fuel power generation, 21. However, ultimately, photovoltaic power generation has a zero fossil load factor, with no other load on the environment, which cannot be said of nuclear power generation and even less so of power generation based on fossil fuel.

21.4 PHOTOTHERMAL CONVERSION

In photothermal conversion, solar radiation energy is absorbed in a conversion device that drives a Carnot engine. The integrated device has been modeled by Müser [9] and has been named the Müser engine. The model is depicted in Figure 21.10. Radiation takes place from an effective sky temperature T_1 (see later) toward a black body at temperature T_3 that absorbs the radiation energy. The Carnot engine, operating between T_3 and T_2, rejects heat to the environment at $T_2 = T_p$, the planet's temperature. This is another example of an endoreversible engine, where the irreversible part of the process is considered to take place outside the engine, namely, in the radiation exchange step.

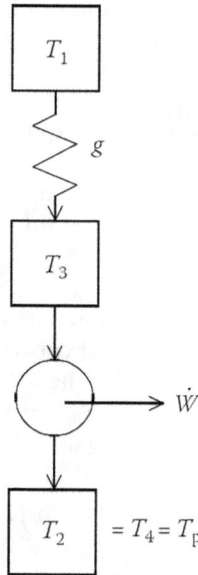

FIGURE 21.10 The Müser engine.

The solar converter is positioned perpendicular to the sun's rays and absorbs heat per m² at a rate.

$$\dot{Q} = f\sigma T_s^4 \tag{21.29}$$

in which f is the dilution factor of Equation 21.10. In addition, the converter absorbs black-body radiation from the environment at temperature T_p:

$$\dot{Q}' = (1-f)\sigma T_p^4 \tag{21.30}$$

Finally, the converter emits black-body radiation at temperature T_3:

$$\dot{Q}'' = \sigma T_3^4 \tag{21.31}$$

assuming that only the upper surface of the converter is an emitter and that the lower surface has a zero emissivity, that is, by applying a certain coating. So the net incident heat flux is

$$\dot{Q}^{net} = \sigma\left[fT_s^4 + (1-f)T_p^4 - T_3^4 \right] \tag{21.32}$$

Defining the effective sky temperature T_1 by

$$T_1^4 = fT_s^4 + (1-f)T_p^4 \tag{21.33}$$

we find $T_1 \cong 5^{1/4}T_p = 1.50\ T_p$. So for $T_p = 288$ K, our planet's effective sky temperature, $T_1 = 432$ K. To establish the maximum power of the Muser engine, we determine the optimal temperature T_3 of the solar converter. For this temperature it is found that the thermal efficiency of the engine is $\eta = 0.20$ and that the solar energy efficiency, based on the heat absorbed from the sun (Equation 21.29), is

$$w = \frac{\dot{W}}{f \sigma T_s^4} = 0.13 \tag{21.34}$$

The *concentration factor C* has been defined as the factor by which the incident solar radiation can be augmented. The ideal concentrator would compensate for the effect of the dilution factor $f = 2.16 \times 10^{-5}$, the value for our planet. The maximum concentration factor therefore is $C_{max} = 1/f = 46{,}300$. For an arbitrary value of C, Equation 21.32 has to be adapted according to

$$\dot{Q}^{net} = \sigma \left[CfT_s^4 + (1 - Cf)T_p^4 - T_3^4 \right] \tag{21.35}$$

Applying the same procedure for determining the maximum power characteristics as before and assuming moderately concentrated sunlight for which $C = 389$ [1], the solar energy efficiency w increases from 13% to somewhat above 60%. For the limit case of $C = C_{max}$—all solar radiation is captured by our planet—w is close to 85%.

Another way to improve on the solar energy efficiency w is to apply selective coating on the converter's absorbing and emitting surface. These coatings oppress the radiation emitted by the converter without affecting the absorption of the solar radiation. This effect can be achieved because the Planck spectra at T_s, the temperature of the sun, and T_3, the temperature of the converter, are more or less positioned next to each other as can be seen in Figure 21.7. This is only true if $T_3 = T_s$ and, thus, if the concentration factor C is not too high. The effect of these so-called bandgap materials can be impressive. Without concentrators, that is, for $C = 1$, $w = 13\%$ without applying these materials. Introducing these materials can enhance w up to 54%. With a moderate concentrator, $C = 389$; this value can even be increased to $w = 70\%$. This should conclude the main features of photothermal conversion.

21.5 PHOTOVOLTAIC ENERGY CONVERSION

Whereas in photothermal energy conversion, radiation creates a temperature difference that is the origin for driving a heat engine, in photovoltaic energy conversion, radiation creates a potential difference that can generate an electric current. Such devices make use of very special materials, called semiconductors. Under certain circumstances a semiconductor can prevail in two states of matter. Within such states, electrons can occupy a range of energy levels; such a range is often named a band. Energy levels between such bands cannot be occupied. The minimum distance between such bands is called a gap, E_g, and this is a material constant. Incident radiation may initiate a transition of an electron from an energy level within one band (e.g., E_1) to an energy level within the other band (e.g., E_2), and thus the corresponding photon frequency g is given by

$$E_2 - E_1 = hv \qquad (21.36)$$

Obviously, E_2-E_1 should be larger than E_g. The energy of electrons within the two bands called the conduction band c and the valence band v refer to different Fermi energies E_F. The distance in Fermi energies E_{Fc} and E_{Fv} is

$$E_{Fc} - E_{Fv} = qV \qquad (21.37)$$

in which

q is the unit of electrical charge
V is the generated potential difference

By tuning the potential difference, the device can emit a certain radiation spectrum; by varying the incident radiation, the device can generate various electrical currents. In this instance, it is important to concentrate on the electron flux $\dot{N}$, just as we did on the heat flux $\dot{Q}$ in photothermal energy conversion. We distinguish between electron fluxes associated with photons absorbed from solar radiation and with photons emitted from the absorbing material. The associated electric current is given by

$$I = q\dot{N} \qquad (21.38)$$

The associated electric voltage V is given by Equation 21.37, and thus the electric power is given by

$$\dot{W} = VI \qquad (21.39)$$

The solar energy efficiency w, as defined before by

$$w = \frac{\dot{W}_{out}}{\dot{Q}_{solar,absorbed}} \qquad (21.40)$$

can now be calculated and its maximum value established. It is found that for maximum power $w = 31\%$ for nonconcentrated sunlight (i.e., for $C = 1$) at an optimal bandgap $E_g^{opt} = 1.30\,\text{eV}$. For moderately concentrated sunlight, with $C = 389$, w rises to 35% for $E_g^{opt} = 1.10\,\text{eV}$. In practice, however, it is difficult to attain values for w larger than 13%.

It has been found that hybrid energy conversion may lead to better values for w than either from photovoltaic or from photothermal conversion. In *hybrid conversion*, solar cells mounted in roof panels, for example, provide the electricity whereas heat is extracted from the panels to possibly drive heat engines. Although theoretical calculations show significant improvements for a hybrid photothermal solar energy convertor, with values of w of nearly 70% compared to those of 50% for photothermal and 30% for photovoltaic conversion for nonconcentrated sunlight, this application has not yet reached the practical stage other than using the liberated heat directly instead of in a heat engine.

Finally, we briefly discuss what is called *multicolor* and *omnicolor* conversion. Instead of using a single solid-state material with one bandgap E_g as characteristic material property, one can select two or more materials, each with its own bandgap. In such systems, according to De Vos, each material is responsible for the conversion of a particular energy band of the incoming solar radiation. Because each band of the light spectrum can be characterized by a particular "color," one speaks of multicolor conversion. Multicolor conversion can refer to photothermal and photovoltaic energy conversion and their hybrid applications. De Vos deals with the nonhybrid cases in his book and shows that for an infinite number of colors, when we speak of omnicolor conversion, the distinction between the two conversion methods disappears and reaches, for nonconcentrated sunlight (i.e., $C = 1$), the thermodynamic limit for solar energy conversion efficiency w of 68.2%. For fully concentrated sunlight, $C = 1/f$, $w = 86.8\%$. This, then, can be considered to be the *exergy* for sunlight for this way of converting sunlight by absorption. When sunlight is used as a heat source at T_s in a heat engine, the exergy reaches the absolute limit of 0.95, which is the Carnot factor between $T_s = 5762$ K and $T_p = 288$ K. For a discussion on the exergy of radiation, we refer to Petela [10] and Lems [11].

21.6 PHOTOSYNTHESIS

The unforgettable Lehninger [12], who in his teaching of biochemistry made extensive use of thermodynamics, called light energy the ultimate source of all biological energy. It is absorbed by photosynthetic cells in the form of chemical energy that is then used to convert CO_2 into glucose. This reaction is usually given by the equation

$$6CO_2 + 6H_2O \rightarrow C_6H_{12}O_6 + 6O_2 \tag{21.41}$$

Writing this equation in the simplified form

$$CO_2 + H_2O \rightarrow CH_2O + O_2 \tag{21.42}$$

and using exergy values for reactants and products (see Chapter 7), we conclude that this reaction requires roughly 480 kJ/mol of CH_2O, an elementary glucose unit, as input of work. This work is furnished by light. Water acts as the donor of hydrogen and electrons, whereas CO_2 acts as their acceptor, which is exemplified by the redox reactions

$$2H_2O \rightarrow 4H^+ + 4e^- + O_2 \tag{21.43}$$

$$CO_2 + 4H^+ + 4e^- \rightarrow CH_2O + H_2O \qquad (21.44)$$

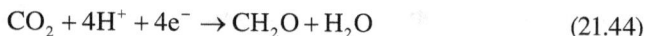

The two oxygen atoms in the liberated oxygen molecule O_2 do not stem from CO_2 but from water, and therefore Equation 21.42 is too simple and should be written as

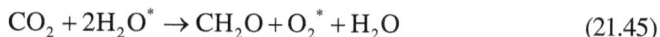

$$CO_2 + 2H_2O^* \rightarrow CH_2O + O_2^* + H_2O \qquad (21.45)$$

and Equation 21.41 should read

$$6CO_2 + 12H_2O^* \rightarrow C_6H_{12}O_6 + 6O_2^* + 6H_2O \qquad (21.46)$$

The asterisk emphasizes that the liberated oxygen originates from reactant water and that CO_2's oxygen is distributed evenly over glucose and product water. Equation 21.44 is representative for how higher plants transform the energy of light into chemical energy. However, purple bacteria absorb light by the simplified reaction

$$CO_2 + 2H_2S \rightarrow CH_2O + 2S + H_2O \qquad (21.47)$$

Using exergy values for reactants and products, this reaction needs only 76 kJ/mol of CH_2O. These bacteria may also absorb light by means of the reaction

$$CO_2 + 2H_2 \rightarrow CH_2O + H_2O \qquad (21.48)$$

and this reaction needs only about 20 kJ of exergy input per mole of the elementary carbohydrate unit. These simple calculations suggest that purple bacteria can deal with scarce light sources, although they have prepared themselves for this by the preliminary synthesis of H_2S and H_2, respectively.

Another interesting aspect of photosynthesis by higher plants is the distinction between so-called light and dark reactions. The *light* reactions absorb light and convert this into chemical energy; the *dark* reactions convert CO_2 into glucose with this chemical energy, but *without* light. In the light reactions, chlorophyl plays an important role in the capture of the energy of light by rearrangement of electrons in its molecules. In the transport of electrons, NADP, or nicotin amide adenine dinucleotide phosphate, plays a crucial role beyond the first light reaction. In the transport of the absorbed energy as chemical energy, the earlier mentioned ATP or adenosine triphosphate, is equally essential (see Chapter 4). To bring these two molecules, NADP and ATP, in proper position for the synthesis of glucose from CO_2, a second light reaction is required that overall generates oxygen while absorbing light. It is interesting to observe that oxygen liberation and CO_2 conversion take place in completely different reactions, not in the same reaction although this is suggested by Equation 21.41 or 21.46.

A final word about the thermodynamic efficiency of photosynthesis. De Vos reports in his book a solar energy conversion efficiency in the order of 2%–2.5% for land plants such as crop, rice, and wheat and values as high as 18% for sea plants such as algae. But he cautions that these results were obtained by agricultural engineers and biologists who tend to take into account only the visible part of solar energy as the input energy. De Vos [1] then shows that a correction factor of 0.368 is required, thus reducing the values to 0.8% and 6.6% for land and sea plants, respectively. In passing, he mentions that 0.1% (!) of the solar flux incident on the earth is converted to chemical energy by photosynthesis, which is an impressive figure indeed.

These thermodynamic efficiencies may be considered to be on the low side, particularly for land plants. But then it must be noted that this photosynthesis is an activity of a *living* system for which photosynthesis, or rather the capture of solar energy and its transformation into chemical energy, is *only* one aspect of life. Supply may exceed demand. Therefore, it may well be that if the *only* purpose of photosynthesis is the capture of light as an energy source, the efficiencies would have been higher. Support for this conjecture can be found in a calculation by Lehninger [12], who shows that under certain laboratory conditions, efficiencies can rise to 36% for green algae or even isolated chloroplasts. Perhaps even higher efficiencies are possible for nonliving systems for the production of exergy carriers. As an aside, we anticipate that light absorption may be enhanced if the "living" system of biomass is skipped and production is focused on a chemical exergy carrier, preferentially liquid hydrocarbons, or even if this carrier is skipped and the focus is on electricity.

21.7 CONCLUDING REMARKS

The sun is an abundant, albeit dilute, source of energy. Some call it the world's most distant and safest nuclear reactor [13]. By the time solar radiation reaches the earth's outer atmosphere, its intensity has been reduced by a factor $f = 2.16 \times 10^{-5}$ due to the earth's distance to the sun. The so-called solar constant $S = 1353 \text{W/m}^2$ is the corresponding flux. A more practical value is $Z = 947 \text{W/m}^2$ in which $Z = (1 - \rho) S$. ρ is the earth's albedo, accounting for the fact that not all solar radiation that reaches the outer atmosphere is absorbed. Solar energy can be naturally converted into wind energy or biomass without the intervention of man. This is in contrast to the conversion into heat for the generation of work as in photothermal conversion, or into electricity as in photovoltaic energy conversion. The efficiency of these conversions is best expressed in terms of the so-called solar energy efficiency w, which quantifies the maximum amount of work that under practical conditions, including irreversibilities, can be extracted from the solar energy absorbed by the planet, which is related to Z. In this way, one can calculate that for the creation of *wind energy* the best value to be attained is $w = 8\%$. Of course wind energy needs to be captured and its conversion into electricity will reduce this efficiency even further.

In *photothermal* conversion, the best value possible seems to be $w = 13\%$. However, by making use of ingenious devices for refraction and reflection, the intensity of captured solar radiation can be enhanced by a concentration factor C close to 400. w can then reach values close to $w = 60\%$. Another way to enhance w is to apply so-called bandgap materials that suppress the emission of absorbed solar energy by capturing this energy with the help of electronic rearrangements in the material. The material is characterized by the so-called bandgap in eV. For nonconcentrated solar radiation, $C = 1$, w can be calculated to reach a value of 0.54, or 54%, for an optimal bandgap of $E_g = 0.90$ eV. For concentrated sunlight at $C = 400$, $w = 70\%$ for a somewhat lower value of $E_g = 0.70$ eV.

Photovoltaic energy conversion also makes use of bandgap materials for the direct conversion of radiation energy into work, electricity. For $C = 1$, the best value for $w = 0.31$, or 31%, at an optimal bandgap $E_g = 1.30$ eV with $w = 25\%$ reached in the laboratory and 15% in industrial production. For $C = 400$, w increases to 35%. For *multicolor* conversion, where many different bandgap materials are applied simultaneously, w can be calculated to increase significantly, namely, to values in the order of 70%–80%. In practice, however, it is difficult to realize values higher than 35%.

In *photosynthesis*,[1] absorbed solar radiation is captured as chemical energy. In this spontaneous natural conversion of absorbed solar radiation, the solar energy efficiency ranges from 0.8% for land plants to 6.6% for sea plants. In laboratory experiments efficiencies above 30% have been observed. Table 21.1 summarizes the results of solar energy conversion, as discussed in this chapter and where the value for wind energy creation is given as reported in [14].

Solar energy is a truly sustainable and ample source of energy, and its conversion still poses many challenges. Meeting these challenges will overcome their obstacles and facilitate the transformation of our society into a sustainable society. It should be clear that there is a great deal of scope for research in the field of solar energy conversion. Practical photovoltaic energy conversions, for example, still fall short of the theoretical limit. To the best of our knowledge, hybrid and multicolor conversions

TABLE 21.1
Summary of Solar Energy Conversion Efficiencies.

Conversion		*w* (%)
Wind energy creation	Theoretical	6–8.5
	Experimental	3
Photothermal	$C = 1$	13
	$C =$ moderate	~60
	$C =$ maximum	85
Photothermal with bandgap material	$C = 1$	54
	$C =$ moderate	70
Photovoltaic	$C = 1$	31
	$C =$ moderate	35
	In practice	15
Hybrid	Theoretical	70
Multicolor	$C = 1$	68
	$C =$ maximum	87
Photosynthesis	Land plants	0.8
	Sea plants	6.6
	Experiment	36

have not gone beyond the experimental stage, even though theoretical conversions are very high. Clearly, the potential benefits of these technologies are great. Advances in the field of biotechnology may enable enhancement of photosynthesis efficiency for the production of electricity or specialty chemicals [15].

NOTE

1 It is interesting to note that researchers at Imperial College in London, United Kingdom, are devising artificial leaves inspired by natural leaves in an attempt to photosynthetically produce chemicals such as hydrogen and methanol [14]. Along a similar vein, it is interesting that Leonardo Da Vinci was inspired by birds when building machines of flight, and was not attempting to build a bird!

REFERENCES

1. De Vos, A. *Thermodynamics of Solar Energy Conversion*, Wiley VCH: Weinheim, Germany, 2008.
2. Chu, S. *Lighting the way. Report of the InterAcademy Council*, 2007. www.interacademycouncil.net/?id=12161
3. Smalley, R.E. Future global energy prosperity: The terawatt challenge. *MRS Bulletin* 2005, *30*, 412–417.
4. Lewis, N.S. Frontiers of research in photoelectrochemical solar energy conversion. *The Journal of Electroanalytical Chemistry* 2001, 508, 1–10.
5. Thekaekara, M. Solar energy outside the Earth's atmosphere. *Solar Energy* 1993, *14*, 109–127.
6. Gordon, J.; Zarmi, Y. Wind energy as a solar-driven heat engine: A thermodynamic approach. *The American Journal of Physics* 1989, *57*, 995–998.
7. De Vos, A.; Flater, G. The maximum efficiency of the conversion of solar energy into wind energy. *The American Journal of Physics* 1991, *59*, 751–754.
8. Gustavson, M. Limits to windpower utilization. *Science* 1979, *204*, 13–17.
9. Muser, H. Behandlung von Elektronenprozessen in Halbleiter Randschichten. *Zeitschrift für Physik* 1957, *148*, 380–390.
10. Petela, R. Exergy of heat radiation. *Journal of Heat and Mass Transfer* 1964, 86, 187–192.
11. Lems, S. Thermodynamic explorations into sustainable energy conversion – Learning from living systems. PhD thesis, Delft University of Technology, Delft, 2009.
12. Lehninger, A.L. *Bio-Energetics*, 2nd edn., W.I. Benjamin: Menlo Park, CA, 1973.
13. Okkerse, C.; van Bekkum, H. Towards a plant based-economy? In *Starch 96—The Book, Carbohydrate Research Foundation*, Zestec: The Hague, 1997.
14. Sahin, A.D.; Dincer, I.; Rosen, M.A. Thermodynamic analysis of wind energy. *International Journal of Energy Research* 2006, *30*, 553–566.
15. Collins, N.; "Artificial leaf" that may help generate clear power on the anvil. *The Daily Telegraph*, August 12, 2009.

22 Hydrogen
Fuel of the Future?

This last chapter will be devoted to what some people call "the fuel of the future," hydrogen (H_2), and to the economy that is associated with it, "the hydrogen economy." Hydrogen is not a resource, but a carrier of energy and its useful component, exergy, and we will discuss its production, transportation, storage, and conversion to its final application, be it heat, work, electricity, fuels, or chemicals. But we will not do this without paying attention to the corresponding thermodynamic efficiency to put things in a proper and realistic perspective. Throughout this chapter, we will try to substantiate the options of a hydrogen economy within the boundary conditions of the laws of thermodynamics, the only way out from what is obviously a labyrinth.

22.1 INTRODUCTION

There used to be much talk in the popular media about hydrogen as the ultimate energy carrier and "energy source," and indeed, a lot of financial resources were dedicated to hydrogen-related research. Proponents of hydrogen (used to) tout it as the silver bullet to provide "green" energy to consumers. Indeed, the talk was about the hydrogen economy. In order to understand this better, it is useful to analyze what exactly was and is meant by the "hydrogen economy."

22.2 THE HYDROGEN ECONOMY

The term *hydrogen economy* was coined by John Bockris during a talk he gave in 1970 at the General Motors (GM) Technical Center [1]. It refers to *a proposed system of meeting energy needs by using hydrogen as a fuel source that could be generated from alternative fuels or other energy sources that do not give off greenhouse gases.* It is this concept that holds promise and potential. Indeed, if hydrogen can be made from sustainable sources and used to store energy or as a transport fuel, this would be a revolutionary development in the way we produce and consume energy.

Hydrogen (H_2) gives off energy when it is combined with oxygen, but the hydrogen has to be produced first, which requires more energy than is released when it is used as a fuel due to inherent process inefficiencies (Figure 22.1). Hydrogen is therefore not really the energy source but simply the carrier, as vast resources of hydrogen are not available terrestrially other than in its oxide form (H_2O). It is interesting to contrast this against the "fossil fuel economy," where the fossil fuels are both the energy source and frequently the carriers as well. The definition after Bockris seems to be forgotten, as the source of hydrogen is frequently omitted when discussing the hydrogen economy. In this spirit, electrolysis of water is frequently mentioned. However, the electricity to achieve this transformation is obtained from nonrenewable-based

DOI: 10.1201/9781003304388-26

Water + Ex_{in} [Ex_{lost1}] $H_2 + \frac{1}{2} O_2$ [Ex_{lost2}]   Water + Ex_{out}

$$Ex_{in} - Ex_{out} = Ex_{lost1} + Ex_{lost2}$$

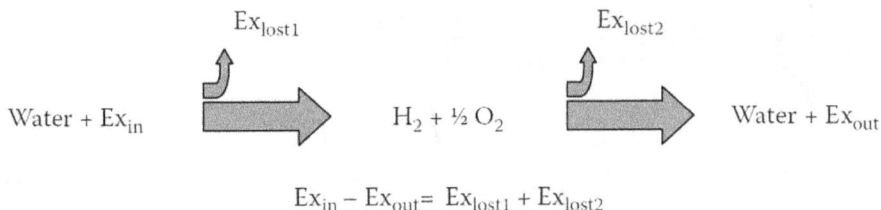

FIGURE 22.1 Water-to-water via hydrogen. Exergy output is less than exergy input.

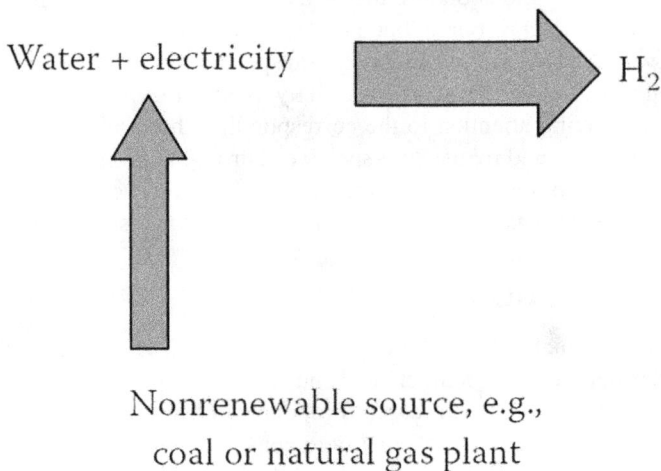

Water + electricity H_2

Nonrenewable source, e.g.,
coal or natural gas plant

FIGURE 22.2 Electrolysis of water requires electricity.

sources (Figure 22.2), and zero emission technology, which incidentally does not exist as of today, would have to be used with nonrenewables. In theory, the same energy input is required to liberate hydrogen as will be obtained when hydrogen is allowed to react with oxygen, but due to inherent thermodynamic inefficiencies, more energy will be necessary (as has been shown elsewhere in this book).

Whereas proponents of the hydrogen economy argue that hydrogen is a clean energy "source" for energy and end users with less CO_2 production and potential pollutants, such as particulate matter, than from fossil sources, the method of production of hydrogen is key. It all depends on where the CO_2 and the pollutants ultimately end up. The energy requirements and efficiency can be calculated for a given process using the tools developed in this book. Critics of the hydrogen economy argue that for many planned applications of hydrogen, direct distribution and use of energy in the form of electricity might accomplish many of the same net goals without requiring the investment in new infrastructure [2]. Hydrogen has been called the least efficient and most expensive possible replacement for gasoline in terms of reducing greenhouse gases [2,3] (see, e.g., Figures 22.3 and 22.4 for a simplified analysis). A comprehensive study of hydrogen in transportation applications has found that "there are major hurdles on the path to achieving the vision of the hydrogen economy; the path will not be simple or straightforward" [4].

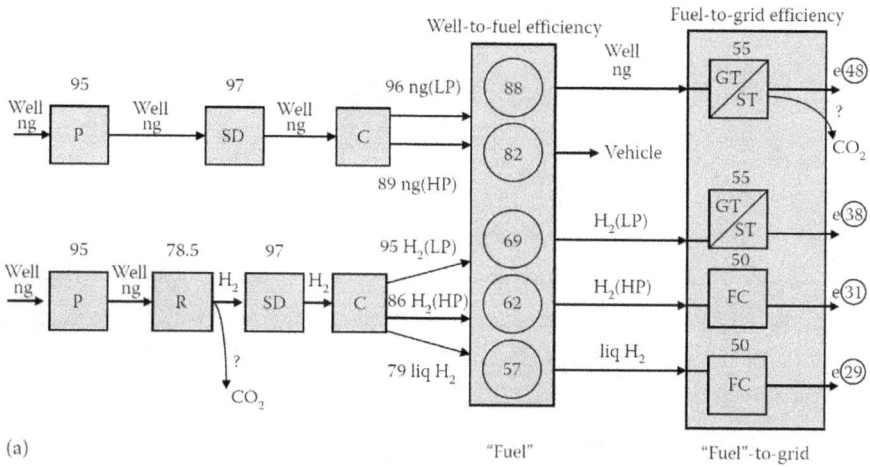

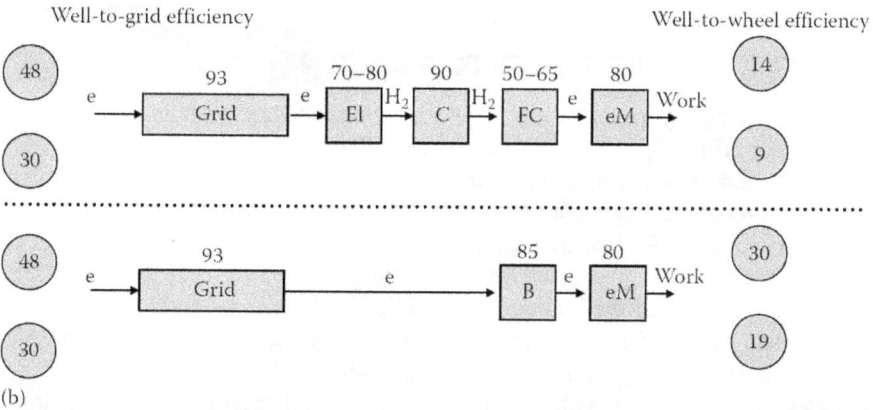

FIGURE 22.3 Well-to-wheel thermodynamic efficiencies for electric transportation with or without H$_2$ as intermediate. (a) Well-to-fuel, fuel-to-grid, and well-to-grid efficiencies for electricity production from natural gas. All routes allow effective CO$_2$ removal but routes including H$_2$ conversion seem most effective in this respect. ng: natural gas; P: production; SD: storage/distribution; C: compression; GT/ST: combined cycle; FC: fuel cell; R: reforming; (b) well-to-wheel efficiencies for the best and worst well-to-grid efficiencies of a. The H$_2$/Fuel option reduces the overall well-to-wheel efficiencies to half their values. El: electrolysis; B: battery; eM: electric motor. (With permission from Kreith, F. and West, R., *J. Energ. Resource Technol.*, 126, 249, 2004.)

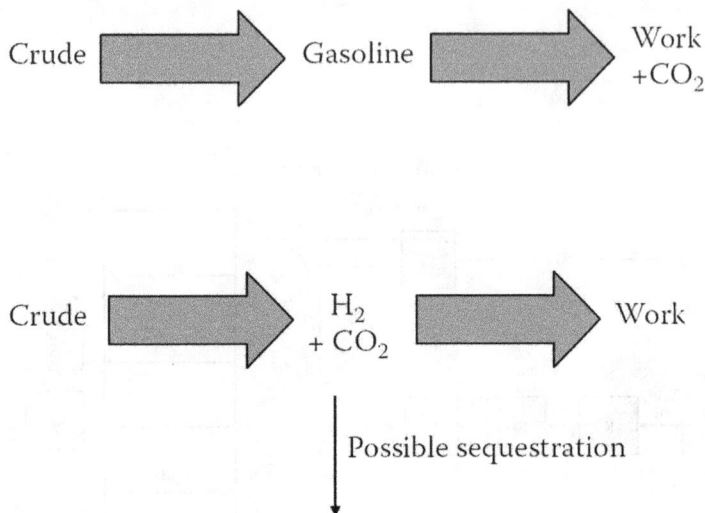

FIGURE 22.4   Crude-to-work via gasoline and crude-to-work via hydrogen. The former is more efficient than the latter, however, the latter will allow for centralized CO_2 capture.

22.3   CURRENT HYDROGEN ECONOMY

The production of hydrogen is a large and growing industry, with an average growth rate of approximately 10%. Approximately 50 million tons of hydrogen [5] or 170 million tons of oil equivalent were produced in 2004. Within the United States alone, 2004 production was approximately 11 million metric tons, or equivalent to 48 GW; this is about 10% of the average electric production in 2003 which was 442 GW.

At present, hydrogen is used in two distinct industries. The first use, which accounts for about half the hydrogen production, is for the production of ammonia, which is then used indirectly as fertilizer. As the world population and agriculture to support this are growing, ammonia demand is up worldwide. The other half of the current production is for hydrogen intensive refinery processes such as hydroprocessing. Hydroprocessing operations include hydrocracking, where heavy petroleum fractions are upgraded to lighter fractions for use as suitable fuels. The scale economies inherent in large-scale oil refining and fertilizer manufacture make possible on-site production and "captive" use. Smaller quantities of "merchant" hydrogen are manufactured and delivered to end users as well.

22.4   CONVENTIONAL HYDROGEN PRODUCTION FROM CONVENTIONAL SOURCES—GRAY, BROWN, AND BLUE HYDROGEN

Currently, global hydrogen production is 48% from natural gas, 30% from oil, and 18% from coal; water electrolysis accounts for only 4%. The distribution of production reflects the effects of thermodynamic constraints on economic choices: of the four processes for obtaining hydrogen, partial combustion of natural gas in a natural gas combined cycle power plant offers the most efficient chemical pathway and the greatest off-take of usable heat energy. If emissions are unabated, these types of hydrogen are termed as gray (natural gas based) or brown (coal) hydrogen. If the hydrogen is captured and sequestered or used otherwise, these types of hydrogen are called blue. If we keep in mind the definition of the hydrogen economy (Section 22.2), then it would seem that only the electrolysis of water which accounts for 4% of the global hydrogen production (green hydrogen) could be considered part of the "hydrogen economy." In practice, this number would most probably be far lower, as the majority of the electricity used for the electrolysis is most likely from nonrenewable sources.

22.5   HYDROGEN FROM RENEWABLES

Hydrogen production from renewable means has been reported and includes the production of H_2 from geothermal energy, fermentative hydrogen production under certain conditions, the production by certain strains of algae, concentrated solar thermal production, and photo-electrochemical water splitting. These are certainly promising leads, but it is hard to predict how these research leads will progress into commercial applications—only time will tell.

Hydrogen can be potentially produced from renewable sources, thus enabling the intermittent and excess power generated *to be stored* for applications in transport, homes, and businesses, thereby making off-grid wind and solar sources economic. However, it must be noted, that commercial applications of these concepts are fairly limited at best or nonexistent at worst, and many decades of research and development may be necessary to solve the technical hurdles for application in a large-scale fashion.

22.6   HYDROGEN AS AN ENERGY CARRIER

In the context of a hydrogen economy, hydrogen is an energy carrier, more specifically an exergy carrier, not a primary energy source. Nevertheless, controversy over the usefulness of a hydrogen economy has been fueled by issues of energy sourcing, including fossil fuel use, CO_2 production, and sustainable energy generation. These are all separate issues, although the hydrogen economy potentially affects them all.

22.7   HYDROGEN AS A TRANSPORTATION FUEL

Some hydrogen proponents promote hydrogen as a potential transportation fuel for cars, boats, airplanes, and to meet the energy needs of buildings and portable electronics. They believe a hydrogen economy could greatly reduce the emission of CO_2 and therefore play a major role in tackling global warming. Countries without oil, such as Iceland, but with renewable energy resources such as geothermal, could use a renewable energy to produce hydrogen instead of fuels derived from petroleum, which are believed to becoming scarcer.

Various options are considered such as direct hydrogen production at retail stations by reforming a fossil derived fuel, electrolysis using solar or wind energy, and so forth. It must be noted that in the former case, CO_2 is potentially released whereas in the latter, this does not occur. When performing these analyses for transportation, the results are frequently compared to equivalents of gasoline and cost per mile. For example, a typical passenger car today has a fuel economy of about 35 miles/gal. Gasoline, excluding taxes, costs about $1/gal. Incidentally, this number is fairly constant throughout the world, and the discrepancy in worldwide gasoline prices is due to taxes. This equates to a price of $1/35 miles = 2.9¢ per mile. Hydrogen-powered fuel cell vehicles are projected to have much higher fuel economies, at the order of 100 miles/gal gasoline equivalent [5]. Here, one gallon gasoline equivalent corresponds to 130.8 MJ (HHV). The delivery cost of hydrogen produced using solar energy in southern California is about $22–$30/GJ. The cost is therefore from $22 × 0.1308 to $30 × 0.1308 = $2.87 to $3.92/gal. The correct metric in this case, however, is the cost per mile, which is $2.87/100–$3.92/100 = 2.9–3.9¢/mile. For hydrogen produced from natural gas it is about twice as low, or 1.4¢–2.0¢/mile. The orders of magnitude of the cost per mile driven for a standard gasoline-powered car and a hydrogen-powered car are about the same. This very simple calculation, subject to the assumptions of fuel economy and cost and not taking into account the energy expended in building the cars and solar panels, etc., indicates that there is no direct showstopper for the use of hydrogen. A noteworthy point, however, is that for these cases, the exergetic and economic cost of dealing with the CO_2 has not been quantified. The production routes are shown pictorially in Figure 22.3.

22.8 EFFICIENCY OF OBTAINING TRANSPORTATION FUELS

It is interesting to examine the exergy inputs to produce hydrogen and conventional transportation fuels. As an example, let us consider the exergy efficiency of pure hydrogen production using state-of-the-art technology [6] by steam reforming of natural gas. The exergetic efficiency of this process is 70% and yields hydrogen at 20 bar and 298 K. Now let us liquefy this using a simple liquefaction process, which yields hydrogen at 21 K and 0.15 MPa [7]. The efficiency is about 57%. If one practices this liquefaction process right after the steam reforming process, one would end up with an efficiency of about 40%. If, on the other hand, compression to about 20 MPa is practiced, the efficiency of the compression step is about 80% [8]. The net efficiency starting with methane would then be 56%. Gasoline, by comparison, requires less energy input per gallon at the refinery, and comparatively little energy is required to transport it and store it owing to its high energy density per gallon at ambient temperatures. Well-to-tank, the supply chain for gasoline is roughly 80% efficient [8,9]. The most efficient distribution, however, is electrical, which is typically 95% efficient. It must be noted that this is solely for the distribution of electricity, and does not reflect the efficiency from source to user. Electric vehicles (EV) are typically two times as efficient as hydrogen powered vehicles (Figure 22.5).

Similarly, a study of the well-to-wheels efficiency of hydrogen vehicles compared to other vehicles in the Norwegian energy system indicates that hydrogen fuel-cell

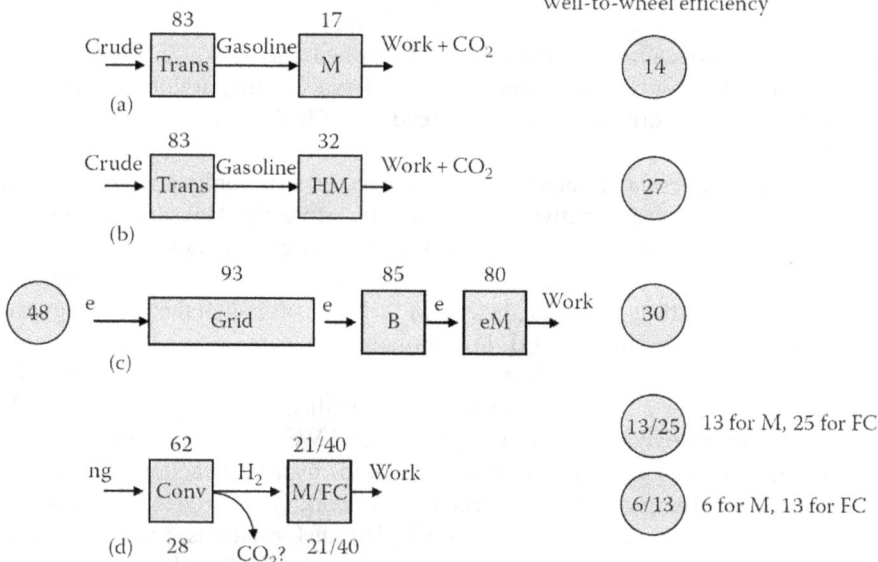

FIGURE 22.5 Well-to-wheel thermodynamic efficiencies for (a) crude to regular vehicle, (b) crude to hybrid vehicle, (c) natural gas to all-electric vehicle (this efficiency increases to 74 respectively 80% for electricity from hydropower or renewable resources), and (d) natural gas conversion to H_2 and natural gas conversion to electricity followed by electrolysis to H_2. Conv: Conversion; M: motor; HM: hybrid; B: battery; eM: electric motor; FC: fuel cell. Routes a and b do not allow for the capture of CO_2. (With permission from Marisvensson, A. et al., *Energy*, 32, 437, 2006.)

vehicles tend to be about 5% less than EVs when electrolysis is used, and even less efficient with hydrogen Internal Combustion Engines (ICE). Even in the case where hydrogen fuel cells get their hydrogen from natural gas reforming rather than electrolysis, and EVs get their power from a natural gas power plant, the EVs still come out ahead. See Figure 22.5 which summarizes the main findings from [10].

22.9 CHALLENGES OF THE HYDROGEN ECONOMY

It seems that three challenges have to be met in order to have widespread use of hydrogen as a sustainable transportation fuel. The first challenge is to produce hydrogen in a sustainable fashion. This can be potentially achieved by using renewable electricity in the electrolysis of water. Examples include using solar, wind, geothermal energy, or nuclear energy, as some have suggested. The challenges regarding spent-fuel processing in the nuclear option still remain, though. Other examples are gasification of biomass and municipal waste as discussed in Chapter 16.

The second challenge is storage of hydrogen, and the third is transportation of hydrogen, which are inseparable. The current infrastructure is designed for liquid fuels and gaseous fuels at moderate pressures; for example, natural gas in Western Europe is transported at pressures at the order of 70 bar. To liquefy hydrogen, low to moderate pressures and low temperatures are necessary,[1] making it a potentially non-viable option or at best a very expensive one. Currently, researchers are looking into developing a hydrogen infrastructure that delivers hydrogen gas at high pressures, at the order of 250 bar or higher. Former President Bush's hydrogen initiative sought to address some of the hurdles facing widespread use of hydrogen [11], by lowering the cost of its application.

Joan Ogden, a research scientist at the Princeton Environmental Institute and a longtime analyst of alternative fuels, testified before the House Committee on Science as part of a hearing on "The Path to a Hydrogen Economy" on March 5, 2003. The hearing was the first formal effort by Congress to respond to Bush's hydrogen initiative. According to Ogden, hydrogen fuel cells, although they are long-term, potentially have a very high payoff. Furthermore, this technology deserves significant government support now. This will act as insurance, if nothing else, that they will be ready in 15 or 20 years if we want to deploy them on a very wide basis [12].

Rigorous thermodynamic analyses of Kreith and West [13] and Bossel et al. [14] raise serious concerns regarding the feasibility of the hydrogen economy. Their analyses indicate that the production of electricity or work via hydrogen as an intermediary exergy carrier is accomplished by multiple transformations, and the upstream production of hydrogen requires large amounts of exergy. Each of these process steps is accompanied by inherent process inefficiencies, which make the direct production of electricity of work more attractive [13,14]. Much of the data used to construct Figures 22.3 and 22.5 came from those sources.

22.10   HYDROGEN PRODUCTION: CENTRALIZED OR DECENTRALIZED?

In a future full hydrogen economy, primary energy sources and feedstock would be used to produce hydrogen gas as stored energy for use in various sectors of the economy. Producing hydrogen from primary energy sources other than coal, oil, and natural gas would result in lower production of the greenhouse gases characteristic of the combustion of these fossil energy resources.

One key feature of a hydrogen economy is that in mobile applications (primarily vehicular transport), energy carrier generation and use is decoupled. The primary energy source need no longer travel with the vehicle, as it currently does with hydrocarbon fuels. Instead of tailpipes creating dispersed emissions, the energy can be generated from point sources such as large-scale, centralized facilities with improved efficiency. This stresses the need for technologies such as carbon capture and sequestration, which are impossible for mobile applications. Alternatively, distributed energy generation schemes (such as small scale renewable energy sources) can be used, possibly associated with hydrogen stations.

Aside from the energy generation, hydrogen production could be centralized, distributed, or a mixture of both. While generating hydrogen at centralized primary energy plants promises higher hydrogen production efficiency, difficulties in high-volume, long-range hydrogen transportation make electrical energy distribution attractive within a hydrogen economy. In such a scenario, small regional plants or even local filling stations could generate hydrogen using energy provided through the electrical distribution grid. This will come at an efficiency penalty (Figure 22.3). While hydrogen generation efficiency is likely to be lower than for centralized hydrogen generation, losses in hydrogen transport can make such a scheme more efficient in terms of the primary energy used per kilogram of hydrogen delivered to the end user.

The proper balance between hydrogen distribution and long-distance electrical distribution is one of the primary questions that arise in the hydrogen economy.

Again, the dilemmas of production sources and transportation of hydrogen can now be overcome using on-site (home, business, or fuel station) generation of hydrogen from off-grid renewable sources.

22.11 INFRASTRUCTURE

While millions of tons of hydrogen are distributed all around the world each year, bringing hydrogen to individual consumers would require an evolution of the fuel infrastructure. The current hydrogen infrastructure consists mainly of industrial hydrogen pipeline transport and hydrogen-equipped filling stations like those found on a hydrogen highway in California. Hydrogen stations which are not situated near a hydrogen pipeline get supply via hydrogen tanks, compressed hydrogen trailers, liquid hydrogen trailers, liquid hydrogen trucks, or dedicated on-site production.

Because of hydrogen embrittlement of steel, natural gas pipes have to be coated on the inside or new pipelines installed, like the over 700 miles of hydrogen pipeline currently in the United States. Although expensive to install, once in place, pipelines are the cheapest way to move hydrogen from point A to B. This can all be potentially avoided, however, with distributed hydrogen production on-site with medium- or small-sized generators ensuring sufficient hydrogen for an entire neighborhood or personal use. In the end, a combination of options is most likely to succeed. The outstanding question remains at what pressure this would become technically and economically feasible.

22.12 HYDROGEN STORAGE

Molecular hydrogen has a very high energy density on a mass basis, but due to its low molecular weight, has a very low energy density by volume at ambient condition. In order for hydrogen to be a practical fuel and compete effectively with current liquid fuels, a sufficient driving range must be available, which can be achieved by increasing its density either by liquefaction or pressurization (Table 22.1). Increasing gas pressure certainly improves the energy density by volume, making for smaller, but not lighter container tanks due to the requirements of the pressure vessels. Achieving higher pressures necessitates greater use of external energy to power the compression. Alternatively, liquid hydrogen may be used which has a higher volumetric energy density. However, liquid hydrogen boils at 20.268 K under standard pressure. Cryogenic storage cuts weight but requires large energy for liquefaction. The liquefaction process, involving pressurizing and cooling steps, is energy intensive. The liquefied hydrogen

TABLE 22.1

Exergy Content of Selected Fuels and Segments.

	kJ/mol	kJ/g	kJ/cc
C-coal	10	34	119 (solid)
CH_4	832	52	2.60 (70 bar)
			9.28 (250 bar)
H_2	236	118	2.63 (250 bar)
			7.37 (700 bar)
			8.38 (liquid)
CH_3OH	718	22.4	20–21 (liquid)
CH_2 (p)~gasoline	658	47	38–45 (liquid)
CH (a)	551	42.4	38–45 (liquid)
CH_2 (c)	652	46.6	38–45 (liquid)
CH_3	748	50	38–45 (liquid)
LPG[a]			~33 (liquid)
CO	275		
CO_2	20		
H_2O	9		

(p) paraffinic; (a) aromatic; (c) cyclic. Standard chemical exergy values reported in this table. See Chapter 7 for more details.

[a] Assumed C_3/C_4 mixture.

Note: CO, CO_2, and H_2O are not fuels, but are used in calculations.

has lower energy density by volume than gasoline by approximately a factor of four, due to the low density of liquid hydrogen—there is actually more hydrogen in a liter of gasoline (116g) than there is in a liter of pure liquid hydrogen (71g). Liquid hydrogen storage tanks must also be well insulated to minimize boil off.

The mass of the tanks needed for compressed hydrogen reduces the fuel economy of the vehicle. Because it is a small, energetic molecule, hydrogen tends to diffuse through any liner material intended to contain it, leading to the embrittlement, or weakening, of its container. Special provisions are therefore required.

Besides storing molecular hydrogen as such, hydrogen can be stored chemically as a hydride or in some other hydrogen-containing compound. Hydrogen gas is reacted with some other materials to produce the hydrogen storage material, which can be transported relatively easily. At the point of use, the hydrogen storage material can be made to decompose, yielding hydrogen gas. Just as mass and volume density problems are associated with molecular hydrogen storage, the high pressure and temperature conditions pose barriers to practical storage systems utilizing hydride formation and hydrogen release. For many potential systems, rates of hydriding and dehydriding and heat management are also issues that need to be addressed.

Another approach is to absorb molecular hydrogen physically into a solid storage material. Unlike in the hydrides mentioned earlier, the hydrogen does not dissociate/

recombine upon charging/discharging the storage system, and hence does not suffer from the kinetic limitations of many hydride storage systems, though absorption and desorption may have their own limitation. Hydrogen densities similar to liquefied hydrogen can be achieved with appropriate absorption media. Some suggested absorbers include metal organic framework, nanostructured carbons such as carbon nanotubes, and clathrate hydrates.

It is interesting to note that the most common method of onboard hydrogen storage in today's demonstration vehicles is as a compressed gas at pressures of roughly 700 bar.

22.13   FUEL CELLS AS A POSSIBLE ALTERNATIVE TO INTERNAL COMBUSTION

One of the main offerings of a hydrogen economy is that *fuel cells* could potentially replace internal combustion engines and turbines as the primary way to convert chemical energy into kinetic or electrical energy. The reason that proponents frequently use is that fuel cells, being electrochemical, are usually (and theoretically) more efficient than heat engines since they are not limited by Carnot efficiencies (see Section 5.2 for more details). Other inefficiencies do occur though, and can be found elsewhere in the fuel cell literature. Currently, fuel cells are more expensive to produce than common internal combustion engines, but are becoming cheaper as new technologies and production systems develop.

All fuel cells can be operated on pure hydrogen, and some are reported to work on methanol and hydrocarbon fuels. In the event of fuel cells becoming price-competitive with turbines, large gas-fired power plants could adopt this technology (Figure 22.5). Such commercialization would be an important step in driving down the cost of fuel cell technology.

Much of the interest in the hydrogen economy concept is focused on the use of fuel cells in cars. The cells are much more efficient than internal combustion engines, and produce no CO_2 emissions at the tailpipe, but CO_2 emissions could be present in the production of H_2. If a practical and engineerable method to store and carry hydrogen is introduced and fuel cells become cheaper, they can be economically viable to power hybrid fuel cell/battery vehicles, or purely fuel cell-driven ones (Figures 22.3 and 22.5). Note that all-electrical vehicles may also be potential alternatives to those having internal combustion engines. However, much research still seems to be necessary, and the issue of having the hydrogen as the fuel available at many locations needs to be resolved.

22.14   COSTS OF THE HYDROGEN ECONOMY

When evaluating costs, fossil fuels are generally used as the cheapest reference, even though the true cost of those fuels is seldom considered. Being fossil fuels—a nonrenewable source of energy—the thousands or millions of years required to be formed inside the Earth seem to mean "no cost" in most calculations and only the production costs are considered. Given such calculated low cost reference, any number of Watts required for hydrogen production seems too much, if those Watts come from a rather opposite—renewable—source of power like the sun. However, this comparison is unfair as simply evaluating a renewable and a nonrenewable on this basis does not take into account the finite nature of the nonrenewable resource.

If a system for hydrogen generation and usage needs to compete with systems that use renewably generated electricity more directly, for example, in trolleybuses, or in battery electric vehicles, it will always be less efficient due to the low efficiency of multiple conversions. From the preceding, hydrogen seems unlikely to be the cheapest carrier of energy over long distances, unless economic bases change. However, it must be emphasized that economics will only change the economic feasibility of options, but the inefficiency of multiple conversions will remain.

Hydrogen pipelines are more expensive [15] than long-distance electric lines. Hydrogen is about three times bulkier in volume than natural gas for the same energy content, and hydrogen accelerates the cracking of steel, which increases maintenance costs, leakage rates which lead to lost hydrogen, and material costs. The difference in cost is likely to expand with newer technology: Wires suspended in air can utilize higher voltage with only marginally increased material costs, but higher pressure pipes require proportionally more material.

Setting up a hydrogen economy would require huge investments in the infrastructure to store and distribute hydrogen to vehicles. In contrast, battery electric vehicles, which are already publicly available, would not necessitate immediate expansion of the existing infrastructure for electricity transmission and distribution, since much of the electricity currently being generated by power plants goes unused at night when the majority of electric vehicles would be recharged. A study conducted by the Pacific Northwest National Laboratory for the U.S. Department of Energy in December 2006 found that the idle off-peak grid capacity in the United States would be sufficient to power 84% of all vehicles in the United States, if they all were immediately replaced with electric vehicles [16].

Different production methods each have different associated investment and marginal costs. The energy and feedstock could originate from a multitude of sources, that is, natural gas, nuclear, solar, wind, biomass, coal, other fossil fuels, and geothermal (Table 22.2). Clearly, the distance from these "resources" can affect the economics.

TABLE 22.2
Cost of Hydrogen from Select Sources in Dollar per Gallon Gasoline Equivalent.

Source	Cost ($/GGE)	Comment
Coal via reforming	1.00	Non-sustainable/high CO_2
Small-scale natural gas	3.00	Non-sustainable/low CO_2
Electrolysis using nuclear electricity	2.50	Nuclear waste
Electrolysis using photovoltaic electricity	9.50	Sustainable
Electrolysis using wind electricity	3.00	Sustainable
Wood reforming	1.90	How far is net $CO_2 = 0$?
Electrolysis using geothermal electricity	1–2.40	Sustainable
Electrolysis using hydro power?	3.00?	

22.15   CONCLUDING REMARKS

In this chapter, we discussed the hydrogen economy, its promise, advantages, and challenges, but also the questions it has raised. Throughout this chapters, we have attempted, however, to substantiate the various claims, suggestions, and reservations with consistent thermodynamic arguments and numbers.

It is likely to take about 40–50 years, according to Ogden and other Princeton researchers, before hydrogen replaces fossil fuels, if it will ever do so. At least 10 years are necessary, according to some experts, before hydrogen finds widespread use in cars or industry. As such, it is useful to take stock of where we are now, and where we want to go. Currently, we are in the so-called fossil fuel age. Our final destination is a sustainable age. This sustainable age could well be the "solar age." Here, we use the term "solar" loosely, since renewables such as solar energy, wind, and biomass all find their origin in the sun. Even hydroelectric energy depends on the flow of rivers, which are fed by precipitation that in turn is partly attributable to the sun driving the water cycle. In the transition period between the fossil fuel and solar age, combinations of renewables and nonrenewables will have to emerge first, since an overnight transformation of energy technology is not possible. As such, natural gas will likely become increasingly important as its high hydrogen-to-carbon ratio makes it a good source of hydrogen, and it has low CO_2 emissions per unit exergy.

NOTE

1   The critical point of hydrogen is at 33 K en 12.4 atm.

REFERENCES

1. *The History of Hydrogen.* www.hydrogenassociation.org/general/factSheet_history.pdf
2. Squatriglia, C. Hydrogen cars won't make a difference for 40 years. *Wired, CondéNet, Inc.*, May 12, 2008. www.wired.com/cars/energy/news/2008/05/hydrogen
3. Boyd, R.S. Hydrogen cars may be a long time coming. *McClatchy Newspapers*, May 15, 2007. www.mcclatchydc.com/staff/robert_boyd/story/16179.html
4. National Research Council and National Academy of Engineering. *The Hydrogen Economy: Opportunities, Costs, Barriers, and R&D Needs*, National Academies Press: Washington, DC, 2004. www.nap.edu/openbook.php?isbn=0309091632
5. *Integrated Hydrogen Production, Purification and Compression System.* www.hydrogen.energy.gov/pdfs/progress08/ii_a_7_tamhankar.pdf
6. Feng, W.; Ji, P.; Tan, T. Efficiency penalty analysis for pure H_2 production processes with CO_2 capture. *AIChE Journal* 2007, *53*(1), 249–261.
7. *Linde Report on Efficiency of Hydrogen Liquefaction Plant.* www.linde-kryotechnik.ch/public/fachberichte/efficiency_of_hydrogen_liquefaction_ plants.pdf
8. The future of the hydrogen economy: Bright or bleak? *Final Report Ulf Bossel.* www.efcf.com/reports/E02_Hydrogen_Economy_Report.pdf
9. Wang, M. Fuel choices for fuel cell vehicles: Well-to-wheels energy and emissions impacts. *The Journal of Power Sources* 2002, *112*, 307–321.
10. Marisvensson, A; Svensson, A.M., Møller-Holst, S.; Glöckner, R.; Maurstad, O. Well-to-wheel study of passenger vehicles in the Norwegian energy system. *Energy* 2006, *32*(4), 437–45.

11. *White House Press Release*, February 6, 2003. www.whitehouse.gov/news/releases/2003/02/20030206-2.html
12. *Princeton Weekly Bulletin* 2003, *92*(21). www.princeton.edu/pr/pwb/03/0331/lb.shtml
13. Kreith, F.; West, R. Fallacies of a hydrogen economy: A critical analysis of hydrogen production and utilization. *The Journal of Energy Resources Technology* 2004, *126*, 249–256.
14. Bossel, U.; Eliasson, B.; Taylor, G. *Hydrogen Economy: Bright or Bleak. Final Report*, February 26, 2005.
15. www.ef.org/documents/NDakotaWindPower.pdf
16. Mileage from Megawatts: Study finds enough electric capacity to 'fill up' plug-in vehicles across much of the nation. *Ascribe*, December 11, 2006. http://newswire.ascribe.org/cgi-bin/behold.pl?ascribeid=20061211.105149&time= 11%2005%20PST&year=2006aaa&public=0

23 Future Trends

The end of the Stone Age was not heralded by a sudden shortage of stones. Similarly, our society's transformation to become sustainable will not be due to a sudden shortage of nonrenewables. Nontraditional sources of energy are becoming increasingly popular, and large investments are being made in a sustainable future. But what will the future hold? A definite answer will only be given once the future truly has arrived, but short of that, we can make some predictions based on current trends and speculate about the future of the chemical and energy industries.

23.1 INTRODUCTION

Experts seem to agree that the reserves of fossil fuels are finite, and one day will run out. However, estimates as to when this will occur vary a great deal. There is a feeling that the use of fossil fuels will lead to elevated levels of the greenhouse gas, CO_2, and that the ensuing global warming will potentially change weather patterns, causing floods and droughts.

The future is by no means gloomy. Efficiency and sustainability are no longer restricted to academic circles. There is a drive to work on alternate energy sources and go "green." The changes in the public mindset are essentially mirrored by what has happened in the research realm, and have sparked the major revisions of the chapters in this book.

In this chapter, we try to offer a glimpse of what the future of the energy and chemical industries may hold and try to point out some of the technological hurdles that have to be overcome to reach the goal, a sustainable society. We point out that we have not discussed nuclear power industries and do not know what role they will play in the future, but we foresee an increased use of nuclear power, perhaps temporary or longer term.

The chemical industry may be poised for changes as well, and different initiatives are in the research stage. The methods of evaluating technological options by reexamining value and financial decisions may also change which options will seem attractive.

23.2 ENERGY INDUSTRIES

Currently, a large proportion of energy comes from nonrenewable fossil fuels (see Chapter 9). Possible viable alternatives are wind and solar energy. In certain cases, for example, Iceland, geothermal energy is viable as well. It is conceivable that large offshore wind farms such as those built off the shore of Denmark may become more commonplace worldwide. The oil industry has considerable experience in operating in hostile offshore environments. Its technological expertise could prove vital in

DOI: 10.1201/9781003304388-27

large-scale offshore projects. A glimpse is already there as wind energy is becoming part of the energy portfolio that major energy companies are offering.

The role of decentralized energy generation is likely to increase in the future. According to the California Energy Commission, the average Californian household requires 6,500 kWh annually, which is less than the national average of 10,400 kWh [1]. Now, commercially available photovoltaic power cells have typical efficiencies of around 14% and can supply about 140 W/m². This means that 10,400 × 1,000/(140 × 8 × 365) = 25 m² of solar panel area are needed per household in the United States for 8 h of daylight. Since the energy requirements per household are generally less outside the United States, this number is likely to be lower in other parts of the world. Note that the new generation of photovoltaic solar panels can function in low light, such as overcast conditions. The implication of this is that using a significant part of the roof area for photovoltaics can take care of a great deal, if not all, of the electricity requirements of an average household. It is therefore conceivable that, in the future, decentralized energy production will supplement large-scale centralized energy production. For example, solar panels mounted on houses will supply a certain amount of electricity. If this supply exceeds the household's demand at that time, the electricity can be sold back to the electricity supplier. Conversely, if the demand is higher than the supply of the solar panels, electricity can be bought from the supplier [2]. Similarly, local waste treatment facilities could generate electricity to serve the local population (see Chapter 20). An interesting point is that not long ago, researchers at the University of California at Berkeley made significant progress in producing cheap *plastic* solar cells [3]. Till date, clean rooms and complex manufacturing procedures are necessary to produce the semiconductors, whereas the researchers only require a laboratory flask at room temperature [4]. Though their development is still in their infancy, such solar cells may prove useful in the future.

Solar energy may prove to be the ultimate renewable energy source, but currently we depend heavily on hydrocarbon fuels. These hydrocarbon fuels, as the name well suggests, consist of hydrogen and carbon, and the hydrogen:carbon ratio of these has interesting consequences. For very small values of this ratio, carbon dominates and the hydrocarbon is essentially solid at room temperature and pressure; coal is a good example. At higher values, the hydrocarbon becomes liquid (e.g., gasoline), which is a mixture of hydrocarbons, and even higher values, a gas (e.g., methane). As the ratio *increases*, the amount of CO_2 emitted per unit exergy *decreases*. Now what happens when the hydrogen:carbon ratio tends to infinity? In that case, we end up with hydrogen, which is no longer a hydrocarbon and has no CO_2 emissions per unit exergy in the ideal case. In the current fossil era, it simply means (as we have shown in Chapter 22) that the production of CO_2 emissions occurs elsewhere in the hydrogen production route.

Nuclear energy will become more important and become a part of the evolving energy mix. It is a proven technology that is low carbon indeed. In view of the climate crisis and recent political events leading to Europe's decision not to import Russian gas, have made discussion of nuclear front and center again.

It is not clear what role hydrogen will play in the future. The current attention has warranted a separate chapter which discussed the pros and cons of the hydrogen economy and the implications from a thermodynamic perspective. We leave the interested readers to draw their own conclusions.

Research is presently being conducted at the Los Alamos National Laboratories [5], among others, on so-called zero-emission power plants using coal. The concept here is to convert water and coal into hydrogen, which in turn is converted into electricity using fuel cells. Hydrogen is produced from water and coal through an intermediate calcium oxide (CaO) conversion into calcium carbonate ($CaCO_3$). In a subsequent step, the calcium carbonate is converted back into calcium oxide and a stream of pure CO_2. The CO_2 is then disposed of through mineral carbonation. It is clear that this technology can be adapted to other fossil fuels or biomass. Technologies such as this could be useful in the transition state between the fossil fuel age and the sustainable age and perhaps, if using biomass, in the sustainable age. Therefore, hydrogen production from fossil fuels will undoubtedly continue, albeit perhaps with sequestration of the CO_2 as mentioned earlier.

On November 20, 2002, Stanford University launched the Global Climate and Energy Project (G-CEP), which is a multimillion-dollar alliance between industry and academia to develop innovative technologies that will meet the world's growing energy needs while protecting the environmental health of the planet [6]. Some of the industrial partners include ExxonMobil, General Electric, and Schlumberger. The sponsors expect to contribute $225 million over the next 10 years. These developments may indicate that a shift in priorities in society is occurring.

In August 2001, researchers at MIT managed to produce hydrogen photocatalytically from water. While not as complete and efficient as photosynthesis, this system comes close to the ideal use of a molecular catalyst as part of a homogeneous reaction, for which scientists have been searching for more than three decades. Even though the process is still in its infancy, improvements of the process could some day allow the energy of the sun to be used to directly convert water into hydrogen and provide a sustainable source of energy fuel [7,8].

One of the trends visible today is to use hydrocarbons as the storage medium for hydrogen, and to use onboard reforming to create the hydrogen on-site. Storage could be possible by physical dissolution in carbon-rich materials or chemically bonded materials. This circumvents many of the problems associated with the storage and transportation of hydrogen, since existing facilities can be used. The carbon in the hydrocarbon could be obtained by renewable means by gasification of biomass to obtain synthesis gas, followed by Fischer-Tropsch synthesis of hydrocarbons. Another, perhaps less cost-effective option would involve the direct decomposition of CO_2 into oxygen and carbon monoxide, followed by Fischer-Tropsch synthesis. Unfortunately, the decomposition reaction is thermodynamically favorable only at very high temperatures. It seems that nature is more efficient at room temperature to fixate the carbon. We mention in passing some of the efforts at Imperial College in London, where researchers are inspired in an effort to mimic photosynthesis with the objective of synthesizing chemicals.

In any case, a storage strategy that involves hydrocarbons as opposed to pure hydrogen may be a useful technology for the transition between the fossil fuel age and the sustainable age and could be implemented without making any significant changes to the infrastructure.

23.3   CHEMICAL INDUSTRIES

Today, much of the chemical industry uses nonrenewable fossil fuels as the raw material. In the future, it is likely that Fischer-Tropsch synthesis will yield many chemicals from renewable hydrogen and carbon, which could come from biomass or even municipal waste. Currently, the proportion of fossil fuel actually used as fuel is much larger than that used as feedstock for the chemical industry. In the future, however, "fossil fuels" will no longer be used for their value as fuel, but could conceivably be used for their interesting chemical properties. It is interesting to note that currently more than 50% of the fossil fuels expended in chemicals production is used in driving the process and does not end up in the product on an exergy basis. This was shown in Chapter 14. The chemical and energy industries are generally quite efficient. Perhaps capitalizing on this, a more prominent trend in the future may be the simultaneous production of energy and chemicals, as shown in Chapter 9.

Needless to say, society is essentially an inert giant, which needs some prodding to change. This change will, of course, require monetary resources. When evaluating the cost of changing, one should not forget that the change may not benefit our generation, but is intended for the generations that come after us, and, hence, the use of standard economic indicators may not be useful. As Jeroen van der Veer, former Chairman of the Royal/Dutch Shell Group, points out in his speech given at the unveiling of the first hydrogen retail station in Reykjavik: "competitiveness must come from the added benefits produced in cleaner air, lower CO_2 emissions, and improved energy supply security." To this he also adds,

> that is why we need a twin track approach that utilizes fuel reformer technology to produce hydrogen from the world's existing hydrocarbons industry, while at the same time supporting the development of an entirely new and commercial hydrogen infrastructure, where commercially and politically feasible. [9]

Standard economic indicators, therefore, may not be appropriate when evaluating the feasibility of sustainable chemical or energy projects. For example, it has been shown [10] that to build a photovoltaic solar energy plant that lasts about 30 years, approximately one-third of the exergy produced over those years is actually necessary to build the plant. This contrasts starkly with the one-twentieth needed for energy plants based on fossil fuels. However, it may be useful to view the current fossil energy economy as a launching pad for a fully renewable economy.

In contrast to the energy industry, where, for example, decentralized production may become more popular, a paradigm shift in the chemical industry leading to radically different production methods is not clear to the authors, and an evolutionary change leading to perhaps different raw materials utilization may be more probable.

23.4   CHANGING OPINIONS ON INVESTMENT

Traditionally, investment decisions have been guided by tools such as the net present value (NPV), which takes into account the time value of money and is formally defined as the difference between the present value of cash inflows and cash outflows using a certain annual discount rate. The NPV is easy to calculate, but, as Dixit and Pyndick have pointed out, it is frequently wrong [11,12] since the NPV analysis is based on faulty assumptions. Either the investment is reversible and can be undone when conditions change, or if it cannot be changed, the investment has to be done now-or-never. This binary approach is not always applicable, and the ability to delay profoundly affects the investment decision. A richer framework is necessary to account for the gray area between the binary possibilities.

Instead of assuming either that investments are reversible or that they cannot be delayed, recent research on investment stresses the fact that companies have opportunities to investment and that they must decide how to exploit those opportunities most effectively, analogous to financial options. A company with an opportunity to invest is holding something akin to a financial call option. When it decides to invest, the call option is exercised.

When a company immediately makes an investment decision, the option is effectively killed, as the possibility to wait for new information is waived. An option value has important implications for managers as they think about their investment decisions. For example, a standard analysis may indicate that an investment is economical right now, but it is often highly desirable to delay an investment decision and wait for more information about market and technological conditions. On the other hand, situations may arise in which uncertainty over future market conditions should call for a company to accelerate certain investments [11,12]. A good example is investment in R&D, since it can lead to patents, which potentially gives the company freedom to operate in future, if it chooses to do so. Research can also potentially act as an "insurance policy" to ensure that companies can obtain experience in certain new developing areas.

Many managers already seem to understand that the NPV rule is too simplistic and that there is value in waiting for more information. In fact, many managers often require that an NPV be more than merely positive. In many cases, they insist that it be positive even when it is calculated using a discount rate that is much higher than their company's cost of capital by setting a hurdle rate. Disinvestment is guided by similar rules. Companies are willing to sometimes take a loss and wait until demand improves to maintain their foothold in the market, rather than close their operations and effectively disinvest. Applying the traditional NPV analysis to funding of long-term research and development projects can also lead to decisions that halt R&D. Research and development gives a company future possibilities, but its value is hard to quantify. High-risk projects may revolutionize the business, but may be very high cost and long term. Low-risk projects may only incrementally improve the business. For example, directed research at slightly improving the yield of a catalyst will only incrementally change the business and be profitable, but changing the process in its entirety may be high-risk and could revolutionize the business by doubling the yield or eliminating various intermediate process steps.

Patenting an invention in a sense creates an option for the company by allowing it to gauge the market potential and giving that company the right to pursue a particular investment.

The acquisition of oil fields is also a good example of decisions where delay is factored in. A company that buys an oil field may develop it immediately or later, depending on market conditions. The company is keeping its options open. Investments in sustainable technology are much the same. These investment decisions may or may not have favorable NPVs, but when market conditions change, it could be a different story. Investments in solar energy, for example, are likely to pay off in the long term. As such, these investments are options to an infinite energy source in the future, and to cash flow. At present, conventional fossil energy sources are less expensive, and as such one could argue that investments should be made in the latter rather than the former. It must be noted that the current price does not take into account the (finite) nature and potential environmental impact of either fossil fuels or renewables. But when fossil energy sources become less abundant, the political or environmental situation is such that the prices go up, so the picture for solar will be much more favorable. That is to say, in the short term, the NPV would give a negative indicator for solar energy, whereas in the long term, when fossil fuels are scarce, would give a positive indicator. However, it is important to realize that as sustainable technology development is a process fraught with technical challenges and its payout is in the long term, the option to use this technology should be pursued now by appropriate research. Changes in economic analysis will therefore also be instrumental in analyzing sustainable technology options.

23.5 TRANSITION

In order to make a transition to a sustainable society, initially, fossil fuels will play an important role. Fossil fuels will be expended in building the first fully sustainable chemical and energy plants (as per the definitions in this book). In the short term, hybrid technologies may be popular, which marks the transition from the "fossil period" to the "solar period," which could be the final sustainable society. Undoubtedly, natural gas will become important in the short term.

We are currently in the fossil fuel age, and our final destination is the sustainable age. Once fossil fuels run out, the fossil fuel age will have to end, and this allows us to estimate the time we have left to make the transition to the sustainable age. We assume that the demand for fossil fuels does not increase and that new reserves are not found. This is a very important assumption in obtaining this estimate, since the population will grow and, as such, with the same energy demands per capita, the demand will grow as well. Furthermore, supply may increase by the discovery of new reserves. Estimates of reserves are difficult to develop, and various sources give different data. We will therefore take the estimate in an order-of-magnitude fashion. The following worldwide statistics are available from the Energy Information Administration [13]:

- Estimated recoverable coal: 929.3 billion tons [14]
- Crude oil reserves: 1243 billion barrels [15]
- Natural gas reserves: 6254 trillion ft^3 [16]
- Coal consumption: 6.74 billion tons/year [14]

- Crude oil consumption: 83 million barrels/day [17,18]
- Natural gas consumption: 104 trillion ft³/year [14]

To obtain the number of years left, we simply divide the reserves by the consumption:

- Coal: 137 years
- Crude oil: 41 years
- Natural gas: 60 years

We note in passing that the consumption of crude oil, natural gas, and coal is expected to increase at 2%–3% [14], and that the reserves typically are replaced at various rates as well. The simple calculation indicates that in about 50 years, the age of fossil fuels will begin to end due to the depletion of crude oil. The fossil fuel age will have to end when the coal reserves run out in about two centuries. This gives us at least 50 years and at most 200 years to make the transition to the sustainable age.[1] In the next 20–30 years, however, fossil fuels will play an important role, as projections of the U.S. Energy Information Administration in the 2010 Energy Outlook indicate [19].

23.6 CONCLUDING REMARKS

The end of the Stone Age did not come because man ran out of stones. But if man runs out of stones, then man must be ready to leave the Stone Age. Similarly, the reserves of fossil fuels such as natural gas, crude oil, tar sands, and coal are expected to last some time, but when this has passed, the alternative options should be ready to go. It is probably best not to wait until there is a real shortage.

Increasing the efficiency of processes will be of utmost importance in the transition period, as well as in the solar age since high-efficiency processes will allow for more to be done with less. For example, it is attractive to use only one solar panel rather than three to get a certain task done. The efficient use of resources will therefore remain important. Undoubtedly, a new type of economics will arise that can accurately quantify and analyze projects leading to sustainable development and accurately inform the consumer of the costs as well.

Some scientists claim that there is no conclusive proof of global warming due to anthropogenic CO_2 emissions and, as such, there is no need to pursue sustainable technology. CO_2 is a greenhouse gas, and levels of this gas will rise if fossil fuels are combusted. At best there is no effect, but at worst it could contribute to global warming. However, sustainable technology is not about global warming. The crux of the matter is that fossil fuels are not renewable, and one day they will be depleted. Sustainability is about preventing such a scenario and, among others, about closing the various elemental cycles so that material and energy resources remain available for future generations and so that quality of life of the latter is not degraded.

In our journey from the fossil age to the sustainable age, we will enter the transition period, where conventional technology will exist side by side with sustainable technology. Hybrid technology will become increasingly more popular. The transition period will most certainly make use of conventional fossil fuel-based processes since this is inevitable. However, when in this transition period, it is important to realize that the path may be hard, but the final destination is well worth it. In any case, many challenges await society, and how well we face these challenges will determine our future.

NOTE

1 These estimates are subject to the assumptions made earlier.

REFERENCES

1. Gorin, T.; Pisor, K. *California's Residential Electricity Consumption, Prices, and Bills, 1980–2005,* September 2007, CEC-200–2007–018. www.energy.ca.gov/2007publications/CEC-200-2007-018/CEC-200-2007-018.PDF
2. Energy Information Administration 1999 data. https://www.eia.gov/totalenergy/data/browser/
3. Huynh, W.U.; Dittmer, J.J.; Alivisatos, A.P. Hybrid nanorod-polymer solar cells. *Science* 2002, *295*(5564), 2425–2427.
4. Sanders, B. Cheap, plastic solar cells may be on the horizon, thanks to new technology developed by UC Berkeley, LBNL chemists. *UC Berkeley Press Release*, March 28, 2002. http://berkeley.edu/news/media/releases/2002/03/28_solar.html (Heeger, A.J. Frontiers of science lecture: Low-cost plastic solar cells, The University of Utah, Salt Lake City, UT, January 29, 2009).
5. Ziock, H.-J.; Harrison, D.P. *Zero Emission Coal Power, a New Concept, 2001 Proceedings.* http://www.netl.doe.gov/publications/proceedings/01/carbon_seq/2b2.pdf
6. *Stanford Report*, November 21, 2002.
7. *MIT News Release*, August 31, 2001. http://web.mit.edu/newsoffice/nr/2001/nocera.html
8. Heyduk, A.F.; Nocera, D.G. Hydrogen produced from hydrohalic acid solutions using a two-electron mixed-valence photocatalyst. *Science* 2001, *293*, 1639.
9. van der Veer, J. *Speech given at the unveiling of the first Shell Hydrogen station in Reybjavik, Iceland,* April 2003.
10. Yoda, S. *Trilemma: Three Major Problems Threatening World Survival,* Central Research Institute of Electric Power Industry: Tokyo, Japan, 1995.
11. Dixit, A.K.; Pindyck, R.S. The options approach to capital investment. *Harvard Business Review*, 1995, 105–115.
12. Dixit, A.K.; Pindyck, R.S. *Investment under Uncertainty,* Princeton University Press: Princeton, NJ, 1994.
13. www.eia.doe.gov
14. International Energy Outlook 2009, Report #: DOE/EIA-0484, 2009.
15. *The Oil & Gas Journal* 2002, *100*(52). https://www.ogj.com/home/article/17233457/worldwide-reserves-increase-as-production-holds-steady
16. *World Oil* 2002, *223*(8). https://worldoil.com/magazine/2002/october-2002/industry-at-a-glance/02-10_worldoilproduction-html-oct-2002/
17. U.S. Energy Information Administration. *International Energy Statistics.* http://tonto.eia.doe.gov/cfapps/ipdbproject/IEDIndex3.cfm?tid=5&pid=54&aid=2
18. U.S. Energy Information Administration. World oil balance, 2005–2009, *International Petroleum Monthly.* www.eia.doe.gov/emeu/ipsr/t21.xls
19. U.S. Energy Information Administration. Annual Energy Outlook 2010 early release overview. *Report#: DOE/EIA-0383(2009)*, December 14, 2009. www.eia.doe.gov/oiaf/aeo/

Index

For Product Safety Concerns and Information please contact our EU
representative GPSR@taylorandfrancis.com
Taylor & Francis Verlag GmbH, Kaufingerstraße 24, 80331 München, Germany

* 9 7 8 1 0 3 2 3 0 3 0 1 7 *